Crime Files Series
General Editor: **Clive Bloom**
Since its invention in the nineteenth century, detective fiction has never been more popular. In novels, short stories, films, radio, television and now in computer games, private detectives and psychopaths, prim poisoners and overworked cops, tommy gun gangsters and cocaine criminals are the very stuff of modern imagination, and their creators one mainstay of popular consciousness. *Crime Files* is a ground-breaking series offering scholars, students and discerning readers a comprehensive set of guides to the world of crime and detective fiction. Every aspect of crime writing, detective fiction, gangster movie, true-crime exposé, police procedural and post-colonial investigation is explored through clear and informative texts offering comprehensive coverage and theoretical sophistication.

Titles include:

Maurizio Ascari
A COUNTER-HISTORY OF CRIME FICTION
Supernatural, Gothic, Sensational

Pamela Bedore
DIME NOVELS AND THE ROOTS OF AMERICAN DETECTIVE FICTION

Hans Bertens and Theo D'haen
CONTEMPORARY AMERICAN CRIME FICTION

Anita Biressi
CRIME, FEAR AND THE LAW IN TRUE CRIME STORIES

Clare Clarke
LATE VICTORIAN CRIME FICTION IN THE SHADOWS OF SHERLOCK

Paul Cobley
THE AMERICAN THRILLER
Generic Innovation and Social Change in the 1970s

Michael Cook
DETECTIVE FICTION AND THE GHOST STORY
The Haunted Text

Michael Cook
NARRATIVES OF ENCLOSURE IN DETECTIVE FICTION
The Locked Room Mystery

Barry Forshaw
BRITISH CRIME FILM
Subverting the Social Order

Barry Forshaw
DEATH IN A COLD CLIMATE
A Guide to Scandinavian Crime Fiction

Emelyne Godfrey
FEMININITY, CRIME AND SELF-DEFENCE IN VICTORIAN LITERATURE AND SOCIETY
From Dagger-Fans to Suffragettes

Emelyne Godfrey
MASCULINITY, CRIME AND SELF-DEFENCE IN VICTORIAN LITERATURE
Duelling with Danger

Beatrix Hesse
THE ENGLISH CRIME PLAY IN THE TWENTIETH CENTURY

Lee Horsley
THE NOIR THRILLER

Merja Makinen
AGATHA CHRISTIE
Investigating Femininity

Fran Mason
AMERICAN GANGSTER CINEMA
From *Little Caesar* to *Pulp Fiction*

Fran Mason
HOLLYWOOD'S DETECTIVES
Crime Series in the 1930s and 1940s from the Whodunnit to Hard-Boiled Noir

Linden Peach
MASQUERADE, CRIME AND FICTION
Criminal Deceptions

Steven Powell (editor)
100 AMERICAN CRIME WRITERS

Alistair Rolls and Deborah Walker
FRENCH AND AMERICAN NOIR
Dark Crossings

Susan Rowland
FROM AGATHA CHRISTIE TO RUTH RENDELL
British Women Writers in Detective and Crime Fiction

Melissa Schaub
MIDDLEBROW FEMINISM IN CLASSIC BRITISH DETECTIVE FICTION
The Female Gentleman

Adrian Schober
POSSESSED CHILD NARRATIVES IN LITERATURE AND FILM
Contrary States

Lucy Sussex
WOMEN WRITERS AND DETECTIVES IN NINETEENTH-CENTURY CRIME FICTION
The Mothers of the Mystery Genre

Heather Worthington
THE RISE OF THE DETECTIVE IN EARLY NINETEENTH-CENTURY POPULAR FICTION

R.A. York
AGATHA CHRISTIE
Power and Illusion

Crime Files Series
Standing Order ISBN 978–0–333–71471–3 (hardback)
978–0–333–93064–9 (paperback)
(*outside North America only*)

You can receive future titles in this series as they are published by placing a standing order. Please contact your bookseller or, in case of difficulty, write to us at the address below with your name and address, the title of the series and the ISBN quoted above.

Customer Services Department, Macmillan Distribution Ltd, Houndmills, Basingstoke, Hampshire RG21 6XS, England

---

The English Crime Play in the Twentieth Century

Beatrix Hesse
Professor of English and Cultural and Media Studies, Universität Passau, Germany

palgrave
macmillan

First published 2015 by
PALGRAVE MACMILLAN

Palgrave Macmillan in the UK is an imprint of Macmillan Publishers Limited, registered in England, company number 785998, of Houndmills, Basingstoke, Hampshire RG21 6XS.

Palgrave Macmillan in the US is a division of St Martin's Press LLC, 175 Fifth Avenue, New York, NY 10010.

Palgrave Macmillan is the global academic imprint of the above companies and has companies and representatives throughout the world.

Palgrave® and Macmillan® are registered trademarks in the United States, the United Kingdom, Europe and other countries.

ISBN 978–1–137–46303–6

This book is printed on paper suitable for recycling and made from fully managed and sustained forest sources. Logging, pulping and manufacturing processes are expected to conform to the environmental regulations of the country of origin.

A catalogue record for this book is available from the British Library.

A catalog record for this book is available from the Library of Congress.

Typeset by MPS Limited, Chennai, India.

Contents

Acknowledgements

The preliminary research for this book was funded by two equal opportunities programmes: the HSP III and HWP (now sadly both extinct), for which I am profoundly grateful. I have also long wanted to thank my academic teacher, Professor Klaus Peter Jochum, who had already retired at the time I was working on this book and still gave generously of his time and expertise. His patience and critical intelligence have been invaluable. Finally, I would like to express my gratitude to Iris Wagner, who proofread the manuscript at very short notice and removed a number of mistakes.

1
The Body in the Library and the Body on Stage

Many of the most popular English stage plays of the last century were concerned with crime, most frequently murder, and its detection. The present study will examine these plays, which in some ways are so like detective fiction and in other respects are very different. The title phrase of this chapter, "the body in the library and the body on stage" plays on the contrast between the popular image of a dead body found in the midst of "fossilized," dead, written language and theatrical performance in which living bodies utter living, spoken language. Twentieth-century English detective fiction created a massive output of stories revolving around the discovery of a dead body and the reconstruction of how that body came to be a dead body. Since drama is concerned with the actions of living bodies one might assume that no comparable genre has evolved in the theatre. This is far from the case.

The term used for this kind of play in this study will be "crime play," since this is a fairly neutral and comprehensive term which does not imply preconceived ideas about the structure or focus of these plays.[1] "Crime play" is defined as a play revolving around a crime and its detection. The crime must be the central topic of the play, which means that, for instance, plays on family conflicts that happen to escalate into violence will not be considered unless the focus is on the illegality of such violence. "Crime" is defined according to the textbooks on criminal law. The definition that states the essential aspects most clearly seems to be the one formulated by the House of Lords in 1957 (*Board of Trade v. Owen*): "an unlawful act or default which is an offence against the public and renders the person guilty of the act liable to legal punishment".[2] The term "unlawful act" already contains several important aspects. "Unlawful" stresses the principle of legality: before an action can be a crime, there must be a law that prohibits it. "Act" refers to the legal principle of "actus reus," meaning a willed, voluntary act committed in "mens rea," a wrongful state of mind, for instance intent – the classic "malice aforethought," – in the case of murder. Besides, the term "unlawful act" draws attention to the difference between legal and moral wrong. A traffic violation may be illegal, though not immoral, while lying or adultery may be immoral but not illegal unless defined as such by the laws of the country. The fact that a crime is considered "an offence against the public" and "liable to legal punishment" differentiates criminal law from civil law. In civil law, a specific person has been offended and the remedy for such an offence is not punishment but compensation.

As the above definition of "crime" suggests, this study will not be primarily concerned with moral wrong, with questions of guilt, responsibility – or Sin – but largely with the aspect of illegality and the threat of punishment resulting from it. The emphasis on this aspect implies that the crime play is understood to be a thoroughly secularized genre that deals with crime and its accompanying intense emotions of aggression and fear in a predominantly rational manner. Hence, treatments of the themes of fear and aggression which present these topics in the context of the operation of occult forces, e.g. Hamilton Deane's *Dracula*, have been excluded from the present study. Plays concerned with terrorism in the context of civil warfare have also been ignored, since they are arguably better discussed together with other plays about war and political conflict.

So much on the aspect of "crime." "Play" in the context of this study means "stage play," excluding TV plays and radio drama, since the conditions of production are vastly different and the three formats are hence not necessarily comparable. Likewise, the present study focuses on full-length plays and largely ignores one-act plays because they present different problems of plot construction. A handful of plays that are less than full-length have nevertheless been considered because they introduced important innovations to the format, for example Tom Stoppard's *The Real Inspector Hound* and Howard Brenton's *Christie in Love*. Since the focus of the present study will be on the development of specific traditions and conventions of the crime play, it turns out to be necessary to limit the selection of plays to those performed in a specific location – in this case, the London West End.[3] Plays that only had a short run in a remote area of the British Isles obviously had little opportunity of establishing a new tradition. In this respect, the study of the crime play necessarily differs from the study of the detective novel: while a novel can be read anywhere, provided the reader is familiar with the language, you can only watch a play in a specific theatre. A historical study of a particular type of play has to be based largely on a study of the published play text. For the present study, only published plays have been considered, in order to render the resulting observations verifiable for the reader.

Finally, only plays of English origin have been considered for the present study, that is, plays usually written by English playwrights that saw their first performance in London or elsewhere in the British Isles. As the study of detective fiction has shown, different countries tend to produce different varieties of crime fiction. Therefore, it seems advisable to limit oneself to one national tradition at a time. For the same reason, plays adapted from foreign detective novels have also been excluded from the present study.

The term "crime play" is not yet established in critical discourse; however, there seems to be a latent awareness that some kind of generic label for the type of play described above is indeed desirable. The website "officiallondontheatre.com" and the regularly published free leaflet "Official London Theatre Guide" offer theatregoers the subgenre of the "thriller" next to established dramatic genres such as "tragedy" and "comedy." John Russell Taylor dubs a play concerned with crime and detection a "whodunit"[4] in imitation of the terminology employed in detective fiction criticism, and Stanley Richards calls his collections of crime plays "Mystery and Suspense Plays."[5] None of these terms, however, is particularly satisfactory, since "thriller" and "whodunit" all by themselves do not clearly announce that they refer to plays, and "mystery play" invites confusion with medieval religious drama. As a neutral and unambiguous term, "crime play" clearly is to be favoured.

As the absence of a uniformly used term suggests, critical discussion of the crime play has so far been virtually non-existent. There is an American dissertation on the subject, Charles LaBorde's *Form and Formula in Detective Drama* (1976), and one monograph that attracted a somewhat wider audience, Marvin Carlson's *Deathtraps* of 1993. Since the number of secondary texts is comparatively small and the texts themselves are not widely known, it may be appropriate to discuss them in some detail.

LaBorde's thesis was subsequently published in *The Armchair Detective* under the somewhat more journalistic title of "Dicks on Stage," but failed to trigger off a lively critical debate on the format. At the outset of his study, LaBorde deplores the lack of secondary criticism on the crime play, remarking: "It remains a major twentieth-century theatre form without detailed scholarly analyses of either its history, development, or structure."[6] While acknowledging the work done by Willson Disher and Frank Rahill in the field of melodrama, LaBorde believes that disproportionately little space is allotted to thrillers and mysteries in their respective works. Besides, he notes that studies of detective novelists like Christie and Wallace also devote but minimal space to the authors' dramatic work. LaBorde limits himself to the discussion of published English and American plays that had substantial runs in London or New York, "substantial" meaning over 70 performances before 1940 and 200 or more after 1940.[7] Besides, he only examines one type of crime play in any detail: the kind he christens the "confined mystery," "a drama in which a group of characters is detained primarily in a single locale until a crime is solved."[8] Although this definition also applies to courtroom drama, this type is explicitly excluded from LaBorde's study.[9]

In his analysis of the structure of crime plays, LaBorde largely relies on Aristotelian dramatic theory and terminology. Like Dorothy Sayers in

"Aristotle on Detective Fiction,"[10] he notes that the crime play admirably obeys the rule of unity of action. Using his favourite type of the "murder-house mystery" as an example, LaBorde describes a typical crime play plot structure as follows:

> Ordinarily the playwright devotes the entirety of the first act of a murder-house mystery to exposition. Despite the rapid insertion of the murder theme, the act moves rather slowly before it culminates in an initial on-stage murder. [...] Thus the end of the act prepares the way for the investigation that occupies most of the remainder of the play.[11]

The investigation characteristically consists of the discovery and interpretation of clues. This part of the play is subject to the "fair-play rule" established by authors of detective fiction, demanding that no important information be withheld from the audience. In the course of his discussion of the fair-play rule, LaBorde introduces a useful distinction between the "functional" and "occupational" detective. While an occupational detective investigates crimes professionally, but may not solve the crime in question (he may even himself be the culprit), the functional detective is whoever takes it on himself to solve the present case. LaBorde strongly disapproves of the practice employed in *The Mousetrap* and *Ten Little Niggers* of presenting the solution in the form of a confession speech delivered by the criminal, since he considers it unsatisfactory for the audience:

> The audience is led to believe that it can solve the problem just as the detective will. Suddenly the murderer, not the detective, reveals the solution. *The audience then realizes that it has preoccupied itself with attempting to solve what is, in fact, an unsolvable mystery.* Another problem with such solutions is that they necessitate a theatrically ineffective final scene.[12]

It is certainly debatable if a confession scene is necessarily "theatrically ineffective" particularly considering the major successes of Henry Irving in his melodramatic parts. The penultimate sentence of the above quotation is more interesting, since it suggests that the crime play, even in its most conventional form, is more sceptical towards positivistic optimism than the traditional detective novel.[13]

The next major part of LaBorde's study is concerned with audience response. Still employing Aristotelian terminology, La Borde deplores the lack of "magnitude" in detective drama occasioned by an insufficient difficulty of the mystery and the lack of seriousness in its treatment. The crime play is said to deal with crime in a "game-like" manner, using sketchily developed characters who "die quickly and their apparent pain or anguish is minimal."[14] Moving from the discussion of plot to the discussion of character, LaBorde observes that a typical crime play cast consists of a detective, a pair of young lovers, a comic servant, a playboy, an older man, and a villain. Character development is rare, with one exception: "One character in each of the plays undergoes a major transformation: that character is the villain."[15] This sudden revelation endangers the Aristotelian virtues in character delineation, which are appropriateness and consistency. The characters' mental processes are almost exclusively centred on deduction, which is mostly done by a single personage, the functional detective. Progressing from characterization to the plays' overall moral message, LaBorde states: "While murder-house mysteries do not make their characters face moral crossroads, the formula posits a basic morality. Stated most tritely, but clearly, that moral viewpoint is 'Crime doesn't pay.'"[16] Not only is the villain punished by arrest or sudden death, but the victims also deserve their fates. Moreover, the "corollary of the 'crime-doesn't-pay' stance states that 'Virtue is rewarded'."[17] Most frequently, the reward of the virtuous takes the shape of the young couple living happily ever after.

After having devoted a very extensive chapter to "murder-house mysteries," LaBorde has considerably less to say about the other formulas he unearths, which include for instance the "police procedural," the "psychological thriller" and the "inverted mystery." The appendix of LaBorde's dissertation contains useful act-by-act synopses of the 34 plays examined. These are the only crime plays listed in the bibliography, with the result that LaBorde's study contributes little to relieving the unsatisfactory lack of a reliable bibliography in the field.

As the subtitle suggests, Marvin Carlson's *Deathtraps: The Postmodern Comedy Thriller* is not concerned with the development of the crime play as a genre but merely with a particular subgenre which evolved between 1970 and 1990. The term "postmodern drama" is usually associated with experimental avant-garde theatre, but Carlson deliberately applies it to commercial mainstream theatre. He defines the type of play with which his study is concerned as follows: "The subject of the play is murder, planned or already committed, and the action is the successful or unsuccessful efforts [sic!] of the murderer to prevent his or her exposure."[18] The earliest instance of a postmodern comedy thriller is Tom Stoppard's *The Real Inspector Hound*, a parody of Christie's *The Mousetrap*, about which Carlson remarks:

> The very solidity and predictability of generic codes in a genre like detective fiction make it a tempting target for the foregrounding or dismantling of those codes that is the basis of parody. [...] Like most of Stoppard's works, however, *The Real Inspector Hound* plays with epistemological and ontological questions that go far beyond its parodic surface and make it a kind of forerunner of a group of equally subversive mystery dramas from the early 1970s onward. In these dramas, the parodic element is much less central and the total effect seems even closer to the enterprise that in other contexts has been characterized as postmodern.[19]

Quoting from Linda Hutcheon's *The Poetics of Postmodernism*, Carlson argues that parody is the perfect postmodern form, "for it paradoxically both incorporates and challenges that which it parodies."[20] The first full-length (and most perfect) example of this type of play is Shaffer's *Sleuth*, "a mystery play of instability and playful self-consciousness,"[21] which ideally conforms to Hutcheon's recipe for creating a postmodern artefact which "uses and abuses, installs and then subverts, the very concepts it challenges."[22] The generic label increasingly used for this type of crime play is "comedy thriller," a classification on which for instance Gerald Moon insists for his play *Corpse!*. Although these plays may be considered "the last mannerist elaborations of a nearly exhausted form,"[23] Carlson believes that they also express common concerns of contemporary culture. He lists the following characteristics of the comedy thriller: a self-referentiality modelled on Pirandello and Brecht, a scepticism towards traditional values and beliefs (such as individual consciousness, objective truth, and the power of reason), and a "simultaneous awareness and playful deferral of death."[24]

Before embarking on an analysis of his main subject, Carlson gives a short overview over the development of the crime play, complaining about the lack of a comprehensive and reliable historical study of the genre:

> The complicated development of the modern thriller has still to be traced and will not be undertaken here [...] [T]he many recent studies of mystery and detective fiction, when they speak of the theatrical thriller at all, tend to consider it as little more than a rather minor derivative form, capitalizing on the success of other detective fiction.[25]

According to Carlson, however, the twentieth-century crime play does not derive from detective fiction but from nineteenth-century melodrama, which differs from later crime plays insofar as there is no mystery about the identity of the villain. According to Carlson, the modern crime play develops in New York:

> In the early years of the twentieth century, a group of New York dramatists, most of them trained in the tradition of melodrama, began to develop a more concentrated type of mystery drama, one in which a murder is committed at or before the beginning of the play and suspicion falls upon a number of characters before the "detective," usually not a policeman, reveals the true murderer at the end of the action, often through some psychological entrapment. There is usually a single setting, either the scene of the crime, in which case the murder is often shown, or a courtroom, portraying the subsequent trial.[26]

As examples of this type of play, Carlson mentions Bayard Veiller's *The Thirteenth Chair* and John Willard's *The Cat and the Canary*. The English crime play, stimulated by the success of Golden Age detective fiction, made use of a different formula:

> This device is to actually show the crime committed early in the play, so that the audience is in no doubt about the murderer's identity, but also to have the murderer develop and perhaps rehearse an elaborate alibi, so that the subsequent tension results from wondering whether the scheme will work.[27]

Carlson devotes the seven main chapters of his study to the analysis of characteristic features of the postmodern comedy thriller. The topics covered include setting (Chapter 2: "The Scene of the Crime"), characters (Chapter 3: "Among Those Present"), plot construction (Chapter 4: "Murder by the Book"), the theme of murder (Chapter 5: "Dead Wrong"), conscious theatricalization (Chapter 6: "Stage Struck"), audience reception (Chapter 7: "The Audience as/for Accomplice") and closure (Chapter 9: "Deathtraps").[28]

With respect to setting, Carlson notes that most postmodern experiments in the theatre have been based on the British variety of the crime play, the type he chooses to call the "cozy."[29] A major difference between the American and the British version of the detective novel is their choice of setting: while the American thriller uses a variety of "open" urban settings, the British detective novel "is most commonly a well-to-do British country home, preferably one [...] isolated by both geography and inclement weather to intensify the aura of entrapment."[30] A subsequent survey of crime plays produced before the period of postmodernism has proved that Carlson is mistaken in assuming that this shift from rural to urban setting is a specific feature of the postmodern comedy thriller. From the outset, crime plays have tended to be situated either in the city or in its immediate vicinity. Plays making use of the traditional country-house setting of Golden Age detective fiction usually are stage adaptations of novels belonging to the said genre, or they are themselves already parodies of this tradition.

Like his chapter on settings, Carlson's chapter on the dramatic characters suffers to some extent from the fact that he compares his postmodern comedy thrillers not to earlier crime plays but to the conventions of classic detective fiction. Starting with George Grella's observation that the cast of the traditional "cozy" represents a microcosm of English upper-class society,[31] Carlson investigates how the comedy thriller departs from this formula:

> [T]he standard modern comedy thriller has replaced the traditional grouping of suspects and detective of the country house murder with a configuration that is quite different but, by this time almost as predictable in these plays as the odd assortment of British upper-class types in the classic cozies. First, we may note that the size of the cast is much reduced;

> most commonly there are only four or five actors and sometimes as few as three or even two. [...] The smaller cast in some measure reflects a general trend in the modern drama, but much more importantly, it reflects a new kind of story being told. Clearly, with a cast of two or even four, a dramatist cannot construct a situation involving a detective, a victim, and a reasonable number of suspects in the traditional manner. What this genre has done is shift from one highly conventionalized pattern to another which, in the theatre, has been equally conventionalized – the French triangle play of amorous intrigue. At the center of almost every thriller of the 1970s and 1980s is found a husband, wife, and male or female lover, with two of them plotting the death of the third. This shift has doubtless been facilitated by the fact that certain conventions of this genre are shared with those of the comedy of manners, whose close relationship to the traditional detective story Grella has noted.[32]

An analysis of earlier crime plays reveals that the reduction of cast size and the substitution of the whodunit formula by the motif of the amorous triangle occur much earlier than 1970, and are more convincingly explained by the difference between detective novel and crime play than by any particularly postmodern concerns.

Applying the structuralist model suggested by Anne Ubersfeld of analysing a dramatic plot in terms of the structural roles played by the characters (i.e. "giver," "receiver," "opponent," etc.) to the comedy thriller, Carlson finds that "actantial roles are cut loose from the actants, so that one never can be sure, from scene to scene, who will be playing any particular actantial role."[33] This even extends to the function of the detective, traditionally considered "an exultation of bourgeois scientific rationalism" and a "guarantor of meaning."[34] In the comedy thriller, the detective is either absent or, if he does appear, "is usually corrupted in some way by the thriller dynamic."[35] Postmodern detectives depart from the codes of their profession in the following ways: they hide the real murderer (in Durbridge's *Deadly Nightcap*); they turn out not to be real policemen but other characters in disguise (in *Sleuth* or Durbridge's *Suddenly at Home*), or they resign without having found the murderer (as in Shaffer's *Whodunnit*). It must be pointed out that – like the shift from rural to urban setting or the shift from whodunit formula to amorous triangle – the absence or corruption of the Great Detective is not a postmodern innovation. Agatha Christie invariably removed her detective in her stage adaptations, and the "bogus policeman" is a standard character in the crime play from the 1920s onwards.

In his analysis of plot structure, Carlson contests Ordean A. Hagen's assumption[36] that crime plays start with murder. He points out that, generally, the crime itself is preceded by some kind of exposition, and the crime then "almost inevitably provides a climax to a scene or an act."[37] As the title of his book suggests, Carlson considers the crime play as predominantly a genre revolving around murder. While the traditional whodunit conceals the identity of the culprit, in the postmodern comedy thriller the identity of the victim may also be in question.[38] "Apart from the questions of whodunit and who-was-it-done-to, comedy thrillers ask the far more unsettling question 'Did it in fact happen?'"[39] Accordingly, comedy thrillers abound in fake death scenes consciously staged by one character or party of characters in order to misguide another. Accordingly, Carlson's main statement in his chapter on audience response is that

> audiences do not come to comedy thrillers planning to play the old and fairly straightforward game of trying to guess the murderer but to play the more complicated game of trying to stay a step ahead of the tricks and misdirections the author presents to them.[40]

To put it differently, the spectator does not attempt to detect the criminal but to detect the playwright, but this is hardly a specifically postmodern phenomenon or one limited exclusively to dramatic performance – even a reader of classic detective fiction feels challenged by the author rather than the criminal.

The last chapter of Carlson's study appropriately deals with closure. Carlson notes that the typical revelation scene of the detective novel in which the detective "assembles the suspects together and, by a physical or verbal reconstruction of the crime, demonstrates which of them committed the destabilizing murder"[41] is usually absent from the comedy thriller. The identification of the culprit may be permanently suspended, as in *Murder by the Book*, or the detective may refrain from making the identity of the criminal public (*Deadly Nightcap*). In most comedy thrillers, the police do not even appear; their arrival is signalled by sirens (*Sleuth, Stage Struck*) or they are being telephoned (*Double, Double*). It must be added that these phenomena are also to be observed in earlier crime plays: in 1932, John van Druten experimented with denying the audience closure in *Somebody Knows*, and in *Rope* (1929), the arrival of the police is merely anticipated by the functional detective's use of his whistle. Generally, the sound of sirens is better understood as a kind of shorthand for the arrest of the culprit rather than as an attempt to undermine the audience's confidence in the triumph of justice. Finally, the detective who allows the culprit to escape appears even in classic detective fiction, for instance in Christie's *Murder on the Orient Express*.

Throughout, Carlson's study suffers from the fact that he tends to compare his postmodern comedy thrillers not to the conventions of the traditional crime play but to those of traditional detective fiction. Besides, his definition of what constitutes a "postmodern" crime play is very generous; while most critics would surely agree that *Sleuth* is a play radically different from, for instance, a product like Christie's *The Hollow*, the plays of Francis Durbridge appear to be firmly rooted in the classic tradition. Carlson's bibliography contains some 160 titles and their authors, ordered according to the dates of the plays' first productions. About half of the plays listed are English, the others American.

The monographs by Carlson and LaBorde have a number of characteristics in common. In their historical outlines, they both stress the crime play's dual indebtedness to the traditions of melodrama and of detective fiction. They both consider the stage whodunit the standard type of the crime play from which all other subgenres deviate – an assumption probably based on the findings in the field of detective fiction studies, though this is never openly acknowledged. This may partly be due to the fact that the authors do not distinguish sufficiently between original stage plays and adaptations of novels. Nor do they attempt a systematic distinction between the British and American traditions in the genre; in this respect, the study of detective fiction has not been imitated.

Apart from the monographs discussed above, a few shorter studies of the crime play have been published, two of which shall be briefly summed up in the following. Clive Barker's and Maggie G. Gale's anthology *British Theatre between the Wars, 1918–1939*[42] contains one article devoted to the stage thriller. John Stokes' essay "Body Parts: The Success of the Thriller in the Inter-War Years" draws attention to the actors' contribution to the development of the format. The article begins by sketching the career of Gerald du Maurier from his success in the title role of *Raffles* in 1906. As Raffles, du Maurier created the twentieth-century stage villain in deliberate contrast to the tormented villain of Victorian melodrama most perfectly embodied by Henry Irving. In *Bulldog Drummond* (1921), du Maurier impersonated a new hero for the "wounded officer class fighting hard to claim its place in the modern world."[43] As an actor, he is also praised for his portrayal of the professional man capable of an extremely efficient deception in Pertwee/Dearden's *Interference* (1927); as a producer, for bringing Edgar Wallace's *The Ringer* and Agatha Christie's *The Murder of Roger Ackroyd* onto the stage. In the context of the Christie adaptation, Stokes explains his basic distinction between "thriller" and "detective story." Although one might assume that he applies "detective story" to narrative and "thriller" to stage plays, he claims that both "thriller" and "detective story" exist both in narrative and dramatic form. The main difference is that in "thrillers" the suspense is directed forwards, towards the future, whereas in the "detective story," the suspense is directed backwards, towards the past.[44] Stokes observes, however, that drama tends to have greater difficulties with flashback than narrative and mentions Milne's *The Fourth Wall* as a successful combination of the formats of "thriller" and "detective story."

After du Maurier, Stokes follows up the career of actor-playwright Emlyn Williams, who created the stage character of the homicidal maniac, first as an actor in Wallace's *The Case of the Frightened Lady* and later, as author and actor, in *Night Must Fall*. Williams' play *A Murder Has Been Arranged* is singled out as an excellent example of theatrical self-referentiality, since the action is set in the theatre. Other examples of self-referential thrillers mentioned by Stokes are Harold Holland's *The Big Drum* and Frank Vosper's *Murder on the Second Floor.* The emphasis on theatricality and self-referentiality in the stage thriller may be indebted to Carlson's argument in *Deathtraps.*

Apart from the careers of actors, playwrights and producers specializing in the format of the thriller, Stokes also discusses various famous true criminal cases of the inter-war period, mentioning the crimes of G. J. Smith, Dr. Crippen, Thompson/Bywaters (as dramatized by Frank Vosper in *People Like Us*) and Leopold/Loeb, which presumably inspired Patrick Hamilton's *Rope.* This concern with the relationship between the stage play and real life is also evident in Stokes' explicit interest in details from his authors' and actors' biographies, particularly their sexual orientation. Stokes sums up his findings, defining the thriller as a somewhat "deficient" genre that never dares to go far enough:

> [I]t flirts with formal or continental experiment, never quite committing itself: it adopts a theatrical self-consciousness of almost Pirandellian complexity, but stops short of any philosophical conclusion; it draws on melodrama's fondness for grating sounds and silent shadows, but never quite makes the transition to full-blown expressionism; it hints at grotesque objects, mutilated bodies and severed heads, but they remain hidden in chests and hat-boxes, preventing any Artaudian visual frisson. This is largely because, as forms of commercial entertainment, the thrillers were ultimately tied to West End methods, to the Lord Chamberlain, and the moral conventions of the time.[45]

The penultimate sentence of this quotation reveals the aesthetic ideals measured against which the thriller appears deficient: the theatre of Pirandello, Artaud and Expressionism. This high regard for Artaud may be related to the taste in theatrical performance dominant in the "nasty" 1990s. The reference to Pirandello is dubious, since self-referentiality in the British stage thriller is probably more closely related to similar experiments in American thrillers of the previous decades, particularly George M. Cohan's wildly successful *Seven Keys to Baldpate.*

Lynn Marian Thomson's recent article on "The Crook Play"[46] discusses some 30 American plays revolving around the character of the criminal that were first produced in the years immediately before World War I, coinciding with the Progressive Era in American politics. Thomson claims that the "crook play" is a distinct literary genre "organized around a character perceived in its time as challenging to both social and dramatic conventions, and a serious, sometimes militant effort to expose destructive social conditions."[47] This character is the "gentleman-thief," first established on the US stage in Paul Armstrong's *Alias Jimmy Valentine,* which was based on the O. Henry short story "The Retrieved Reformation" and first performed in 1910. Thomson's second major example is Bayard Veiller's *Within the Law* (1912) about a shop-girl turned criminal, which already indicates that the "gentleman-thief" may also be a working-class woman.

The crook play is a hybrid genre evolving from various strands of popular American theatre, particularly farce, melodrama and detective play, but develops its own set of conventions. It dispenses with the classical three unities, employing multiple settings and expanding plots governed by the twin principles of reversal and surprise. Instead of aspiring to unity, the crook play exploits opposition and contrast in its juxtaposition of different social milieus and interactions. There tends to be a happy ending, but no poetic justice – since the hero is a criminal, the two would presumably be mutually exclusive. Taking her cue from Carlson, Thomson notes a frequent blurring of identities, as crooks disguise themselves as detectives and vice versa. The crime in a crook play is often theft, frequently that of a pearl necklace. An important stylistic feature of the crook play is its humour, generated by the use of well-placed slang expressions and the typically American linguistic trope of the wisecrack.

A crook play may also create humour by parodying melodrama, for although the genre is to some extent related to melodrama, it markedly departs from the conventions of the older genre: heroines are not necessarily pure, and the guilty are not punished. Melodrama as an inherently conservative genre prohibits cross-class marriage, whereas in crook plays, cross-class marriages abound. The differences between the crook play and the detective play are even more substantial. According to Thomson, detective plays consider "universal subjects,"[48] while crook plays are consciously topical. In a detective play, a well-ordered society is disrupted by crime, whereas in the crook play, the crime is but a symptom of a basic disorder in society. Accordingly, while the detective play casts out the guilty, the crook play contests the simplistic opposition between Good and Evil. In her description of the detective play, Thomson largely draws on W. H. Auden's influential essay "The Guilty Vicarage,"[49] suggesting that Auden's theme is detective drama, while the essay itself explicitly speaks of "reading" and "stories," that is, of narrative.[50] Probably Thomson refers back to Auden for lack of a suitable study of the detective play to consult; moreover, she apparently intends to establish an opposition between a conservative English genre and a politically radical American one,[51] rather than an opposition between narrative and drama. Thomson's sole example of a detective play, however, is American: Bayard Veiller's *The Thirteenth Chair*.

> *The Thirteenth Chair* functions exactly according to this map, with all that is unknown made known, all injustice put to right, the good rewarded and the evil punished; mystery is explained, and the social system, once purged of corruption, is reasserted.[52]

Although Thomson's article does not discuss English crime plays in any detail, it contains a number of suggestions helpful for the purposes of the present study, particularly in respect of the conventions operating in the genre. Concerning the crook play's tendency towards political criticism and social reform, however, Thomson appears a little over-enthusiastic.

Apart from the articles surveyed above and the books by LaBorde and Carlson which were interested in the development of the crime play genre as a whole, there are a few monographs dealing with adjacent subjects that also make important observations about individual crime plays. Under the title of *Agatha Christie: Murder in Four Acts*, Peter Haining has published a full-length history of Christie's dramatic works presented on the stage, in films, on radio and TV. The book contains numerous photographs and illustrations and a certain amount of useful background information. Being intended for the general reader, however, it does not attempt any theoretical discussion of generic characteristics and developments. Studies of English melodrama, for instance Maurice Willson Disher, *Melodrama – Plots That Thrilled* (1954) and Frank Rahill, *The World of Melodrama* (1967) have also devoted some space to the discussion of mystery melodrama.

Rahill in particular presents an outline of the history of the twentieth-century murder mystery that has influenced later critics of the crime play. He dates the development of the genre from *The Argyle Case* of 1912, which "begins after the crime has been committed (a new approach),"[53] though he considers the play old-fashioned in several respects; in particular, he criticizes its episodic structure. Rahill thus establishes two critical traditions: considering the whodunit as the standard type of crime play and judging a crime play's quality by the degree to which it adheres to the Aristotelian unities. Rahill introduces the subcategories of "mystery farce" (*Seven Keys to Baldpate*), "fright-wig melodrama" (*The Cat and the Canary*), "detective story," "murder mystery," "thriller" and "study of criminological verisimilitude" and praises the improvement brought about by British examples of the genre appearing in the 1930s, in particular *The Fourth Wall* and *Rope*: "A whole new dimension in criminal types and motivations is introduced; and the dialogue has grace and wit, qualities which are sadly lacking in the mechanically slick productions of our Broadway artisans."[54] Discussing the plays of Anthony Armstrong, Rahill notes that the English practitioners of the genre departed from the whodunit structure – a decision of which he approves since the "thrill of the chase is no less keen for the quarry's being known, and the opportunities for characterization are vastly enlarged."[55] He identifies a new phase in the history of the British murder mystery beginning with *Night Must Fall* and *Gas Light*, plays that feature "homicidal maniacs." The dominance of the psychological case study leads directly to the demise of the genre: "These villains – and villainesses – are not wicked but neurologically ill, in the patter of the psychiatrist, and hence not responsible; the warrant of the moral law does not run here."[56] This view is also shared by James Smith: "When villains are pitiable delinquents, and Divine Retribution leads directly to the psychologist's couch, melodrama is dead."[57] The end of this development, according to Rahill, is the parody of the homicidal maniac play, as achieved in the phenomenally successful *Arsenic and Old Lace*. Rahill's observations on the development of the twentieth-century crime play have clearly been influential; however, the studies of Rahill, Disher and Michael R. Booth[58] are even more helpful, even indispensable, when discussing the genre's prehistory. Another very comprehensive study of crime plays before 1900 is offered by Chance Newton's *Crime and the Drama*, a monograph concentrating on versions of famous true crimes presented on the London stage, mainly in the Victorian period.

Some of the plays discussed in the present study have received critical attention not as crime plays or melodramas but in other contexts. Since the crime play is closely related to the well-made play, a number of crime plays are discussed in John Russell Taylor's *The Rise and Fall of the Well-Made Play*, for instance Galsworthy's *Loyalties*, Maugham's *The Letter* and *The Sacred Flame*, Lonsdale's *The Last of Mrs Cheyney*, Rattigan's *The Winslow Boy*, and the works of Joe Orton. However, Taylor is eager to distinguish the well-made play from the "whodunit" by remarking that "all whodunits should be well-made, but that does not necessarily mean that all well-made plays need stick at being whodunits."[59] In order to save the genre of the well-made play from critical contempt, Taylor clearly distinguishes it from the adjacent genre of the crime play, which he dismisses as a genre not worthy of serious criticism.

Not all crime plays have met with the same critical contempt. Seven of the plays discussed in this study were included in the NT2000 list of the 100 most important plays of the century.[60] The crime plays elected were *Plunder* by Ben Travers, *Night Must Fall* by Emlyn Williams, *Gas Light* by Patrick Hamilton, *An Inspector Calls* by J. B. Priestley, *The Mousetrap* by Agatha Christie, *Loot* by Joe Orton and *Sleuth* by Anthony Shaffer. Crime plays written by critically acclaimed authors like Priestley, Orton and Stoppard have received critical attention in monographs devoted to their authors.

Finally, a number of crime plays discussed in the present study belong to the movement of "in-yer-face theatre" observable in late twentieth-century British drama. The term "in-yer-face theatre" was coined by *Tribune* drama critic Aleks Sierz and became the title of his seminal monograph on the topic. The book, the term and the plays in question will be discussed in greater detail in Chapter 6.

The scarcity of secondary criticism about the crime play also extends to the field of bibliographies. In his bibliography of "detective, mystery and suspense fiction" *Who Done It?*, Ordean A. Hagen lists some 120 crime plays in his chapter on "Mystery Plays." The plays are listed alphabetically according to title, and the name of the author and publisher are given. Neither the date of publication nor the date of first production are mentioned, but the author provides a one-sentence commentary and an indication of the number of male and female actors required. This bibliography is evidently intended for (amateur) theatre companies. About 40 of the crime plays in Hagen are of English origin, but a handful of them can only be considered crime plays if applying a very liberal interpretation of the term, for instance Graham Greene's *The Potting Shed* and Michael Redgrave's *The Aspern Papers*, a stage adaptation of the tale by Henry James. Jacques Barzun's *A Catalogue of Crime* of 1971 lists two crime play collections and about seven individual crime plays according to no recognizable system whatsoever. At present, the most helpful tool for the discovery of crime plays is Allen J. Hubin's excellent, extremely comprehensive bibliography *Crime Fiction*. Novels and plays are listed together, alphabetically by author, with plays marked "play" in brackets. Hubin's bibliography is also "comprehensive" insofar as it applies very liberal standards to the definition of "crime fiction," including for instance versions of the Dracula motif. Some of the approximately 2,000 items selected from Hubin's bibliography have therefore been excluded from the present study on grounds of content.

The most recent contribution to the study of the crime play is Amnon Kabatchnik's enormous four-volume work *Blood on the Stage*,[61] evidently a labour of love. The individual volumes cover the periods of 1900–1925, 1925–1950, 1950–1975 and 1975–2000 respectively. Kabatchnik's collection provides a wealth of material on a large number of crime plays of diverse provenance, describing the plays' content, structure and performance history. There is no attempt, however, to relate the individual entries to general cultural shifts and trends either in drama or detective fiction. *Blood on the Stage* is therefore best regarded as an extremely helpful annotated bibliography of the crime play without any particular interest in generic development. As in the case of Hagen and Hubin, Kabatchnik's tastes are extremely catholic, and some of the plays he describes would not qualify as "crime plays" in the sense of the present study.

As Marvin Carlson has pointed out, detective fiction studies have largely ignored the crime play, considering it a minor derivative form. Examples might be listed at will; suffice it to say that the major collections of essays on detective fiction do not discuss stage plays at all. Reynolds/Trembley's essay collection *It's a Print* which discusses film versions of detective fiction also ignores drama entirely, and Harriet Hawkins in *Classics and Trash* analyses the novel and the film version of *Rebecca* without mentioning the stage adaptation written by Daphne du Maurier herself. Besides, critics often insufficiently distinguish between detective novel and crime play. Martin Priestman, for instance, discusses the history of detective fiction and, reaching the 1970s, suddenly comes up with examples from drama (David Hare's *Knuckle*) and television (Dennis Potter's *The Singing Detective*).[62] And German critic Ulrich Suerbaum explains the characteristic features of the detective novel by comparing it to Shakespeare's *Macbeth*, without references to the generic differences between novel and drama.[63]

In spite of this "blind spot" of detective fiction studies, detective fiction and its criticism will serve as an important point of reference in the present study. An excellent comprehensive monograph on the academic study of detective fiction is Heta Pyrhönen's *Murder from an Academic Angle*.[64] Pyrhönen notes that until the 1940s, criticism of detective fiction was mainly written by practitioners of the genre (such as G. K. Chesterton, Dorothy Sayers, S. S. van Dine and Nicholas Blake), who attempted to make it respectable by pointing out its moral message and clearly defined genre rules. In the 1940s and 1950s, there followed analyses of detective fiction on the basis of psychoanalytic theory (Pederson-Krag,[65] Lacan[66]) and discussions of detective fiction in relation to the social norms of the society that produces and consumes it (Orwell,[67] Aydelotte[68]). Since the 1960s, two main strands of detective fiction criticism have been established: studies that relate detective fiction to literature in general, and studies that relate it to specific themes and cultural preoccupations. The first strand has produced structuralist and narratologist criticism that uses the detective story as a test case for investigating narrative in general.[69] It has also brought forth investigations of the relationship between popular detective fiction and "serious" literature.[70] The second strand has produced studies on detective fiction in relation to psychology,[71] to hermeneutics and semiotics,[72] and as a reflection of the values and norms of a society.[73] This last aspect gained momentum with the inception of cultural studies in the US in the 1970s. John G. Cawelti's *Adventure, Mystery, and Romance* (1975) marks this point in the development of the critical discussion, examining the function of "formulaic fiction" for social stability. Other representatives of the cultural-studies type of detective fiction criticism are Jerry Palmer's *Thrillers: Genesis and Structure of a Popular Genre* (1978) and Stephen Knight's *Form and Ideology in Crime Fiction* (1980). From the late 1980s onwards, the cultural study of detective fiction has tended to concentrate on aspects of gender[74] and race,[75] a trend that continues to the present day.

As the above survey reveals, detective fiction studies may profit from an analysis of the crime play. While detective fiction criticism with a background in cultural studies supposes that recurrent motifs in detective novels are to be explained by certain cultural preoccupations, structuralist and narratologist analyses tend to relate standard patterns to the specific demands of narrative. A study of the crime play will permit a clear distinction between conventions imposed by dominant cultural norms, which should be shared by detective narrative and the crime play, and conventions imposed by the necessities of narrative, which should appear in detective stories though not in crime plays.

A few words need to be said about the structure of the present study. It consists of three main parts. The first main part, consisting of Chapters 2 to 6, is a history of the crime play, which is largely descriptive. This part is mainly intended to introduce the reader to a body of works virtually unknown by giving short plot synopses and tentatively relating the plays to cultural shifts – such as the emergence of television – to trends in drama and detective fiction, as well as to political upheavals. It is for instance remarkable that the crime play always saw a marked increase in productivity about ten years after an armed conflict. This first part admittedly is of interest mainly to scholars intending to do further research in the field. Readers mainly interested in the crime play as a genre and its poetics are advised to skim through these chapters and focus on Part II (made up of Chapters 7 to 11), which is more analytically inclined, seeking to isolate characteristic recurrent features and conventions of the plays surveyed in the first part. These characteristic traits are discussed in relation to the basic elements of dramatic performance – set, plot, and characters – and to the treatment of the central theme of crime and punishment. Chapter 11, the last chapter of the second part, will address the question of whether the crime play is a genre and examine its relationship to various other dramatic genres, such as melodrama and farce, and eventually propose a typology of crime plays based on the categories of central character and mode. The third and last main part will discuss the relationship between the crime play and detective fiction on the basis of a comparison between detective novels (and some short stories) and their respective stage adaptations. Part III consists of Chapters 12 and 13, the first of which is devoted to stage adaptations of narratives by Agatha Christie, while the second one examines stage adaptations of novels and stories that depart from the whodunit formula to some extent. The study is rounded off by a brief "Coda," summing up the previous results from the perspective of the twenty-first century. The bibliography lists all plays discussed in the present study together with the date and site of their first West End performance. Detective novels and plays mentioned but not analysed in depth will be listed together with secondary criticism.

Finally, a brief note on political correctness may be in order. In some cases, terms which may be found offensive by some readers have been employed for the sake of historical accuracy. The most blatant example clearly is Agatha Christie's *Ten Little Niggers*, which will be referred to by its original title rather than the cautious "And Then There Were None" of later editions, since nothing is to be gained by pretending that discrimination – racial or other – never occurred. According to the same principle, terms like "foolish female" or "invalid" have been allowed to stand where they help to reveal the underlying ideological assumptions of certain texts. This usage in no way reflects the convictions and habits of the author.

Part I
A History of the Crime Play

2
From Victorian Melodrama to the End of the Twenties

An important cultural shift in the first decades of the twentieth century is the decline and eventual demise of Victorian melodrama. In order to account for this development, critics have often related the decline of melodrama to the simultaneous rise of cinema. However, detective fiction also established itself as a major popular art form during the same period. While melodrama lost its attractive feature of spectacle to the newly evolving medium of film, it lost its attractive theme of crime and punishment to detective fiction. The dominant form of drama that was to replace Victorian melodrama was the well-made play. Within this framework of cultural shifts and trends, the crime play began to form, influenced by Victorian melodrama and detective fiction as well as by the well-made play.

In the course of its history, Victorian melodrama had developed specific conventions for representing the theme of crime and punishment, and a number of these conventions were taken up by the twentieth-century crime play. To begin with, Victorian melodrama had frequently adapted true criminal cases for the stage. The Chelsea Trunk Murder, the Wainewright Poisoning Case and the misdeeds of Dr Crippen and of course Jack the Ripper inspired successful plays. Owen Davis recalls that "if a particularly horrible murder excited the public, we had it dramatized and on the stage before anyone knew who had been guilty of the crime."[1] In the crime play of the early twentieth century, for instance Frank Vosper (*People Like Us*) and Patrick Hamilton (*Rope*) used true criminal cases as an inspiration for their plays.

The creation of the criminal protagonist of melodrama was strongly indebted to the dramatic art of Henry Irving. Irving scored his first major success playing Mathias in Leopold Lewis's *The Bells*, a melodrama of guilt and remorse, at the Lyceum in 1871. In the following years, he played Eugene Aram (1873) and Sir Edward Mortimer in his revival of George Colman's *The Iron Chest* (based on William Godwin's novel *Things as They Are; or, The Adventures of Caleb Williams*). All these plays follow "the conventional melodramatic pattern – the expiation of guilt followed by a kind of secularized apotheosis."[2] Irving's repertory comprised both villains and saints, and in Reade's adaptation of *The Lyons Mail* (1877), he played both the hero Lesurques and his criminal double Dubosc. While Irving specialized in the portrayal of tormented "tragic" villains, a new type of elegant, gentlemanly villain appeared on the West End stage in the 1880s. Sir Charles Young's *Jim the Penman* (Haymarket, 1886) showed an elegant City gentleman leading a double life as an international jewel thief, and the villain of *The Silver King* by Henry Arthur Jones and Henry Herman (Princess's, 1882) cracked safes in evening dress. On the twentieth-century stage, the character of the gentleman crook was most satisfactorily embodied by actor-manager Gerald du Maurier, who appeared as the title character in *Raffles*, based on the stories by E. W. Hornung, at the Comedy in 1906 and as Arsène Lupin at the Duke of York's in 1909. In the later history of the crime play, the character of the tragic villain tortured by remorse and the glib international criminal continued to exist side by side.

Rather than using the criminal as the play's protagonist, melodrama soon hit on the even more attractive possibility of making the hero a man accused of a crime he did not commit. In Buckstone's *Presumptive Evidence* (Adelphi, 1828), a sailor is mistakenly believed to be a murderer, because his Trafalgar medal has been found next to the body of a murdered man. The innocent accused continued to be one of the genre's favourite heroes. Peter Brooks explains this preference as follows: "melodrama typically not only employs virtue persecuted as a source of its dramaturgy, but also tends to become the dramaturgy of virtue misprized and eventually recognized. It is about virtue made visible and acknowledged, the drama of a recognition."[3] The American crime play soon used the motif of the innocent accused, particularly in courtroom drama. English crime playwrights were more reluctant to pick up the theme, but for instance in Walter Hackett's *The Barton Mystery* (Savoy, 1916), additional urgency is lent to the investigation because an innocent man is about to be executed for a crime he did not commit. The same mechanism is in operation in Percy Robinson's *To What Red Hell* (Wyndham's, 1928), which shifts between scenes from the life of the innocent prime suspect and the true culprit. Due to the popularity of the motif of the innocent suspect, a lot of melodramas contained scenes set in prison. It was not until Charles Reade's *It's Never Too Late to Mend* (Princess's, 1865), though, that a dramatist made an attempt to portray prison life realistically. Booth writes about the scenes set in prison that

> complete with treadmill, punishment jacket, and the suicide by hanging of a prisoner who had been tortured, [they] were so unpleasantly realistic that they called forth loud protests from the first-night audience. Although heroes spend much of their time in prison dramatists had previously been contented with a stock cell.[4]

Reade's play therefore is the earliest example of a treatment of crime and punishment in the form of what James L. Smith has called "protest melodrama," a trend continuing in twentieth-century prison plays. Among twentieth-century crime plays, Galsworthy's *Justice* (Duke of York's, 1910) contained a particularly impressive prison scene.

While most crime plays revolved around the character of the criminal or the innocent suspect, the detective as hero first emerged on stage in Jerrold's *Vidocq, the French Police Spy* (Surrey, 1829). Vidocq's main skill resides not in his superior powers of observation and deduction but in his talent for disguise and impersonation; in the course of the play, he variously appears as a German, a beggar and a monk. Hawkshaw, the detective of Tom Taylor's influential Victorian crime melodrama *The Ticket-of-Leave Man* (Olympic, 1863) also is mainly a master of disguise. Rahill characterizes him as follows:

> Hawkshaw's principal resources were a mastery of disguise and an impressive, if largely unexplained, ubiquity. There was nothing brilliant, dashing, subtle, or exotic about him; he was a plain, honest, and rather kindly policeman who tracked down wrongdoers, snapped handcuffs on them in the last act, and marched them off with a few well-chosen words at the final curtain. He owed his immense and lasting popularity to his very commonplaceness.[5]

The play culminated in a revelation scene, in which a drunken navvy, who has offered to help the innocently accused protagonist, drops his disguise and reveals himself to be no other than Hawkshaw himself. In the twentieth-century crime play, for instance Edgar Wallace reused the motif of demasking not only the villain but also the detective in his plays *The Terror*, *The Squeaker* and *The Flying Squad*.

The archetypal Great Detective of detective fiction, Sherlock Holmes, was put on the stage by William Gillette in 1899. Doyle himself had produced a Sherlock Holmes play for Herbert Beerbohm Tree, who turned it down. Gillette rewrote the play with Doyle's permission and starred as the title character. Since he felt that a play without a love interest is incomplete, he famously cabled to Doyle "May I marry Holmes?" and obtained the author's permission. *Sherlock Holmes* was so popular that Gillette spent the rest of his professional life performing it, and it was revived in 1974 by the Royal Shakespeare Company. In plays that were adapted from novels, the injection of an additional love interest for the Great Detective continued to be considered necessary; Michael Morton for example introduced a love interest for Hercule Poirot into his play *Alibi* (Prince of Wales's, 1928), a dramatization of Agatha Christie's *The Murder of Roger Ackroyd*.

The additional love interest is not the only change to be observed between stage adaptations of detective fiction and their respective novels. More importantly, dramatists tended to dispense with the element of mystery, following the model of William Wilkie Collins in his stage adaptation of *The Moonstone* (Olympic, 1877). "A critic (Dutton Crook) agreed with the change, since while it is allowable to perplex and mystify a reader to almost any extent 'it is found advisable to enlighten a spectator concerning the secrets of a plot at the earliest possible opportunity.'"[6] This belief in the different poetics of detective novel and the crime play was widely shared and Collins' decision to eliminate the element of mystery was imitated for example by A. E. W. Mason when he adapted his novel *At the Villa Rose* for the stage.

In the first three decades of the twentieth century, the detective novel experienced a rapid development and an unprecedented popularity, hence the 1920s have frequently been called the "Golden Age" of detective fiction. Main practitioners of the newly developing genre were A. E. W. Mason, Agatha Christie, Dorothy Sayers, Anthony Berkeley Cox, Freeman Wills Crofts and, in America, S. S. van Dine. In order to make their work respectable and eligible for literary criticism, some of these authors deliberately and explicitly laid down the rules of the new genre. To the same end, the Detection Club was founded in 1928.

Unsurprisingly, successful detective novels were soon adapted for the stage. *At the Villa Rose* and *Alibi* have already been mentioned above, to which might be added Sapper's *Bulldog Drummond*, Levy's *A Man with Red Hair* (after the novel by Hugh Walpole) and *The Ringer*, *The Squeaker* and *The Flying Squad* by Edgar Wallace. Arthur Conan Doyle dramatized his short story "The Speckled Band" and W. Somerset Maugham adapted his short story "The Letter" for the stage. Due to the large number of adaptations, the crime play has frequently been accused of being merely derivative. However, even in the first three decades of the century, original crime plays already outnumber those adapted from detective fiction.

The shift from a type of crime play still largely modelled on Victorian melodrama to a type imitating the "whodunit" structures established by Golden Age detective fiction is best demonstrated by comparing two individual plays, *Grumpy* of 1914 and *The Creaking Chair* of 1924. *Grumpy* by Horace Hodges and Thomas Wigney Percival is still strongly influenced by the conventions and plot configurations of Victorian melodrama: an innocent girl (aptly named Virginia) is in danger of being seduced by the villain and consequently losing the affection of her decent fiancé. Virginia's fiancé happens to be a courier transporting a large diamond from South Africa to London, which is stolen from him by his rival in love. The affections of the heroine thus are symbolically represented by the large exotic jewel, a motif probably derived from Collins' *The Moonstone*. "Grumpy" is the nickname of the heroine's irritable, seemingly senile (though actually rather shrewd) grandfather, a retired lawyer, who makes sure the crime is discovered and the right couple get married. The emergence of the detective character from the comic man of Victorian melodrama is still much in evidence, and the crime element is clearly secondary in importance to the love interest.

In Allene Tupper Wilkes/Roland Pertwee, *The Creaking Chair*, by contrast, the detective interest is paramount. The action revolves around the theft of an Egyptian headdress, with accompanying murder. This is the first real English crime play with a "whodunit" structure, in which everybody comes under suspicion in turn. In its subtitle, *The Creaking Chair* identifies itself as "a farcical mystery play," though most of its humour probably needs an experienced cast of actors to carry it off. Charles LaBorde has shown persuasively that Wilkes/Pertwee imitated American models, particularly of the "murder-house mystery" variety, which is hardly surprising since Pertwee also anglicized a number of American crime plays for the consumption by English audiences.

A concern shared by the detective fiction of the period and the contemporary crime play is to define the boundary between the ghost story and the tale of horror. From Poe onwards, one of detective fiction's central concerns had been to distinguish itself from the story of horror and of the supernatural. Poe himself, as will be remembered, wrote "tales of mystery and imagination," which deal with the supernatural, and "tales of ratiocination," in which mysterious and apparently supernatural events are finally given a perfectly rational explanation. Doyle's *The Hound of the Baskervilles* is a particularly good example of a detective story that approaches the borderline towards the supernatural tale of horror. The story safely returns to the realm of detective fiction by reassuring the reader that the supposedly mythical hound of the title is a real dog, heavily made up with phosphorescent paint. This is also the mechanism adopted by Arnold Ridley in his play *The Ghost Train* (St. Martin's, 1925). The "ghost train" of the title has reportedly been haunting the area – a remote Cornish railway station – ever since a disastrous train accident years ago. It is eventually revealed that it is a real train after all, used by smugglers who therefore have an interest in keeping the legend alive. *The Ghost Train* was received unenthusiastically at Brighton, but after a major advertising campaign, ran for 600 performances in London. For his next crime play, *The Wrecker*, Ridley removed the suggestion of the supernatural but retained the setting in a railway station and had his play end in a spectacular train crash accomplished with the help of a drum, a thundersheet, a cylinder of compressed air and a garden roller.[7]

In order to negotiate the demarcation line between the detective plot and the horror plot, a number of crime plays of the period made use of the contemporary craze for Spiritualism. In A. E. W. Mason's *At the Villa Rose*, an elderly woman is murdered in the course of a séance, and a bogus medium, who has been sponging off her, consequently is suspected of being the murderess. In Hackett's *The Barton Mystery*, the medium Mr Beverley assumes the part of functional detective, but it is left open if he really has psychic powers or is merely a clever charlatan. Apart from the suggestion of the supernatural, Hackett introduced yet another plot element that undermines the audience's expectations of stage realism: nearly 20 per cent of the play consists of the extended dream of one of the characters. The effect of this device is particularly destabilizing since it is never formally announced that it is a dream we are watching; the dream sequence merely starts with the dreamer falling asleep. A third play making use of the motif of Spiritualism is Bayard Veiller's *The Thirteenth Chair*, which was imported from New York in 1917 and saw 245 performances at the Duke of York's. In this play, the medium successfully summons spirits to her aid in order to clear her daughter from a murder charge.

Veiller's play was not the only American import to influence the development of the English crime play. On the contrary, the American crime play developed significantly earlier than its English counterpart, and some of its innovations were only adopted by English playwrights in the following decades. The subgenre of courtroom drama, for instance, which had brought forth plays like Elmer Rice's *On Trial* (1914) and Veiller's *The Trial of Mary Dugan* (1917) in the United States, only appeared on the London stage in the 1930s and never became quite as popular as in America.

A type of American crime play that proved more successful on the London stage was the subgenre disrespectfully dubbed "frightwig melodrama" by George Nathan, represented for instance by Rinehart/Hopwood, *The Bat* (based on the novel *The Circular Staircase* by Mary Roberts Rinehart; 1920) and John Willard's *The Cat and the Canary* (1922). Both plays transferred to London in 1922 and had runs of 327 and 181 performances respectively. These plays belong to the subgenre called "murder-house mystery" by LaBorde, that is, plays involving "an ongoing series of murders among a group of people who were confined in a secluded house or apartment."[8] "The Bat" is the nickname given to a mysterious murderer prowling the neighbourhood while "the cat and the canary" refers to the situation of the heiress in the latter play: like a canary in a cage she cannot actually be harmed by the villain but is in danger of dying of sheer fright. In both plays, the criminals deliberately try to frighten the heroines, the creation of maximum fear is more important than the quiet, unobtrusive elimination of the victim – not only to the authors but also to their villains. The subjection of the heroine to a rapid succession of situations of extreme horror thus becomes credible and justified in terms of dramatic logic. The formula of the "murder-house mystery" was later imitated by Agatha Christie in *Ten Little Niggers* and *The Mousetrap*, while the attempt to drive the heroine insane became the main motivating force in Patrick Hamilton's *Gas Light*.

As mentioned above, both American and English plays had problems crossing the Atlantic. George M. Cohan's mystery farce *Seven Keys to Baldpate*, for instance, ran for over 300 performances in New York but closed after some 50 performances in London. What is remarkable about this play is that it consciously highlights the artificiality of the conventions of the genre. The plot revolves around a writer of detective fiction who bets that he can write a mystery novel within 24 hours. In order to remain undisturbed, he withdraws to Baldpate Inn, to which he holds the only key. As the title suggests, however, six more keys are in circulation, and the writer receives several unwelcome visits – from a gangster, a hermit posing as a ghost, some corrupt city officials and from a lovely lady journalist. At the end of the second act it is revealed that all the events of the night were staged by actors to prevent the writer from finishing his novel. This final reversal is subjected to a further reversal when it turns out, in the epilogue, that everything that was shown on stage was nothing but the novel the author has been writing, thereby winning his bet. The device of framing the play's main action by a subplot in which the main plot is invented by the playwright or rehearsed by the director was imitated on the London stage by Frank Vosper in *Murder on the Second Floor* (Lyric, 1929) and A. A. Milne with *The Man in the Bowler Hat* (Prince of Wales's, 1925). It is perhaps surprising that in this early phase of the formation of the genre playwrights should already be beginning to undermine generic conventions by pointing out their artificiality. However, something very similar happened in the field of detective fiction. E. C. Bentley's *Trent's Last Case* (1913, like *Seven Keys to Baldpate*) acquired the status of one of the earliest classics of the "Golden Age" English detective novel, in spite of the fact that it demolished generic conventions rather than helping to establish them.

As stated at the outset of this chapter, with the demise of melodrama, the well-made play became the dominant popular genre on the London stage. The "well-made play," however, is an umbrella term covering a variety of different subgenres, most importantly the problem play, drawing-room comedy and farce. In each of these subgenres, dramatists successfully experimented with the possibilities opened up by the crime play.

The most substantial body of "well-made" crime plays in the above period was produced by John Galsworthy, who between 1906 and 1926 wrote four problem plays about questions of justice and the Law. The earliest of these plays, *The Silver Box* (Royal Court, 1906), which Galsworthy ironically subtitled "a comedy," is a social melodrama dealing with legal injustice as a result of social position. The play only had a short run but was a success with the critics.

Galsworthy's next drama, subtitled "a tragedy," was *Justice* (Duke of York's, 1910). Falder, a junior clerk in a solicitors' office, forges a cheque, but his crime is soon discovered by his employers and senior clerk Cokeson. The second act of the play shows the trial against Falder, in which his young and inexperienced defence counsel unsuccessfully tries to get him off on grounds of temporary insanity. Act III is set in prison, and the silent scene III.iii, showing Falder in solitary confinement, is reported to have been almost unbearably moving. The last act is set in the solicitors' office again: Falder has been released from prison and comes to his former employers for help. They are sympathetic, but when Falder finds himself threatened with renewed arrest for failing to report to his parole officer, he jumps to his death. *Justice* was the rare case of a play that actually brought about legal reform: Winston Churchill was so impressed by it that he initiated a long overdue reform of penal practice. Critics have mainly praised the play's scrupulous fairness:

> Galsworthy is not trapped into a cheap antithesis; his prison officers are decent chaps with the best intentions, and his convict is a moral weakling sent down for petty forgery. The wordless scene of Falder's solitary confinement harrows not because he is a victim of grotesque injustice but because we feel no penal system ought to treat a man this way.[9]

John Russell Taylor considers Galsworthy's fairness as an aspect of the "well-madeness" of his plays: "Galsworthy believed passionately in truth: in the scrupulously fair presentation of both sides in any conflict, the undeviating arguing-out of any issue to its logical conclusion. [...] His is above all a theatre of decency, where fair-play and other gentlemanly virtues reign uppermost."[10] In spite of all this highly praised fairness, there is no doubt that audience sympathies are with Falder and that the character of Cokeson is deliberately established as a model of humane behaviour.

Galsworthy's third play on crime and punishment has also been praised as an example of supreme fairness: in *Loyalties* (St. Martin's, 1922), "he balances his sympathies so neatly that there can be no suspicion that he has rigged things to favour one view or another. If we ask where our own sympathies should lie the only possible answer is, with everybody, suffering as they do in a fallen world."[11] The action begins with a weekend in the country. One member of the country-house party in a rather daring stunt robs a rich social outsider – a Jew – and is at first protected by the other members of his set – until the trial, when suddenly the professional ethos of the lawyers' caste gains the upper hand over class loyalty.

The fourth and last play is *Escape* (Ambassadors, 1926), an episodic play describing the progress of an escaped convict through the country, revealing the attitudes of the diverse members of society he encounters. James Gindin recognizes a veiled self-portrait of Galsworthy the social reformer in the character of the protagonist, summing up the play as follows:

> The play becomes, in tone, something of a picaresque comedy, and its resonance has much less to do with implicit commentary on the law or society than it does with the half-mocking and half-admiring tone in which the gentle knight-errant, who will lose rather than violate his own standards, is depicted.[12]

The possibility of a successful combination of drawing-room comedy with crime play elements was best demonstrated by Frederick Lonsdale's *The*

Last of Mrs Cheyney (St. James's, 1925). The title consciously alludes to late nineteenth-century problem plays around the "Woman with a Past," as for instance in Pinero's *The Second Mrs Tanqueray*, while the name of the protagonist may be a deliberate echo of Wilde's Mrs Cheveley in *An Ideal Husband*. Mrs Cheyney poses as a respectable Australian widow, but she and her servants are really a gang of international jewel thieves. However, they are so smitten by the charm of the British upper classes that they start feeling scruples about robbing them. Lord Dilling, a disreputable but brilliant man-about-town, recognizes Mrs Cheyney's "butler" Charles as a notorious criminal, but since he has fallen in love with Mrs Cheyney, he simply crosses the gang's plans by exchanging rooms with their intended victim. Dilling gives Mrs Cheyney "the choice of sleeping with him or being handed over to the police. She chooses the only course a middlebrow audience could approve – to be turned over to the police."[13] The gang escapes arrest because they are in possession of a letter written by Lord Elton, a pompous ass, that reveals his opinion about his class. When the gentry attempt to buy the letter back from Mrs Cheyney, she reveals that she is not a blackmailer: she has already destroyed the letter, which makes her eligible for marriage to Lord Dilling.

The attempt to relieve the idle rich of their superfluous jewellery is also at the centre of *Plunder* (1928), one of the "Aldwych farces" produced by Ben Travers between 1922 and 1933, always with the same team of actors.

> Travers's theory of farce revolves around his belief that to be funny, characters must be recognizable to the audience as types with whom the audience can identify. The humour arises from outrageous, unfamiliar, and sometimes dangerous situations in which the characters find themselves. Travers's plays are *farces of suspense*.[14]

Plunder starts with a gentleman crook and his supposed "sister" who manage to persuade a boyhood friend of his to take part in a jewel robbery. The victim has no claim on the audience's sympathies since she herself is a ruthless social climber who married a much older man for his money, becoming a wealthy widow soon after the wedding. Since the crooks stick together in the end, Scotland Yard has no means of pinning the crime on them. To Travers as to several later practitioners of farce, the crime plot must have meant first of all a welcome change from the ubiquitous theme of adultery.

As pointed out above, the rise of the well-made play coincides with the rise of the cinema. But since films were silent before the late 20s, they did not endanger for instance the genre of drawing-room comedy, which relies mainly on sparkling dialogue. Since the crime play also largely relies on dialogue, it was comparatively safe from the competition of the film industry, even though crime and detection is obviously an attractive theme for cinema. Of the crime plays produced before 1930, about 50 per cent were later made into films, but they only became available as raw material for the film industry with the advent of talking films in 1929. 1929 saw the release of the first of several film versions of *Bulldog Drummond, The Last of Mrs Cheyney* and *The Letter* respectively. All these films were American; the only British film based on a crime play released in 1929 was Alfred Hitchcock's *Blackmail,* based on a play by Charles Bennett. The play, first performed at the Globe in 1928, was only moderately successful, but the film version is still occasionally shown due to the reputation its director acquired in the course of his career.

The plot of the play runs as follows. A shop girl stabs an artist who has attempted to rape her, and the murder is covered up by her policeman boyfriend. The two lovers are then blackmailed by a witness, but by a stroke of poetic justice, suspicion of the murder alights on the blackmailer himself. The girl considers giving herself up to the police, but it then conveniently turns out that the victim died not of the wounds inflicted by her knife but of heart failure. In the film version, there are no expiatory circumstances: the girl is still guilty of manslaughter but will not be convicted, since the blackmailer has fallen to his death during a police chase through the British Museum. Her policeman boyfriend will marry her, a scene filmed as if he were arresting her. The film is famous for its shots of the chase through the British Museum and its use of expressionistic sound effects, but to those interested in the development of the English crime play, it is above all a significant document of the acting style of the period. After *Blackmail,* Bennett went on to write screenplays for Hitchcock but produced no more crime plays for the stage.

Crime plays that were the basis of successful films frequently became entirely overlaid in the public memory by their respective film versions. This happened for instance to Patrick Hamilton's *Rope* (Strand, 1929), which was also filmed by Alfred Hitchcock and in today's criticism is customarily referred to as "Hitchcock's *Rope,*" with authorship being tacitly attributed to the director, not the playwright.[15]

The invention of film, and the advent of talking film in particular, clearly had an important impact on the genre. The main political event of the period presently under survey, however, the Great War of 1914–1918, had comparatively little direct influence on the crime play, with the exception that, as J. C. Trewin put it,

> German spies began to arrive in play after play, and one hugely successful piece, *The Man Who Stayed at Home* (1914), by Lechmere Worral and J. E. Harold Terry, proved to be as good in a schoolboy's-annual fashion as anything of its kind during the war. It had plenty of rivals, Austin Pages's *By Pigeon Post* (1918) for one.[16]

The Man Who Stayed at Home is set in a small hotel in East Anglia that is also the headquarters of a group of German spies. The detective is a young man of the "silly ass" type who turns out to be unexpectedly smart and, what is more, a member of the Intelligence Department, comparable to an earlier fictional hero, the Scarlet Pimpernel of the novels by Baroness Orczy. The Germans speak broken English among themselves, and gadgets include a fireplace containing a wiring station – a very attractive feature which may have prompted Trewin's comparison with schoolboy annuals.

The main impact of the Great War, however, became evident some ten years after its end, when the genre of the crime play experienced a sudden marked peak in productivity. As a look at the chronology of crime plays reveals, two thirds of the crime plays written before 1930 were produced between 1926 and 1929. After the end of World War II, a similar phenomenon is to be observed: in the postwar period, the crime play once again became extremely popular. The increasing interest in plays about crime and detection in the postwar periods may perhaps be accounted for by the population's longing for the re-establishment of "law and order," and of pre-war standards of propriety and safety.

The increase of productivity in the genre from 1926 onwards may also be related to the emergence of Edgar Wallace as a playwright. As a playwright, just as previously as a novelist, Wallace was prodigiously productive, writing the crime plays *The Ringer, The Terror, The Squeaker, The Flying Squad, The Calendar, On the Spot, The Case of the Frightened Lady* and *The Green Pack* all in the six years immediately preceding his death. Three of these plays are stage adaptations of novels he had previously written: *The Ringer, The Squeaker,* and *The Flying Squad*. Of Wallace's original crime plays, *The Case of the Frightened Lady* (Wyndham's, 1931) is the most interesting. The play transferred to New York under the title of *Criminal at Large* where it had 161 performances, 30 fewer than in London. The criminal is the helpless and charming young Lord Lebanon, who turns out to be a homicidal maniac strangling his nearest and dearest. His mother does her best to protect him from the consequences of his actions. She wants him to marry Ailsa, a distant relation and the "frightened lady" of the original title, in order to continue the family line. There is a complete reversal at the end of the play when the apparent victim turns out to be the criminal and his two menacing footmen are revealed to be his carers. Instead of a sudden surprise revelation of the identity of the culprit, Wallace chose to have the truth gradually dawn on the audience, as LaBorde has pointed out:

> Wallace provides many fair but not immediately transparent clues at first. Obviousness of clues then increases in the later scenes. Similarly, the murderer's on-stage behavior gradually alters until his insanity becomes apparent. Thus, by the moment the unwitting characters realize the solution to the mystery, it has been clear to the audience for some time. The author allows his audience the unique (for mystery plays) pleasure of discovering the identity of the villain for itself.[17]

And, it may be added, Wallace thus manages to combine the pleasures of the whodunit with those of the suspense thriller.

Another playwright who made a significant contribution to the genre of the crime play within a comparatively short period of time was Frank Vosper, who produced three classic crime plays within seven years, *Murder on the Second Floor* and *People Like Us* in 1929, and *Love from a Stranger*, an adaptation of Agatha Christie's short story "Philomel Cottage," in 1936. *Murder on the Second Floor* is set in a Bloomsbury boarding house, which serves as a cover for a dope-smuggling ring. The rather sensational plot contains a suicide and a particularly gruesome murder – one of the criminals is stabbed in the throat by a jealous husband – but the violence of the play is contained by a framing device: a playwright resident in the boarding house is trying to invent a play on the spur of the moment, an idea obviously borrowed from *Seven Keys to Baldpate*, though Vosper's play is both more neatly constructed and funnier.

People Like Us likewise shows imagination and reality becoming intertwined, contrasting the drab ordinary life of its lower middle-class characters with the world of lurid prefabricated romance. The play is considerably more serious than its predecessor, demonstrating how popular role models and masterplots of popular fiction may shape people's perception of their own lives. Vosper based the plot of this play on a true criminal case, the Thompson/Bywaters trial, but since Thompson's relatives objected to a dramatization of the case, the play was banned by the Lord Chamberlain and saw a mere 18 performances at the Arts Theatre Club. In the marriage of the Thompsons in *People Like Us*, a disturbing note of sado-masochistic mutual dependence may be discerned, an element even more in evidence in the marriage of Bruce and Cecily in Vosper's later play *Love from a Stranger*. Whether Vosper took his cue from Cohan, from true crime, or from Christie, his own dominant interests and concerns always asserted themselves in the finished plays.

Apart from the plays by Wallace and Vosper, the 1920s saw the production of two more crime plays that have elicited univocal critical praise. A. A. Milne's *The Fourth Wall* (Haymarket, 1928) has mainly been appreciated for its lightness of touch and the elegance of its style, yet it contains a particularly brutal and cold-blooded murder committed on the open stage – a potentially disturbing combination. Years ago, Arthur Ludgrove helped to send two criminals to prison for life. They have now returned to get revenged on Ludgrove, who fails to recognize them. One of the two men attacks Ludgrove, while his accomplice – a close acquaintance and house guest of Ludgrove's – asks the victim to call the police and then suddenly forces him to shoot himself with the police as earwitness at the other end of the line. Ludgrove's ward Susan serves as the functional detective and solves the murder case. While in the first act we see the murderer manipulating his victim, in the third act we see Susan manipulate the murderer. LaBorde notices the major reversal characterizing the plot structure:

> In the Milne drama Laverick and the less obviously villainous Carter lay their trap for Ludgrove, the victim. After they catch and kill their prey, the direction of the plot fully reverses itself so that the former pursuers, Laverick and Carter, become the pursued, the objects of the detectives' search.[18]

The cat-and-mouse game between villain and female detective recalls Wilkie Collins' Fosco and Marian, a parallel recognized by Frank Rahill, who praised Milne's murderer Carter as "not unworthy of standing alongside that other fat villain, Count Fosco of *The Woman in White*."[19] The play's title is explained in the initial stage direction, which informs us that the set is "Arthur Ludgrove's private sitting-room at Heron Place, Sussex, through the fourth wall of which we see what happened,"[20] a self-consciously metatheatrical reference.[21]

When Ordean A. Hagen included *The Fourth Wall* in his annotated bibliography of crime fiction *Who Done It?*, he listed the play under its American title of *The Perfect Alibi* and added the one-sentence commentary: "We know at once whodunit, but how will they be caught?"[22] This short description of the mechanics of plotting and suspense would also apply to Patrick Hamilton's *Rope* (Strand, 1929). Like Milne, Hamilton revealed the identity of the culprits at the outset, creating suspense by making the audience wonder if (and how) they will be caught. Hagen describes *Rope* as "a thriller based on the Loeb-Leopold case,"[23] but Hamilton claimed that he had not heard about this famous murder trial until later. Two young homosexual men murder a fellow student for the sheer excitement of it and, for an extra thrill, invite the victim's family to have dinner with them from a chest containing the corpse. Another fellow student, formerly himself a staunch advocate of Nietzschean categories of supermen and subhumans, discovers the murder and, to their surprise, hands the criminals over to the police. In his preface, Hamilton calls *Rope* a "horror play" rather than a "thriller" and discusses the dynamics of disgust. Throughout his work in the detective genre, Hamilton specialized in the portrayal of cases of insanity, as in his later play *Gas Light* and his excellent novel of schizophrenia, *Hangover Square*. Hamilton's choice of homicidal maniacs as his culprits and his decision to reveal the identity of the criminals early in the play contributed to establishing two major conventions of the crime play genre. In the following decades, for instance Emlyn Williams successfully employed the mentally deranged murderer as protagonist in *Night Must Fall*, and Anthony Armstrong presented both the crime and its investigation in their chronological order in *Ten-Minute Alibi*.

3
From the 1930s to the End of World War II

In the field of detective fiction, the 1930s marked the beginning of the second phase of the "Golden Age." The genre had become respectable in the previous decade; now it was becoming increasingly literate as writers like Michael Innes (Oxford don J. I. M. Stewart) and Nicholas Blake (poet Cecil Day Lewis) started to contribute, albeit under pseudonyms. Symons, however, also notes that a reaction against the established conventions set in. The most violent attack against the established generic rules was carried out by the American writers who had come in from the pulps and whose cause Raymond Chandler eloquently championed in his essay "The Simple Art of Murder." Another American writer, though of a somewhat more conservative persuasion, John Dickson Carr, reacted to the established conventions by allowing his characters to self-consciously discuss the generic rules of the fictional universe in which they moved. A particularly good and much quoted example is Dr Gideon Fell's lecture on locked-room mysteries in *The Hollow Man* (1935). The detective novelist Anthony Berkeley Cox assumed the pseudonym of Francis Iles and established what has become known as "inverted" detective fiction: the criminal is known from the start, and the reader follows his actions rather than the investigations of the detective. As has been pointed out in the preceding chapter, however, generic self-consciousness (as in *Murder on the Second Floor*) and a criminal revealed to the audience at the outset (as in *Rope*) were nothing new in the format of the crime play.

While Anthony Berkeley Cox took up a new pen name to experiment with the formula of detective fiction, Agatha Christie was mainly concerned with perfecting the technique she had developed so far – the famous "Christie sleight of hand." In 1930, she tried her hand at the original crime play *Black Coffee*, which was performed at the St. Martin's. Hercule Poirot investigates the case of a "purloined letter" and the murder of an eminent scientist. Christie was not satisfied with the way the character of Poirot had been presented on stage and consequently eliminated him from her later plays. Ten years later, Arnold Ridley adapted Christie's novel *Peril at End House* for the stage (Vaudeville, 1940), with moderate success. Christie herself dramatized *Ten Little Niggers*, which opened at the St. James's in 1943 and had a most satisfactory run of 260 performances, even though it had to be moved to the Cambridge for a few months since the St. James's had been hit by a bomb. For her next effort, a dramatization of *Appointment with Death* (Piccadilly, 1945), Christie reworked her material thoroughly, even to the point of changing the identity of the murderer. In spite of her efforts, this play turned out to be noticeably less successful than *Ten Little Niggers*.

While Christie continued to write in her chosen genre, Dorothy Sayers' series of detective novels featuring her aristocratic detective Lord Peter Wimsey reached its natural end in her Great Detective's marriage to his some time Watson, Harriet Vane – marriage being notoriously fatal to serial detectives. However, Lord Peter made his last appearance not in a novel but in a stage play which was later adapted as a novel. *Busman's Honeymoon*, co-scripted by Muriel St. Clare Byrne, opened at the Comedy in 1936. Both the novel and the play were published by Gollancz in 1937. Harriet and Peter are on their honeymoon and move into a cottage whose owner has disappeared. The body is found in the cellar with the skull broken by a blunt instrument. Through reconstruction they manage to prove that the old man was killed by an ingenious trap consisting of a large brass pot with a cactus hauled up by a fishing line. This means of course that the murder weapon is in sight throughout the entire play. The trap is actually shown in operation, almost bringing down the criminal. He is caught after having lost audience sympathies by his cruel treatment of his victim's niece, an ageing spinster he has promised to marry for financial gain. What is perhaps as interesting as the play itself is the "Authors' Note" included in the published version, in which Sayers and Byrne distinguish between the three subgenres of "the true detective problem," "the thriller" and "the psychological crime story" and insist on observing the "fair play rule," which was gradually being established in detection narrative during the 1930s.[1] Sayers/Byrne experiment with the fair play formula on the stage, where, as they claim, it has so far been strangely lacking. "It was necessary to invent a technique to express this formula, since the novelist's approach by argument and explanation is clearly unsuited to the stage."[2] The clues are largely visual ones, appearing in Act I, while the dialogue serves mainly to distract the spectators' attention. In Act II, the inquiry is directed towards the question of motive, whereas the information actually imparted to the audience concerns opportunity. Act III conveys information as to motive, while the audience's attention is directed to the emotional implications of the situation. As the subtitle of the novel version, "a love story with detective interruptions," suggests, the love interest is central to the plot. In this context, Sayers/Byrne observe that the influence of the comedy of manners has already invaded detective fiction. In the final sentence of the Authors' Note, Sayers and Byrne express the hope that *Busman's Honeymoon* may start off a new trend in the genre.

"The future development of the detective play may lie in this direction, [since] neither sensation without thought nor argument without emotion can ever provide the basis for any permanent artistic structure."[3] However, the play failed to establish a new tradition, partly because Sayers herself lost interest in the genre.

In the context of the relationship between detective fiction and the crime play, a few other stage adaptations of detective fiction ought to be mentioned. In 1931, Jeffrey Dell adapted C. S. Forester's *Payment Deferred*, an inverted detective novel, for the St. James's, and in 1940, Daphne du Maurier dramatized her neo-Gothic novel of terror, *Rebecca*. Evadne Price, a lesser known author, wrote a crime play called *The Phantom Light* for the Haymarket in 1937, based on her own novel *The Haunted Light*. The principle is the same as in Ridley's *The Ghost Train*: a local superstition, here surrounding a lighthouse, is used as a cover for illegal activities. The functional detective is an emergency lighthouse keeper called Sam, a comic character prone to malapropisms. This part was played by Gordon Harker, who had created the Cockney petty criminals in Edgar Wallace's crime plays. Price's next crime play *Once a Crook* (co-written by Kenneth Attiwell, Aldwych, 1940) was also intended as a vehicle for Harker. This is the rare case of an English gangster play, and the influence of American models of crime fiction and film is much in evidence. Most conspicuously, the play's master criminal "The Duke" is an American, described in the stage directions as looking like the actor Edward G. Robinson. This indicates that the hardboiled school of American writing had influenced English writers and their audiences mainly via the medium of cinema.

It has been stated above that since generic rules in detective fiction were now firmly established, certain authors began to question and undermine these rules. A similar trend may be observed in the genre of the crime play: a number of 1930s' crime plays questioned the values and traditions of the genre altogether by violating basic genre rules. Three plays which attack the genre's conventions in different ways are Mordaunt Shairp's *The Crime at Blossoms*, John van Druten's *Somebody Knows* and *Murder in Motley* by Ingram d'Abbes and Fenn Sherie.

Shairp's *The Crime at Blossoms* (Embassy, 1931) criticizes the public's morbid fascination with "true crime." Because they have run into debt, Valerie and Christopher Merryman have had to let their Elizabethan cottage "Blossoms" to a stranger. In their absence their tenant is murdered, and Valerie soon realizes that they can benefit from the crime financially by showing the scene of the crime to sensation-hungry visitors. She even goes so far as to perform a ghastly "reconstruction" of the murder, getting the essentials of the matter all wrong. This is revealed when the real culprit comes forward and confesses the truth. Shairp attacks the conventions of the sensational type of the crime play while at the same time employing its devices: in *The Crime at Blossoms*, suspense is created by keeping the protagonists and the audience in the dark about the identity of the criminal and the events that led to the tenant's death for a considerable time.

In John van Druten's *Somebody Knows* (St. Martin's, 1932), by contrast, the identity of the criminal is never revealed with complete certainty. Lily, a streetwalker, has been taken into the community of a boarding house as the fiancée of the unsuccessful actor and musician Lance Perkins. When she is murdered, Lance is the prime suspect, but he is acquitted on grounds of insufficient evidence. Instead of focusing on the criminal, the play concentrates on the responses of the people around him, and van Druten has been praised in particular for the creation of the female characters surrounding Perkins. Commercially the play was not a success, since the audience apparently felt duped by being led to expect a murder mystery and then denied a conventional solution. *Somebody Knows* undermined one of the crime play's – and detective fiction's – basic assumptions, namely that absolute truth exists and may be discovered by quasi-scientific means.

The "comedy thriller" *Murder in Motley* by Ingram d'Abbes and Fenn Sherie (Royalty, 1934; revived Fortune, 1952) is based on the principle of Cohan's *Seven Keys to Baldpate* and set in actual place and time: the theatre on the evening of performance, as the initial stage direction specifies: "[T]he true names of the actors and actresses taking part in the production should be used. [...] Similarly the name of the theatre in which the play is being given, and the names of local buildings and streets should be substituted where necessary."[4] In the course of the first act of a drawing-room comedy the leading lady's understudy is shot, apparently from the auditorium. The police are called in, and the investigation gets increasingly complicated when the body vanishes[5] at the end of Act I and the stage manager O'Bryan starts receiving ominous phone calls. In the end, the corpse returns alive and well, and it is revealed that the whole murder mystery was nothing but a clever publicity stunt invented by O'Bryan who had realized that the drawing-room comedy they had on their hands was too boring ever to be a success. This resolution, which may delight critics because of its Pirandellian repercussions, may actually infuriate crime play addicts who, in adherence to the "fair play rule," expect to find the loose ends of the plot cleared up.

Another tendency observable in the crime play of the 1930s is an increased interest in questions of justice and the law. Previously, as will be remembered, only Galsworthy had shown a pronounced interest in the theme, possibly because of his own legal training, and the genre of courtroom drama, which flourished in America, had not been imitated by English playwrights. In 1931, however, Michael Barringer's *Inquest* opened the new Windmill Theatre. *Inquest* is a genuine mystery drama, which observes the unities of place, time and action and has a surprise ending providing a complete reversal of audience expectations. The play is set in an improvised coroner's court in a village school, and its plot is based on an exchange of identities: the battered dead body is not that of the presumed victim but that of the supposed culprit. Edward Wooll's courtroom play *Libel!* (Playhouse, 1934) likewise revolves around an exchange of identities. Sir Mark Loddon has to prove that he is indeed Sir Mark, who has returned home from World War I suffering from severe amnesia, and not a daring impostor who has done away with the real Sir Mark. The central idea of the plot was probably derived from the case of Martin Guerre reported in Pitaval's *Causes célèbres*. What makes this plot so appealing is the underlying debate of the problem of identity: how does a man prove that he is himself? In *Libel!*, two different concepts of identity combine: the nineteenth-century concept, according to which for instance Sir Percival Glyde in *The Woman in White* is "not himself" in the sense of having no claim to the title, and a modern concept of psychological integrity. Sir Mark has lost his sense of self through the trauma of war, shell shock has deprived him of his memories. It is only when he is confronted with another potential self – a character called merely "Number 15," having lost all claim to humanity through being reduced to mere animal functions – that Sir Mark regains his sense of identity. In both plays, the truth about the innocent and the guilty is revealed at the end of the courtroom action.

A number of playwrights of the 1930s, however, seemed more sceptical concerning the ability of the legal institutions to actually dispense justice and preferred to show individuals taking the law into their own hands: Edgar Wallace in *The Green Pack*, James Parish in *Distinguished Gathering*, and Alec Coppel in *I Killed the Count!*.[6] The title of *The Green Pack* (Wyndham's, 1932) refers to a pack of cards used in a rather sinister lottery: three men who have recently discovered a gold mine draw lots to determine which of them is to kill the man who tried to rob them of their earnings. In *Distinguished Gathering* (St. Martin's, 1935), a publisher gives his guests the opportunity to stab one of his authors in the blacked-out living-room, in order to prevent the publication of a book that would disclose disreputable secrets about each of them. From play to play, plot complexity increases: the murder in *Distinguished Gathering* is more ingeniously planned than the murder in *The Green Pack*; and the murder of Count Mattoni in *I Killed the Count!* (Whitehall, 1937), is very intricately plotted indeed. After Mattoni has been found murdered in his flat, Detective Raines finds abundant clues that point to three different suspects – each of them, unfortunately, confesses to having killed the Count. It turns out that the three suspects conspired to kill Mattoni and deliberately left behind clues implicating each of them, because "two or more persons cannot be charged as principals with a crime known to have been committed by only one person." They drew lots to decide who was to do the actual killing, but a fourth person interfered: Mattoni's wife, who really did murder the Count, but the ingenious device invented by the three conspirators protects her as well. Mattoni is a typical thoroughly unpleasant victim, and since the play employs flashbacks to illustrate each of the suspects' versions of how they killed the Count, he is murdered over and over again on stage.

The most significant individual author of crime plays in the period presently under scrutiny is J. B. Priestley, whom Courtney considers "by far the most important serious dramatist working in the commercial theater."[7] Priestley, an immensely productive "middlebrow" author, wrote in a variety of dramatic genres and his crime plays *Dangerous Corner* (Lyric, 1932) and *Laburnum Grove* (Duchess, 1933) as well as the later *An Inspector Calls* are among his most enduring successes and were revived later to popular and critical acclaim. What is most remarkable about Priestley as a crime playwright is his stout refusal to fasten guilt on one individual character who may then be cast out, in the manner of a scapegoat, to purge society from sin. And as W. H. Auden and Nicholas Blake have argued, this is precisely what detective fiction normally does. Auden's famous formula for detective fiction, laid down in his seminal essay "The Guilty Vicarage" reads: "a murder occurs; many are suspected; all but one suspect, who is the murderer, are eliminated; the murderer is arrested or dies."[8] In Priestley's serious crime plays, *Dangerous Corner* and *An Inspector Calls,* this formula is subverted. A suicide occurs; nobody is suspected, but gradually it is revealed that everyone around is guilty to some extent; nobody is arrested or dies; life goes back to normal.

Dangerous Corner, for instance, begins with an after-dinner conversation. A closely knit group of relatives and friends happen to hit on the topic of a recent suicide in their midst. It has been taken for granted that the deceased, Martin Chatfield, killed himself because he had embezzled money from his brother's firm. Now it turns out that the money was stolen by one of the company's managing directors, who cleverly tricked Martin and Robert Chatfield into suspecting each other. The logical conclusion that Martin killed himself because he could not bear to discover that his adored brother was a thief, however, is mistaken – there are further revelations in store. Olwen, a secretary, confesses that Martin did not shoot himself but that she killed him in self-defence when he attempted to rape her. She also reveals that Martin was a drug addict. Unable to cope with the shattering news, Robert rushes to his room intending to kill himself. The catastrophe is averted, however, because at this point of the action a "time-loop" is introduced: the action reverts to the beginning of the first act, where this time a chance remark in the conversation does not lead to further discussion of Martin's death; a "dangerous corner" has been passed unharmed. The experiment with the play's time structure relates *Dangerous Corner* to other plays by Priestley, most notably *Time and the Conways* and *I Have Been Here Before,* both of 1937.

In contrast to *Dangerous Corner, Laburnum Grove* is a comedy, "an immoral comedy," so the subtitle specifies. George Radfern, a respectable, if somewhat boring, middle-class businessman, actually runs a successful counterfeiting ring. He confesses the truth to his family and escapes police detection in the nick of time. However, some doubt remains as to whether he really is a criminal or just invented his criminal career in order to impress his family and study their reactions. These doubts do not affect audience sympathies: even if Radfern is a forger, audience sympathies are with him rather than with the police investigating the case. This may be due to the nature of the crime: forgery is a crime sufficiently abstract not to alienate middle-class audiences. Priestley later claimed that the choice of counterfeiting as the central crime also carried a political message: "At the time I wrote it, when I was also gathering material for *English Journey*, I was very suspicious about our financial system, if only because the banks appeared to flourish when industry was failing,"[9] he writes in the preface to his collected plays. Retrospectively linking *Laburnum Grove* to *English Journey*, Priestley invests it with a more acid social criticism than is discernable when reading the play. In spite of its author's attempt to associate *Laburnum Grove* with the more overtly political *English Journey*, it remains set cosily in its suburban living-room, as befits a well-made play.[10]

Apart from J.B. Priestley, who is usually considered a successful commercial author with a few crime plays to his name, the 1930s also saw the emergence of a number of crime playwrights whose reputation today rests mainly on their crime plays: Anthony Armstrong, Emlyn Williams, Barré Lyndon, Roland Pertwee, and the team consisting of Edward Percy and Reginald Denham. All these authors experimented with the genre of the crime play for some time but produced only one major success.

In the case of Anthony Armstrong, the major success was his *Ten-Minute Alibi* (Haymarket, 1933), which ran for over 800 performances and was later adapted by its author as a novel and made into a film. In his previous crime play *Well-Caught* (Strand, 1929), Armstrong had experimented with the whodunit structure and the genre of the "criminal comedy," as the subtitle calls it. For *Ten-Minute Alibi*, Armstrong largely dismissed the element of comedy and, more importantly, also eliminated the element of mystery. The identity of the culprit is revealed from the start, as for instance in the earlier crime plays *The Fourth Wall* and *Rope*, or in the inverted detective novels of Francis Iles that had just appeared. Trussler comments on the play: "Anthony Armstrong's *Ten-Minute Alibi* (1933) had taken the dramatic thriller about as far towards applied mathematics as it could be expected to go."[11] By "applied mathematics" Trussler is referring to the construction of the criminal's alibi, which is based on fiddling with a clock and in the third act leads to a lot of involved arguing about tampering with the exact time of the murder. A mathematically inclined mind is also in evidence in the neat construction of the play: the first act shows the planning of the crime, the second its execution, and the third act the investigation. The crime that is committed is a murder the audience is meant to approve of: the victim is a seducer and pimp involved in white slave traffic, and the murderer is the lover of an innocent girl who is in the seducer's thrall. Audience sympathies are manipulated in such a manner that, by Act III, spectators are torn between the desire to see the police discover the truth and the hope of seeing the murderer escape. Both wishes are gratified: the police do discover the truth, but the criminal is allowed to get away with his crime nevertheless.

Armstrong wrote two more crime plays, which, however, did not achieve the same success as *Ten-Minute Alibi*. In *Without Witness* (Duke of York's, 1933), which was co-scripted by Harold Simpson, the main interest resides in the investigation method: a re-enactment of the crime which brings out the truth. *Mile-Away Murder* (Duchess, 1937), by contrast, is attractive because of the ingenious murder method involved: the culprit murders his brother by poisoning the back of a stamp, which will be posted back to him so he can destroy the evidence. Judging from the comparative lack of success of the two later plays – they clocked up runs of 72 and 42 performances respectively – neither an ingenious murder plan nor a clever investigation technique is in itself sufficiently interesting to hold audiences spellbound; the two must be combined as in *Ten-Minute Alibi*.

While Armstrong's plays mainly appeal to their spectators' intellectual faculties, the crime plays of Emlyn Williams aim at the audience's gut reactions – they set out to create maximum terror. Williams's first attempt in the genre, *A Murder Has Been Arranged* (Strand, 1930) relies on supernatural intervention for its solution: the ghost of the murdered man returns to the scene of the crime to avenge his death. More interestingly, the play is set in the actual place and time, the theatre on the evening of performance. Williams probably borrowed this device from American plays like *Cock Robin* and *The Spider* and in turn inspired Ingram d'Abbes and Fenn Sherie's *Murder in Motley*.

Williams had started his theatrical career as an actor and played for instance the homicidal young lord in Wallace's *The Case of the Frightened Lady*. He recreated the part of the boyishly charming psychopath in his own play *Night Must Fall* (Duchess, 1935). Williams played the character of Dan, a bellhop who has murdered a hotel guest and carries the victim's head around in a hat box. The play begins with a prologue in court, where Dan's appeal for a reprieve has just been dismissed, and then, in a flashback, opens on a domestic scene with Olivia, a spinsterish paid companion, reading *East Lynne* to her rich hypochondriac old aunt Mrs Bramson. Enter Dan, a baby-faced youngster, who quickly manages to worm his way into Mrs Bramson's favour, although his homicidal leanings are pretty obvious to everyone. At least they are pretty obvious to Olivia, who nevertheless shields Dan out of a romantic misconception about the attraction of evil. The seminal idea for this play, according to Williams, was provided when a fellow passenger on a train suddenly recognized a distant acquaintance in the photograph of a notorious matricide in a newspaper. In Williams' play, however, Dan does not murder his mother, since this might have seriously damaged his appeal to audiences – indeed, matricide and infanticide remain largely taboo in the genre until as late as the 1990s. *Night Must Fall* proved immensely popular: it saw over 400 London performances and afterwards transferred to New York and numerous other cities and was filmed in 1937. Williams continued to write plays with attractive parts for himself, catering to his audience's varying tastes, and produced another major crime play about 20 years after *Night Must Fall*: *Someone Waiting* (Globe, 1953).

The most successful crime play by Barré Lyndon was *The Amazing Dr. Clitterhouse*, performed at the Haymarket in 1936 with Ralph Richardson as the title character. It ran for nearly 500 performances and was later made into a film. Clitterhouse is a respectable citizen brought into contact with crime through professional curiosity. He is a well-established GP, who attempts to study the nervous reactions of criminals during the act and hence begins to lead a double life as a criminal. When he has finished his research and wants to give up his criminal career he is identified and blackmailed by his fence Kellermann. He murders Kellermann in the belief that a chapter on murder will complete his book. However, he commits a few mistakes and is found out by the police, but a barrister friend of his trusts that he will get Clitterhouse off on grounds of insanity. It is the murder of Kellermann that shifts the balance and finally makes the audience realize that Clitterhouse is a dangerous lunatic and ought to be arrested. In the doctor's gradual decline from scientist to murderer the influence of R. L. Stevenson's Dr Jekyll may be discernible. One of the most attractive features of the play, however, is the contrast between two professional milieus, the medical and the criminal. In *They Came by Night* (Globe, 1937) Lyndon reused the theme of a respectable professional finding himself drawn into the criminal milieu: this time the hero is a jeweller forced to collaborate in a bank robbery because he has once – unwittingly – taken stolen goods into commission. For *The Man in Half Moon Street* (New, 1939), Lyndon once again employed the theme of the mad scientist, this time in order to pursue the question of whether eternal life is desirable. The play has a strong science fiction element, and like most science fiction, it has not aged well. Neither of the two plays reached the popularity of *The Amazing Dr. Clitterhouse*.

As will have been noticed, all the three major successes discussed above – *Ten-Minute Alibi*, *Night Must Fall* and *The Amazing Dr. Clitterhouse* – chose the criminal as their protagonist. The plays Edward Percy and Reginald Denham produced between 1937 and 1967 also invariably have the culprit as their central character. The earliest play of the Percy/Denham team (then writing under the pseudonym of Rex Judd) was *Suspect* (St. Martin's, 1937), which revolves around a peculiarity of the Scottish judicial system, the verdict of "not proven." Thirty years ago, Margaret Wisehart was accused of having killed her father and stepmother with an axe and judged "not proven" by a Scottish jury. She has since lived quietly in the South of England under the name of Smith. When her son's fiancée discovers her true identity and resumes the investigation into the old murder case, Wisehart manages to dispel her suspicions, but the audience learns, right at the end of the play, that she was indeed guilty of the crime of which she was suspected. In spite of the gruesome nature of the crime, audiences are asked to side with Wisehart against the snooping amateur investigator.

In *Trunk Crime* (first performed at the Comedy in 1937, under the title of *The Last Straw*), Percy/Denham expected audiences to sympathize with a murderer who plans to bury his victim alive. Wolfe Guldeford, a rich, hypersensitive young man, has his college rooms ransacked by a group of drunken cads led by Richard Steele. When Guldeford's fiancée Pamela deserts him to marry Steele, Wolfe drugs his rival and puts him in a large trunk, planning to dump him in the swamp behind his cottage in the marshes. A courageous young woman intervenes; Steele survives and Guldeford is sent to a sanatorium to recover.

Ladies in Retirement (St. James's, 1939), probably the finest play by Percy/Denham, is set in Victorian England and was based on an actual French murder case of 1882. Two of the "ladies in retirement" are Leonora Fiske, a retired prostitute, and her companion Ellen Creed. Ellen is in charge of her two younger sisters, who are mentally retarded and hence entirely dependent on her. The sisters come to stay with Ellen and Leonora, but after a few months start getting on Leonora's nerves, so she orders Ellen to send them away. Ellen, who feels desperately protective towards her sisters, strangles Leonora and walls up the body in an old oven that has formerly been used as a safe by Leonora herself. The three sisters live happily together until their nephew Albert turns up looking for a hiding place after having committed a theft. Albert finds out about the murder and with the help of the maid Lucy stages a ghostly apparition scene, in which Lucy dresses up as Leonora. Ellen faints, but on recovering persists in her intention to send Albert away to Canada. By now Lucy has discovered the truth and calls the police. Ellen, who has been tormented with guilt at the thought of having murdered her friend, gladly surrenders. Her sisters are blissfully unaware of the impending change in their lives. Critics praised the play for its "unity of effect" and compared it to the works of Poe and Emily Brontë. It also deserves praise for its believable and yet insoluble central conflict – the audience cannot help sympathizing with both Leonora and Ellen – and for its pervading atmosphere of claustrophobia. Thus, the walling up of Leonora's body is but the most claustrophobic image in a wholly claustrophobic play.

Percy's greatest popular success *The Shop at Sly Corner* (St. Martin's, 1945) was written without Denham's assistance. Here the criminal protagonist is an antique dealer with a sideline in stolen goods. He murders his shop assistant, who has discovered the truth and begun to blackmail him, and is brought to bay by an act of poetic justice. In 1953, Percy returned to the genre with *The Man with Expensive Tastes* (Vaudeville, co-written by Lilian Denham), which revolves around an eminent graphologist who is also a successful forger. Even later, in 1967, Denham produced his last crime play *Minor Murder* (Savoy, co-written by Mary Orr) about two teenage murderesses in an Australian outback. In spite of a wide variety of themes, settings and the influence of different collaborators, Percy and Denham invariably stuck to the formula of putting the criminal at the centre and allowing him/her at least a modicum of audience sympathies.

In the crime plays of Armstrong, Lyndon and Percy/Denham, some kind of general "recipe" is discernible. This is not the case in the work of Roland Pertwee. In the 1920s, he had collaborated with Allene Tupper Wilkes on the whodunit *The Creaking Chair* and with Harold Dearden on *Interference* (St. James's, 1927), an adaptation of his own novel, based on the character configuration of two interlocking amorous triangles. In the period currently under survey, he produced another whodunit play with an array of assorted suspects (*To Kill a Cat*, Aldwych, 1939), co-written again by Harold Dearden, and another play in which crime plot and amorous intrigue intermingle (*Honours Easy*, St. Martin's, 1930). His greatest success, *Pink String and Sealing Wax* (Duke of York's, 1943) abandoned the whodunit structure and made use of the amorous triangle in a modified manner. The play is set in the Victorian era among the rising middle classes and shows the attempts of the family's children to liberate themselves from their father's absolute authority. The father is a chemist and a forensic expert; his eldest son Albert, however, would rather have been an engineer than follow in the family profession. Albert gets involved with a lower-class loose woman and brags about his medical knowledge. The woman uses the information thus obtained and a bottle of strychnine stolen from the laboratory table to do away with her brutal husband. She relies on Albert's father to forge the results of the *post mortem* for his son's sake, but since all members of the family stick together for once, the woman's attempts to escape discovery are defeated, and she confesses and drowns herself. Although absolute patriarchal authority is questioned, the play is an unqualified eulogy of middle-class morals and aggressively denounces members of the lower classes as drunkards, whores and murderers. Reviewers were delighted with the play, and *The Evening News* particularly praised the domestic scenes showing "the authentic family history of real people;"[12] "real people" obviously referring to the middle classes only. Comparing *Honours Easy* to *Pink String and Sealing Wax*, one cannot help noting a distinct change from a comparatively liberal attitude towards crime and adultery towards a stricter moral code. This change may be indicative of a general shift in the system of values that took place between the 1930s and 1940s.

Apart from the work of particular authors specializing in the genre, a few individual crime plays of the period deserve closer scrutiny because they established enduring trends in the genre. Two of these plays, Martin Vale's[13] *The Two Mrs Carrolls* and Patrick Hamilton's *Gas Light*, revolve around a murderous husband and his victimized wife, a character configuration that was soon to become excessively popular. *The Two Mrs Carrolls* opened at the St. Martin's in 1935 and soon successfully transferred to the United States where it was also made into a film. Geoffrey Carroll, the type of the "immoral artist," is married for the second time. Before his divorce, he attempted to kill his first wife by poisoning her nightcap of hot milk. The first Mrs Carroll returns just in time to warn Geoffrey's second wife, who is already showing symptoms of lead poisoning, that her Bluebeard husband has recently fallen in love with another woman. Together, the two Mrs Carrolls finally manage to confront Geoffrey, who is allowed the gentlemanly exit of poisoning himself. In *The Two Mrs Carrolls*, suspense is created mainly through the danger of an earlier crime repeating itself. This formula is obviously very attractive for a crime play, since it allows audience expectations to be both specific and vague: we know what is going to happen, but maybe history will not repeat itself and the victim will escape this time. This mechanism is in evidence in all serial-killer plays, as for example in *Love from a Stranger* and *The Mousetrap*.

In *Gas Light* (Apollo, 1939), Patrick Hamilton employs a similar technique by linking a crime in the past – a jewel robbery that occurred 20 years earlier – and a crime in the present: Mr Manningham's attempt to slowly and methodically drive his wife insane. As the title suggests, the play strongly relies on lighting effects for the creation of horror: "Perhaps the chief moment of horror came when a fading gas jet revealed the presence of the murderer in the room overhead."[14] The title as well as the subtitle "A Victorian Thriller" allude to the atmosphere of terror and gloom usually associated with Victorian melodrama. What is just as terrifying as the effect of the gaslight dimming, however, is the characters' obsessive behaviour – Mrs Manningham's hysteria, and her husband's compulsive ransacking of the house for the jewels he was unable to find 20 years ago. His antagonist Rough, who acts as compulsively as the villain, is a retired policeman who assists Mrs Manningham in regaining confidence in her own observation and judgment. *Gas Light* was extremely successful with 1,293 performances altogether and two different film versions[15] following each other in rapid succession.

J. Lee Thompson's *Murder without Crime* (Comedy, 1942), a considerably weaker play, is of interest mainly insofar as it introduced a number of plot elements that soon became stock motifs of the genre: the love triangle, the unintentional killing, the "busy body," mysterious occurrences that seem to point to the activities of a poltergeist, the return of a character presumed dead and the framing of an innocent man for a crime he did not commit. The play's plot is extremely intricate, and in the following murder plays triggered off by marital infidelity, plot complexity was even going to increase. In this respect, *Murder without Crime* may already be considered the prototype of the "postmodern comedy thriller" of the 1970s.

The end of World War II has been chosen somewhat arbitrarily as the end of an era, for – as Maack has pointed out – World War II is not really a watershed in English drama.[16] The war mainly affected drama in general by forcing theatres to close and by interrupting runs of successful plays when their theatres were destroyed by bombs. Of the plays discussed in this chapter, *Rebecca* terminated its run on 7 September 1940, because the Queen's had been hit by a bomb, but the play saw another successful season in a new production at the Strand and the Lyric in 1942. And as mentioned above, *Ten Little Niggers* had to transfer to the St. James's on 24 February 1944 and only returned to its original venue in May. Generally, crime plays survived bomb scares comparatively well, as did farces. Courtney remarks: "As the bomb menace receded, some theatres played farces and thrillers."[17] And Trussler comments: "Ironically, two of the established successes which continued their runs regardless of the rockets were farces concerned, as were their audiences, with outfacing the ultimate taboo of death – Noël Coward's *Blithe Spirit*, and that masterpiece of the comic macabre, Joseph Kesselring's *Arsenic and Old Lace*."[18] *Arsenic and Old Lace* had been imported from New York in 1942 and saw 1,337 London performances. It had a lasting impact on the development of the English crime play by inspiring a number of black farces that addressed "the ultimate taboo of death" in a playful manner in the following decades. The explanation ventured by Trussler to account for the success of *Arsenic and Old Lace* sounds equally convincing when applied to *Ten Little Niggers*. Moreover, the basic situation of waking up in a country-house to find another of your fellow guests murdered may seem less outrageous to spectators who may well wake up to find another block of flats in their neighbourhood gone. However popular the genre of the crime play may be in wartime, it tends to flourish even more in the era of reconstruction following an armed conflict. In the decade following World War II, the crime play was once again to see a peak in productivity comparable to the one reached ten years after the end of World War I.

And as in World War I, plays about espionage began to emerge in the course of World War II. However, while World War I's *The Man Who Stayed at Home* had been relatively close to the spy novels of its era, the spy plays of World War II began to divorce themselves from contemporary spy fiction. The spy thrillers of the 1940s, most notably those by Eric Ambler and Graham Greene, were decidedly bleak in outlook. Contemporary spy plays like Geoffrey Kerr's *Cottage to Let* (Wyndham's, 1940) or Vernon Sylvaine's *Warn That Man!* (Garrick, 1941) are treating the subject light-heartedly. Kerr's German spies present no real threat, since they are eventually outwitted by a teenage boy. The representatives of British Intelligence, on the other hand, are not to be taken too seriously either: an undercover agent posing as a butler struggles severely with his domestic chores and thus provides some comic relief. In *Warn That Man!* a group of German agents have invaded an English country-house and replaced the entire household by doubles in order to entrap an important visitor expected over the weekend – Winston Churchill, as it turns out at the end.[19] Their plans are crossed by a group of unemployed actors, whose theatre has just been bombed. Since espionage seems mainly a game of impersonation and deceit, the professional actors are clearly superior to their adversaries, who may be professional spies but are obviously mere amateurs when it comes to acting. The spy play and the spy novel were to continue in their separate directions in the following decades.

4

From the End of World War II to 1955

In the first decade following the end of World War II, Britain saw thorough political changes: the independence of India and Pakistan signalled the end of the British Empire. The newly elected Labour Government initiated a number of social reforms, but general living conditions were poor: the pound was devalued, and until 1954, consumer goods remained rationed. To the conservatively inclined readers of detective fiction, the destabilization of the formerly rigid class structure presented a considerable threat. In her novel *A Murder Is Announced*, Agatha Christie had Miss Marple complain that you no longer knew who people *were* – numerous people had become dislocated by the war, and family ties and institutions that had formerly helped to place people socially had broken away:

> Every village and small country place is full of people who've just come and settled there without any ties to bring them. The big houses have been sold, and the cottages have been converted and changed. And people just come – and all you know about them is what they say of themselves.[1]

In Christie's most famous play *The Mousetrap* (Ambassadors, 1952), a number of isolated individuals meet in a snowbound boarding house, and they may all very well be something other than they pretend. According to Athanason, the experience of dislocation and the destabilization of the class system was essential for the play's initial success[2]: "Agatha Christie dramatizes the malaise of cultural displacement that many English people experienced during the 1950s, when formerly established life styles had to be drastically modified to adapt to an England that no longer had a stable empire but a precarious welfare state."[3]

The killer in *The Mousetrap* is a young man from a working-class background, who has been severely damaged by his experiences during the war. This apparently was the type on whom middle-class audiences could most easily project their general anxieties. However, the choice of a young male criminal protagonist may be more than a mere projection and allude to the urgent social problem of what to do with demobbed servicemen who – like Bulldog Drummond in World War I – may "find peace dull." Other, more serious crime plays also took up the theme of male juvenile delinquency: Reginald Beckwith, *Boys in Brown* (Arts, 1947), Michael Hutton, *Power without Glory* (Fortune, 1947), Gordon Glennon, *Gathering Storm* (St. Martin's, 1948) and Ludovic Kennedy, *Murder Story* (Cambridge, 1954). The "boys in brown" are inmates of a Borstal institution for young offenders, and Beckwith put a considerable amount of preparatory research into his comprehensive study of the various forms of male juvenile delinquency. The play was first performed at the Gate in 1940 and revived after the war.

While Beckwith had presented a number of different case studies, the three other plays each concentrate on an individual case of a young working-class man who commits murder. In *Power without Glory*, Cliff, the younger son of a shopkeeper's family, murders his pregnant girlfriend, and the action of the play revolves around the reactions of the members of his family, who in different ways try to protect him. Ned in *Gathering Storm* murders his grandmother in order to inherit and sell her farm, since this would enable him to win the affections of Myrtle, a city girl who dreams of opening her own hairdressing salon in London. Cliff and Ned are young men driven by sexual desire as well as a desire to break free from their stiflingly close-knit family circles. They may thus be considered "angry young men" *avant la lettre*; and in fact, *Power without Glory* and *Gathering Storm* achieved for the crime play what the drama of the angry young men was to accomplish in the following decade: an escape from the eternal drawing-rooms of the well-made play. *Power without Glory* is set "in the living-room behind an East End provision shop,"[4] and *Gathering Storm* "in the kitchen-living-room of the Hardlestone's farmhouse."[5]

In *Murder Story*, which was first performed six years after *Gathering Storm*, a new theme has been added to the plot motif of the young working-class man with no prospects. *Murder Story* is a deliberate and obvious issue play, demanding the abolition of the death penalty. Jim Tanner, aged 19 and mentally retarded, lives with his working-class family. He gets involved with a gang of juvenile delinquents, and in the course of a burglary one of the gang, Clift, shoots a policeman. Jim is arrested and found guilty of murder. In prison he learns to read and is introduced to the basics of the Christian faith. When both his appeal and reprieve fail, his family produces a petition signed by 20,000 people, including the victim's widow. Nevertheless, Jim is executed. Kennedy himself remarked later that he had "made the mistake of not allowing the audience to feel sufficiently the desire for revenge in the theatre"[6] by making Jim the central character instead of Clift. Nor does he show the death of the policeman. The technical circumstances of capital punishment, by contrast, are described again and again in great detail throughout the play. The play is unashamedly biased, and, in contrast to

Robinson's *To What Red Hell*, for instance, the audience is denied the comfort of a happy ending. Kennedy's "Epilogue on Legal Killing" gives the rational arguments and data to reinforce the emotional effect of the play. Kennedy claims that his play was a reaction to the decision to ban public executions but continue them behind closed doors, thus enabling the public to ignore a practice they felt to be revolting.

> By putting hanging out of sight they also put it out of mind; by relegating it to a place where it could neither be seen nor heard they rid themselves of their guilt and transferred to a handful of people the responsibility of what they could no longer tolerate themselves.[7]

The example of *Murder Story* shows that the crime play, like detective fiction, does not have to be politically conservative and escapist but may be used as a tool in the fight for social reform. However, only very few plays in each period used the format for such purposes.[8]

Apart from *The Mousetrap* and the angry-young-men plays listed above, a few more crime plays of the period immediately reflected the living conditions in postwar England. Gerald Anstruther's *The Third Visitor* (Duke of York's, 1949), for instance, uses the motif of dislocation through war – a young husband does not know his cheerful little wife Steffy was once an inmate in a concentration camp – and employs an "evil German" as its villain, who used to be the commandant of that camp.[9] *The White Sheep of the Family* (Piccadilly, 1951) by L. Du Garde Peach and Ian Hay employs topical references to the black market and the rationing of meat and eggs. More importantly, as the title suggests, the play presents a complete reversal of (pre-war?) moral values: a family of successful criminals worries about their only son, who has suddenly decided to go straight. The son recovers from his lapse into virtue in time, and they all live happily ever after. The play systematically undermines the doctrine common to plays of this type that "Crime Pays No Dividends" – the title of a film the maid Janet goes to see in the first act. And as in Priestley's earlier *Laburnum Grove*, the family maintains a respectable façade – the father for instance is a member of several prestigious clubs and a churchwarden. The facile assumption that there exists a clear-cut distinction between the respectable citizen and the hardened criminal is thus subverted.

In spite of the general social changes alluded to above, drama as an art form did not undergo any substantial changes during the first decade after World War II. The stage was still dominated by playwrights who had been active before the war, of whom the commercially most successful was Terence Rattigan. Rattigan had produced a number of comedies in the 1930s and continued to write plays well into the 1970s. In 1946 he wrote the crime play *The Winslow Boy* (Lyric), which is not merely set in the Edwardian period but also imitates the dramatic forms of that era – David Mamet, who filmed the play in 1999, praised its melodramatic quality, and Christopher Innes noted its similarity to Galsworthy's *Justice* and Pinero's *Benefit of the Doubt* of 1895.[10] Rattigan took his cue from a famous true criminal case, the Archer-Shee trial of 1910. The *Winslow Boy* questions the relationship between (moral) Right and (legal) Justice. A 14-year old naval cadet is accused of having stolen a classmate's five-shilling postal order. The boy's father engages the help of famous barrister Sir Robert Morton, and the interrogation scene, in which Morton questions the boy and is finally convinced of his innocence, is the climax of the whole play. The Winslow family makes substantial financial and personal sacrifices in order to see the boy cleared of suspicion. Sir Robert wins the case, and besides reveals the humanity hidden behind his fish-like exterior. As in *Libel!*, the trial is conceived as an ordeal that brings about character revelation and self-discovery to those concerned. The play was extremely popular and is revived frequently.

Priestley's almost exactly contemporary crime play *An Inspector Calls*[11] is also set in the period before World War I. The year is 1912. The celebrations of the engagement between Sheila Birling and Gerald Croft are interrupted by the appearance of mysterious Inspector Goole, who has come to investigate the chain of events that led to the suicide of a young woman. It turns out that all those present were in some way involved: Arthur Birling sacked the girl from his company, then Sheila caused her dismissal from her next job at a dressmaker's, whereupon the girl was taken up by Gerald as his mistress. After Gerald had dumped her, she became the mistress of Sheila's brother Eric. When the girl became pregnant, Eric embezzled some money from his father's firm to help her, but she refused to see him again and appealed to a charitable organization for single mothers instead. However, Mrs Birling, the head of the committee, refused her any assistance. The Inspector forces the members of the family to confess their guilt, but once he is gone, the older generation go back to normal. Gerald discovers that the visitor was not a real inspector after all, and that they probably all encountered different young working-class women in the course of the past year. The play ends with a call from the local police department: they are sending round an inspector to inquire into the suicide of a young woman. In contrast to *Dangerous Corner*, the characters of *An Inspector Calls* do not qualify for a second chance. The play makes clever use of some of the stock elements of the conventional crime play such as everybody's guilty secrets and the characters of the omniscient detective and the "bogus policeman," here fused into one. It is, however, rather more melodramatic in attitude than a traditional whodunit, delivering its message that "we don't live alone. We are members of one body."[12] World War I is seen as the inevitable result of a rigid class structure and people's inability to repent, a scourge of God. The fact that the play is a period piece set over 30 years before its first performance, however, allows the audience to distance themselves comfortably from the play's moral appeal. *An Inspector Calls* has proved consistently popular. It was last revived in London in the 1990s, and the National Theatre's survey of 2000 elected it one of the most important plays of the century. Its unambiguously voiced moral message has made it a favourite for reading in the classroom.

However, *An Inspector Calls* is difficult to classify. Is it "a taut, well-made murder mystery"[13] or a "modern morality,"[14] as John Gassner suggests? The answer depends on who we think the Inspector is. He is clearly no ordinary policeman, but is he a celestial agent of justice, or is he – as Shepherd and Womack have argued – an author surrogate?

> [I]n *An Inspector Calls* (1946), the whole action is motivated by an inquisitive intruder who appears to be a police inspector, but who ultimately turns out not to be [...]: in his omniscience, his didactic authority, and the cunning manipulation by which he advances the dramatic action, he is clearly the playwright.[15]

A few lines down, the authors summarize their observations: "round its discontented edges, realism tends to generate author surrogates,"[16] giving the arms manufacturer in Shaw's *Major Barbara* and the doctor in Eliot's *The Cocktail Party* as examples. What is a sign of the exhaustion of naturalistic drama for Shepherd and Womack is considered a sign of Priestley's conscious experimentation with non-naturalistic styles by Innes. Innes groups Priestley under the heading of "Poetic drama," together with Yeats, Eliot and Beckett and claims:

> Priestley's stylistic experimentation is comparable to Pirandello in its overt theatricality. [...] Despite its flaws, Priestley's work marks the loosening of Shaw's dominance over mainstream theatre. His achievement was to bring experimental techniques, symbolism, and existential themes onto the West End stage.[17]

Among crime plays, *An Inspector Calls* most closely resembles Emlyn Williams' *A Murder Has Been Arranged*, which also begins like a conventional crime play and only allows the metaphysical element to intrude at the point of closure. However, while Williams uses the appearance of the ghost as a *deus ex machina* device for solving the mystery (as well as for horror effects), Priestley deliberately employs the "supernatural twist" in order to demonstrate that the topic of moral responsibility cannot adequately be dealt with in the format of the traditional crime play.

After *An Inspector Calls*, Priestley wrote another, more conventional crime play entitled *Bright Shadow* (1950), based on the principle of Vera Caspary's *Laura* (1944), which had been extremely successful as novel, play and film. After the death of a beautiful young woman, a detective reconstructs her life and her personality from the divergent testimonies of people who knew her.[18] Peter Warton, the functional detective of *Bright Shadow*, is a typical representative of the classless postwar society dreaded so much by Miss Marple:

> He speaks good English without any suggestion of any class and social background, very much a contemporary type, without any "good public school manner." He clearly belongs to a different social order and world from those of the Colonel.[19]

The Colonel as a stock character of Golden Age detective fiction and as the representative of pre-war society eventually turns out to be the murderer, motivated by fear that his comfortable life-style may be disrupted. After *An Inspector Calls*, *Bright Shadow* surprises by its conventionality. The guiding questions structuring the reception of the crime play plot,[20] usually submerged in the text in detective fiction and the "well-made" variety of the crime play, become very conspicuous (not to say obtrusive) as Peter recites his list:

> I want to know what it was that took her from the top of the world down to a suicide's grave. I want to know if Mrs Probus was right about her, or her husband was right about her. Was she unfaithful to him or wasn't she? Was this other fellow – Kemp – her lover, and, if so, what happened when she went to him? Why did she leave him and come back here? Why did she commit suicide – *if* she did commit suicide? What sort of woman was she? Was she what we thought she was, or what her old nanny thought she was, or what her husband thought she was, or somebody quite different?[21]

While the detective in *An Inspector Calls* had been an author surrogate, the detective of *Bright Shadow* is a spectator surrogate, or, arguably, a *reader* surrogate, since the plot of *Bright Shadow* seems much better suited to narrative than to the stage, consisting entirely of stories revolving around a character already dead: a body in the library. The fundamentally non-dramatic nature of this central device was proved by the fact that *Bright Shadow* was only moderately successful and never reached the West End.

The highly acclaimed plays of Rattigan and Priestley had little impact on the development of the crime play as a genre. In the period under survey, the crime play as a specific dramatic genre, comparable to the detective novel, was dominated by the work of Agatha Christie. In the decade after World War II, Christie was the undisputed "Queen of Crime" on the stage. She adapted her novel *Death on the Nile* as *Murder on the Nile* for the Ambassadors in 1946 and *The Hollow* for the Cambridge in 1951. In between, she allowed her novel *The Murder at the Vicarage* to be dramatized by Moie Charles and Barbara Toy (Playhouse, 1949). The early 1950s saw her two greatest successes on the stage, *The Mousetrap* (Ambassadors, 1952) and *Witness for the Prosecution* (Winter Garden, 1953). *Witness for the Prosecution* was adapted from a previously published short story, while *The Mousetrap* had first been a radio play of half an hour in length, commissioned by the BBC in honour of Queen Mary's 80th birthday entitled *Three Blind Mice*. In 1950, Christie turned it into a novella of over 70 pages and had it published in the United States only in the volume *Three Blind Mice and Other Stories*. Hence, the narrative version was not freely accessible in Great Britain, which may have been essential for the continued success of the play. Although Christie adapted the stage version from the novella, the play still in places betrays its origin as a radio play: the exposition is accomplished very economically by a short audiodrama relating the discovery of the initial murder while the stage is in total darkness. When the lights come on, the radio motif is continued: additional information concerning the crime is imparted via a radio announcement on the wireless. The central murder of the play is committed while the wireless is turned up loud and the stage is once again plunged into perfect darkness. Though the turning off of the lights serves first of all to conceal the identity of the culprit, the result is that the stage play invariably reverts to its earlier stage as a radio play during the traumatic scenes of murder.

As in her previous success *Ten Little Niggers*, Christie bases the plot on the principle of the "murder-house mystery" and has the serial killings follow the pattern of a nursery rhyme, though in this case, there are obviously fewer victims. *The Mousetrap* is set in a snowbound guesthouse – the converted country-house of earlier detective fiction – where a homicidal maniac is at large, about to kill two people who were responsible for his traumatic childhood and his brother's death. The culprit eventually turns out to be the "Inspector" purportedly sent by Headquarters to solve the initial murder and prevent two impending crimes. The mentally deranged killer is neither killed nor handed over to the police but given a sedative and sent to bed, watched over by his elder sister, with whom he has finally been reunited. Apart from this slightly unconventional ending, *The Mousetrap* is impeccably classic in construction, strictly adhering to unity of plot,[22] combining a crime in the past and a crime in the present, and using the time-honoured plot motifs of the homicidal maniac, the bogus policeman and the murder-house. The assorted suspects include the charming young couple running the guesthouse, an overbearing elderly lady who makes an ideal victim, a retired major (who at first seems thrown in for good measure, but finally turns out to be the real policeman), an elderly racketeer, an effeminate young man and a masculine young woman.[23]

Since the title "Three Blind Mice" had already been used for another West End play, Christie's crime play was rechristened "The Mousetrap." The *Hamlet*-reference of the new title suggests a strong metadramatic element, since in *Hamlet*, "The Mousetrap" refers to the play within the play. And indeed, *The Mousetrap* is more self-consciously parodic than Christie's other plays, and, in fact, than most other crime plays. Paravicini's closing remarks in I.i are a particularly good example of this type of self-referential genre parody:

> "Who am I? You do not know. Where do I come from? You do not know. Me, I am the man of mystery." (He laughs) [...] "I complete the picture. From now on there will be no more arrivals. And no departures either. By tomorrow – perhaps even already – we are cut off from civilization. No butcher, no baker, no milkman, no postman, no daily papers – nobody and nothing but ourselves. That is admirable – admirable. It could not suit me better."[24]

Paravicini, who is first introduced as a slightly revised edition of Poirot to mislead the audience, here assumes the part of the stage manager, openly discussing plot design with the audience. Since *The Mousetrap* constantly parodies itself, it is largely immune to parody from outside. This self-ironic streak may help to account for the play's phenomenal success; particularly since the playfully parodic note is associated with a theme evoking profound childhood fears. The story of the three little children abandoned at a remote farmhouse recalls fairy-tale motifs, and the recurrent nursery rhyme likewise strikes a note between childhood terror and playfulness. *The Mousetrap* opened at the Ambassadors in 1952, and in 1974 transferred to the St. Martin's where it is currently running, having long beaten all records for the longest running play ever. Since 1960, the cast has been changed annually to prevent the effects of fatigue. Interestingly, *The Mousetrap* was but reasonably successful in New York.

Of the six Christie plays running in the London West End between 1945 and 1955, only one was an original play: *Spider's Web* (Savoy, 1945), tailor-made for the actress Margaret Lockwood. Lockwood played Clarissa Hailsham-Brown, a very imaginative young woman, who finds a corpse in her living-room the very evening her husband is bringing an important foreign politician to his house. Clarissa manages to persuade three of her men friends to dispose of the body for her, but they are interrupted by the arrival of the police and Clarissa must improvise. This highly self-conscious play contains the positively last appearance of secret passages and sliding panels, as well as the not quite final appearance of the "busy body." The play was filmed by Godfrey Grayson in 1960, but by then, the play was already slightly dated, as a contemporary review suggests: "It'll amuse, though hardly scare, average audiences and youngsters. [...] The mystery element seldom thrills, but the twist ending is theatrically effective, and Technicolor tarts up the interiors"[25] is the overall verdict. Due to this change in audience tastes, the period of Christie's great stage successes was definitely over by 1960.

Apart from stage plays adapted from Christie's detective fiction, a number of dramatizations of novels and short stories of detection by other writers appeared on the West End stage. Aldous Huxley adapted his short story "The Gioconda Smile" (New, 1948); Rodney Ackland transformed Somerset Maugham's short story "Before the Party" into a "comedy of expediency" (St. Martin's, 1949), and detective novelist Gerald Verner dramatized Peter Cheyney's *The Urgent Hangman* under the title of *Meet Mr Callaghan* (Garrick, 1952). *Meet Mr Callaghan* will be discussed in some detail in Chapter 13. *Before the Party* is a very cynical crime comedy about moral hypocrisy. Laura Skinner has killed her alcoholic husband, and her family is intent on hushing the matter up in order to get her married off again, to David Marshall. Only Laura's spinsterish sister Kathleen insists that Laura must tell David the truth. David is still prepared to go through with the marriage, but the formula of "living happily ever after" is subverted by the sly suggestion that David, too, may be a closet alcoholic. The theme of a group of people going back to normal after murder was obviously particularly appropriate for the postwar period. Rodney Ackland apparently found the format of the crime play so attractive that he later wrote a successful original crime play called *A Dead Secret.*

Apart from the influence of crime fiction in evidence in the number of plays adapted from novels and short stories, the English crime play of the period was inspired by foreign imports, to be more exact, by the model of Joseph Kesselring's *Arsenic and Old Lace*. Due to the popularity of the Frank Capra film version, the plot is fairly well-known, therefore it may suffice to say that the play revolves around two charming old ladies who have made a favourite pastime out of "mercy killings" of lonely elderly men. In R. C. Sherriff's *Miss Mabel* (Duchess, 1948), the title character is an equally nice elderly lady who poisons her selfish twin sister for the common good. Other playwrights soon realized that the crime of murder is not necessarily incompatible with comedy, though – in contrast to *Arsenic and Old Lace* and *Miss Mabel* – murder may be suspected or intended but is never shown to be executed successfully. Kenneth Horne's *A Lady Mislaid* (St. Martin's, 1950) revolves around an imaginative woman who falls in love with an otherwise unremarkable man because he reputedly murdered his wife. However, the presumed victim turns out to be still alive, and the promising bones the police dug up in the chicken run turn out not to be human remains after all. Reviewers differed as to whether the search for a dismembered murder victim makes suitable light comedy fare but generally praised the author's tact, delicacy and humour. Kenneth Horne's subsequent comedy *Trial and Error* (Vaudeville, 1953) also derives much of its suspense from the suggestion that the heroine may have attempted to murder her first husband. While Horne's plays revolve around murders that have not been committed, Ian Stuart Black's *We Must Kill Toni* is concerned with an intended murder that never materializes. Toni is a beautiful, only seemingly naïve young woman who inherits a country estate and thus becomes the target of several murder attempts by her charming but villainous cousins Francis and Douglas. Toni escapes unharmed, and the cousins commit suicide.

Meet a Body by Frank Launder and Sidney Gilliat likewise revolves around a series of thwarted murder attempts. The involved plot deals with the attempts of Hawkins and Montague, the one a paid assassin, the other a muddle-headed communist, to kill politician Sir Gregory by means of a time-bomb concealed in a radio. Sir Gregory's faithful secretary Winifred discovers their murder plan and tries to warn her employer. Consequently, Winifred herself becomes the target of several murder attempts; first she is hit over the head, believed to be dead and hidden in a piano, and later in the course of the action she gets chloroformed. In between, however, she manages to impart the details of the planned assassination of Sir Gregory to William Blake, a vacuum cleaner salesman, who unwittingly butted in on the first attempt to dispose of Winifred. Blake saves Sir Gregory by throwing the radio out of the window in the nick of time, bringing down most of the surrounding scenery in the ensuing explosion. The most ingenious idea in the play is the plan for the elimination of Sir Gregory: in order to make sure that he is close to the radio when the bomb goes off, the villains have chosen the exact moment one of his speeches is broadcast on radio. Twice, the audience is subjected to Sir Gregory's speech: in Act II, when Hawkins listens to it in order to time the explosion, and again in Act III; therefore the audience knows exactly when to expect the detonation. *Meet a Body* was subtitled "an improbable adventure," and this label appropriately characterizes the play's pervasive mode of burlesque and of frantic activity. The work for which Launder/Gilliat are best known today is the screenplay for the early Hitchcock film *The Lady Vanishes*. Like *The Lady Vanishes*, *Meet a Body* attempts to combine a genuine threat and high-spirited comedy, plus a – somewhat minor – love interest.

As mentioned in the previous chapter, murder in crime plays was increasingly motivated by marital tensions (*The Two Mrs Carrolls*, *Gas Light*) or extramarital relations (*Murder without Crime*). In the decade after World War II, this trend continued, and adultery became the most common motivation for murder in the crime play, twice as common for instance as gain or revenge. This notable increase in the (violent) breaking up of marriages mirrored a general trend in society: "The war had, in many instances, disrupted marriages and family life. Divorces reached a peak of 60,000 in 1947, ten times the pre-war figure."[26]

Three of the plays based on novels by Agatha Christie are triggered off by an adulterous relationship (*Murder on the Nile*, *Murder at the Vicarage* and *The Hollow*), as well as Huxley's *The Gioconda Smile*. The following seven plays by various authors were also based on the character configuration of the homicidal amorous triangle: St. John Clowes, *Dear Murderer* (Aldwych, 1946), Terence de Marney and Percy Robinson, *The Crime of*

Margaret Foley (Comedy, 1947), William Dinner and William Morum, *The Late Edwina Black* (Ambassadors, 1949), Frederick Knott, *Dial "M" for Murder* (Westminster, 1952), Emlyn Williams, *Someone Waiting* (Globe, 1953), Philip Mackie, *The Whole Truth* (Aldwych, 1955) and Jack Popplewell, *Dead on Nine* (Westminster, 1955).

The theme of the battle of the sexes is most evident in *Dear Murderer*, a play extremely typical of its kind, using a number of plot elements that were soon to become stock devices of the crime play: a game of cat-and-mouse enacted by the murderer and his victim, an innovative murder method (here: gas poisoning with the help of a gas mask), a wife turning the tables on her villainous husband, a police inspector reduced to attempted blackmail, and a forged suicide note, a plot motif here realized twice in order to bring about poetic justice. *Dear Murderer* is unusual only insofar as it explicitly addresses changed gender roles and relations: in her final feminist speech, the heroine declares that generations of oppression have made women intellectually superior to men. The part of the murderous husband in *Dear Murderer* was played by Terence de Marney, who had starred in *Ten Little Niggers* and later played the hardboiled detective in *Meet Mr Callaghan*. He also co-scripted *The Crime of Margaret Foley*, in which the initial set description already informs us that Margaret Foley has married beneath her and despises her husband.

The Late Edwina Black seems to be consciously imitating successful earlier models: like *Gas Light*, it is set in the Victorian period; like *The Ghost Train*, it flirts with the supernatural; like *The Gioconda Smile*, it revolves around the murder of an invalid wife; and as in the drama version of *Appointment with Death*, the victim eventually turns out to have poisoned herself. The play eventually achieved a run of 158 performances, which shows that a mere mixture of approved ingredients does not yet ensure popular success.

That it is impossible to predict a play's success accurately is proved by the example of *Dial "M" for Murder*. The play at first failed to find a theatrical producer and was hence broadcast on TV instead. After its phenomenal success as a TV play, it transferred to the stage of the Westminster. Today, the play is widely known because of the popularity of the film version by Alfred Hitchcock and a considerably weaker remake of the 1990s. Yet this very ingeniously plotted play deserves some closer attention on its own merits, particularly its use of character configuration, of reversals of fortune and of favourite crime play props like the telephone and the latchkey.

Ex-tennis-champion Tony Wendice is alarmed when his rich wife Margot shows signs of falling in love with Max Halliday, a writer of TV crime plays. At first Tony is content with stealing a letter from Max to his wife and then anonymously blackmailing her, but then he decides to have Margot killed by a paid assassin. He contacts Lesgate, a former schoolmate, who has since turned criminal, and blackmails him into cooperation. Lesgate is to enter the apartment by means of a latchkey left on the stairs by Tony and to kill

Margot the very moment Tony phones her from a faraway location. This is intended to secure Tony's alibi, but there is an undercurrent of Tony wanting to be in at the kill. Tony's plan miscarries, for Margot manages to defend herself by stabbing Lesgate with a pair of scissors. Tony has to change his plans and now makes it appear as if Margot had been blackmailed by Lesgate and killed him in cold blood. He removes Lesgate's murder weapon, returns the latchkey from Lesgate's pocket to Margot's handbag and plants Max's love letter on Lesgate. His plan is successful, and Margot is tried and sentenced to death. The third act, however, shows a major reversal: independently, Max and police inspector Hubbard have hit on the real sequence of events. Hubbard has even realized that the latchkey in Margot's handbag was Lesgate's own key, and that Lesgate returned the Wendice key to its place on the stairs before entering the apartment. It is proved that Margot knows nothing of the latchkey on the stairs, but Tony realizes his mistake when he finds he cannot get into his flat. He picks up the fitting latchkey from the stairs and lets himself into his apartment where he encounters Hubbard, Max, and Margot. The traditional motif of the murderous husband is here varied and developed by revealing a certain sadistic strain in Tony: he does not simply try to get rid of his wife but chooses a murder method of prolonged torture – public trial and execution – echoing his earlier blackmailing of her. His delight in inflicting pressure on others is also shown in his interactions with Lesgate, who, however, is a thoroughly unpleasant character and therefore the ideal murder victim.

Both in *Dear Murderer* and *Dial "M" for Murder* the major part of the action revolves around the attempt to frame an innocent character for a murder (s)he did not commit. The plot motif of framing is also essential for Emlyn Williams' last crime play *Someone Waiting*. Here the plot motif of adultery leading to murder only serves to set the action in motion. A man strangles his mistress; an innocent man is convicted and executed in his place, and the play's main action now revolves around the attempts of the father of the executed man to get revenged on the true culprit by framing him for another murder that he did not commit. The two plays of 1955, *The Whole Truth* and *Dead on Nine*, also employ both the theme of adultery leading to murder and the motif of framing an innocent character for a crime (s)he did not commit. In *Dead on Nine*, even the representative of the police fabricates a fictional "murder story" to entrap the guilty. Against this abundance of invented stories of crime, only "the whole truth" can save the protagonists: both plays express a pronounced belief in the healing power of confession. The involved plots of *The Whole Truth* and *Dead on Nine* are characterized by several reversals of fortune and strategy, including the return of a person presumed dead. This type of plot construction was later brought to perfection (and exhaustion) in the crime plays of Francis Durbridge.

Although these crime plays reflect an increasing destabilization of the institution of marriage, they do not choose the facile option of collectively blaming women for the break-up of marriages, as for instance the Hollywood *film noir* of the 1940s revolving around the character of the *femme fatale*. The English crime plays of the period are not simply plays about unfaithful wives, since the amorous triangle of wife-husband-mistress is quite as frequent as the one consisting of husband, wife and lover. Nor are the women in these plays inordinately aggressive: in five of the above plays, the aggressor is male; in a mere two, female. Moreover, the plays revolving around a murderess are adaptations of narratives, hence the motif of the murderous female cannot be said to be a peculiarity of the crime play.[27] However, women do kill even in the original crime plays, but usually only in self-defence or revenge after a previous act of male aggression. The terrible woman of Hollywood cinema is not a favourite character on the postwar English stage.

In the context of changing gender relations it is perhaps interesting to consider the character of the woman detective. In the previous periods, woman detectives had occasionally appeared, for instance in *The Wrecker, The Fourth Wall* or *Trunk Crime*, but more often than not women "discovered" (rather than investigated) crimes committed by their husbands, as for instance in *The Two Mrs Carrolls, Love from a Stranger, Gas Light* or *Rebecca*. In the postwar period, we encounter a number of woman detectives who deliberately set out to track down a murderer: Steffy in *The Third Visitor* (see above), Pinkie in *Grand National Night* (see below) and Miss Marple in *Murder at the Vicarage*. To these might be added Sister Mary Bonaventure in Charlotte Hastings' *Bonaventure* (Vaudeville, 1949) and Dora in Janet Green's *Murder Mistaken* (Ambassadors, 1952). Sister Mary Bonaventure serves as the play's functional detective and solves a murder case during a flood at Denzil St. David, which has brought the inhabitants of the nearby village to the high ground of the convent and hospital. Among the refugees is a young woman wrongly convicted for the murder of her brother, and Sister Mary proves her innocence and identifies the true culprit. As befits a female detective, her detection process is linked to domestic chores – an important clue is provided by newspapers used for lining a cupboard. In *Bonaventure*, Hastings successfully combines character study – Bonaventure's struggle with convent discipline – and an ingenious puzzle.

Janet Green, by contrast, dispensed with the puzzle element in *Murder Mistaken*: Edward Bare murders his wealthy wife Monica by staging an accident with the gas fire. It is only after Monica's funeral that he learns that the bulk of her fortune will go to her sister Dora in Jamaica. Bare takes refuge in another advantageous marriage, and soon afterwards meets an attractive young schoolteacher, who also seems ready to succumb to his charms. This woman, however, is no other than Monica's sister Dora, who has come over to investigate the circumstances of her sister's death. When Bare discovers the truth, he plans an accident for Dora, but she survives and Bare escapes punishment by committing suicide with the help of his loyal wife. The most interesting aspect of the play is Edward's genuine attachment to Monica, with whom he still conducts – albeit rather one-sided – conversations even after he has killed her.

Murder Mistaken and *Bonaventure* belong to two different types of crime play: *Murder Mistaken* follows the formula of inverted detective fiction, *Bonaventure* conforms to the model of the classic puzzle. It is interesting to note that different types of detective were considered suitable for the two types of play. In the "inverted" play, the detective is a member of the family and hence personally involved in the outcome of the case. In *Bonaventure*, Hastings thought it more appropriate to create a detective who is an impartial investigator, and made her somewhat "impersonal" by turning her into a nun – possibly in imitation of Chesterton's Father Brown; but then all great detectives have tended to be curiously sexless.

In the postwar period, a handful of new writers of the crime play emerged who shall be briefly introduced in the following. The team of Dorothy and Campbell Christie[28] had already produced a successful crime play in 1935: *Someone at the Door* (St. Martin's), in which a group of bright young things plan to stage a fake murder mystery and suddenly find themselves involved with real crooks. The Christies' two next plays, *Grand National Night* (Apollo, 1946) and *Carrington, V.C.* (Westminster, 1953) both revolve around a gallant hero encumbered with an insufferable wife. Gerald Coates in *Grand National Night* accidentally kills his alcoholic wife. Since he is used to taking risks both as an amateur jockey and a brilliant bridge player, he decides to bluff his way out and indeed manages to dispel the suspicions of the police and of his wife's sister, who serves as the functional detective. The title character of the Christies' next play, *Carrington V.C.* (Westminster, 1953), is a character as reckless as the protagonist of *Grand National Night* and, incidentally, like the latter, an expert jockey. Carrington, wearer of the Victoria Cross, serving under an incompetent superior, Henniker, and married to a hysterical wife, Val, commits a number of strategic mistakes: he has an affair with Captain Alison Graham, and when his overdue pay does not come through, he helps himself from the battery safe and bets on his own horse at the Sandown races. He wins a race but is injured in the following race, and while he is still in hospital, Henniker discovers that the money is missing. Carrington is court-martialled for fraudulent misapplication of money, absence without leave and for entertaining a female officer – Alison – at his barracks. He defends himself, but just when he is about to prove that he took the money openly as a gesture of defiance to the Paymaster, Val finds out about the affair with Alison and refuses to corroborate his statements: she even makes it sound as if he had coached her to give false evidence. Carrington is discharged from the army and Val leaves him, but he finds some comfort in Alison's love, and even more in the support of his men. The Christies' last crime play was the considerably weaker *The Touch of Fear* (Aldwych, 1956), which shows a young wife investigating the murder of a nursery governess and increasingly growing afraid that her husband may be to blame. Like the "bright young things" of their play from the 1930s, the Christies' dashing gentlemanly heroes were quickly becoming outdated, as well as their insistent distinction between respectable and "loose" women.

In histories of twentieth-century British drama, R. C. Sherriff is mainly mentioned for his play *Journey's End*, "the only British play of note successfully to attempt a depiction, however diluted, of the reality of the trenches."[29] *Journey's End*, which premiered in 1928, remained its author's greatest success, and it was nearly 20 years later that Sherriff, now in his fifties, began to experiment with the crime play. His first crime play *Miss Mabel* has already been mentioned above as a probable offshoot of *Arsenic and Old Lace*. Like *Miss Mabel*, Sherriff's next two crime plays, *Home at Seven* (Wyndham's, 1950) and *A Shred of Evidence* (Duchess, 1960) deal with "ordinary" middle-class people who come into contact with crime, and with the effect this experience has on their lives.

The plot of *Home at Seven* revolves around a case of amnesia: When David Preston returns home one Monday evening he finds that he has "lost" a whole day: it is Tuesday evening instead and he has no recollection of the previous 24 hours. Things are made worse by the fact that money is missing from the safe of the social club he is running, and the club steward, who was first suspected of having taken it, is found dead with his head bashed in. Preston even starts suspecting himself, although his wife and family doctor firmly believe in his innocence. It turns out that the steward was killed by his accomplice and that Preston spent Monday night and most of Tuesday in a pub he had a habit of visiting on his way home, without ever telling his wife. In the pub he suddenly lost his sense of time and place and believed himself to be back in World War II, during which he had had a traumatic experience. The pub landlord allowed him to sleep out his attack, partly because he himself would have been in for a fine for selling alcohol outside licensing hours.

What is interesting about the play is the way in which the uncanny – theft, murder, loss of memory – intrudes into a comfortable middle-class home characterized by respectability and routine. The initial scene of Preston returning home to his wife, who is still in a panic, while he is unaware of anything extraordinary going on, is particularly effective. In the production at the Wyndham's, David Preston was played by Ralph Richardson, who had previously created the character of Dr. Clitterhouse, another middle-class professional drawn into crime.

It is significant that Preston's amnesia is triggered off by his experience of being transported back into World War II. It is his traumatic war experience that brings about his subsequent feeling of guilt, which – in an act of transference – he attaches to the crime that has just been committed in his neighbourhood. This psychological mechanism of the transference of feelings of guilt from the war experience to contemporary crime may offer a feasible explanation for the popularity of the crime play genre in the two postwar periods. The appearance of *Journey's End* ten years after World War I suggests that ten years may be the usual "incubation period" until the first attempts to deal with the war experience on stage are to be expected. And this is precisely when the crime play saw its great peaks of productivity.

The protagonist of *A Shred of Evidence* also suspects himself of being responsible for the death of a man but does his best to destroy the evidence of his involvement. In this case the victim is a cyclist who met with a fatal accident, possibly after being hit by the protagonist's car. Once the hero has been cleared of suspicion, he quickly recovers and goes back to normal. While the rapid return to ordinary life had been criticized in *An Inspector Calls* and Ackland's *Before the Party*, Sherriff does not seem to intend his heroes' quick recovery to be morally questionable.

Like Sherriff, Lesley Storm showed respectable middle-class people coming into contact with crime. In *Black Chiffon* (Westminster, 1949), Flora Robson starred as Alicia Christie, a middle-aged woman who is caught shoplifting, stealing of all things a black chiffon nightdress she would never be able to wear. A psychiatrist consulted by the defence finds out that there is a strong mutual attachment between Alicia and her son Roy, who is about to get married to Louise in a few days' time. The black chiffon nightdress is just the type of dress Louise would wear. Alicia refuses to have a Freudian reading of her motives delivered in court; she prefers to be sent to prison for three months: "I'd rather be tried for a petty crime and accept their judgment on it than have the secrets of my heart explored in public."[30] What is most remarkable about this play is its ambiguous attitude towards psychoanalysis: while the basic tenets of Freudian psychology are accepted, Storm maintains that the cure may be worse than the disease and that, in order to make life in a civilized community possible, subconscious desires had better not be brought to light. *Black Chiffon* was followed by a comparatively minor effort, *The Day's Mischief* (Duke of York's, 1951), which revolves around a schoolteacher who comes under suspicion of being responsible for the disappearance of one of his pupils. The characterization in *The Day's Mischief* is considerably weaker than in *Black Chiffon* – while the study of Alicia Christie had been psychologically convincing, the later play abounds in female characters who are more or less hysterical. Storm did not produce any more crime plays in the following decade.

5
From 1956 to the End of the Sixties

The years of 1955 and 1956 are generally considered a watershed in twentieth-century English drama. In 1955, *Waiting for Godot* was performed at the Arts, directed by Peter Hall, and 1956 saw the visit of Brecht's Berliner Ensemble to London and John Osborne's *Look Back in Anger* at the Royal Court. In their different ways, these productions seemed to offer ways of escape from the stifling routines of the well-made play. While Osborne's play, considered revolutionary at its time, is now considered "fairly orthodox,"[1] the impact of absurd drama and epic theatre was more profound and enduring. *Waiting for Godot* was a liberating experience for playwrights because it broke all the rules: there is no specified place or time, the characters do not seem to have a history or any psychological "depth" and, perhaps most importantly, the plot does not follow the traditional formula of "rising action – climax – falling action," but after some time starts to repeat itself. Brecht and the Berliner Ensemble, by contrast, served as a model for a pronouncedly political, radical theatre, and on the level of theatrical style, suggested an escape from the confines of the well-made play by smashing the "fourth wall."

These different trends in drama affected the crime play in various ways and to a varying extent. As noted in the previous chapter, the crime play had experimented with the theme of "angry young men" soon after World War II, but the genre was no longer particularly interested in the character type after *Look Back in Anger* – on the contrary, there is only one angry-young-man crime play to be found in the post-Osborne era: Philip King's *How Are You, Johnnie?* (Vaudeville, 1963), which contributed nothing very innovative. The example of Brecht inspired the evolving theatre workshops and a young generation of left-wing political playwrights, for instance Howard Brenton and David Hare. Both Hare and Brenton also experimented with the crime play; two of Brenton's plays in the genre will be discussed below, Hare's *Knuckle* of 1974 in the following chapter. As far as the impact of absurd drama is concerned, the plays of Harold Pinter and Tom Stoppard have been claimed for the absurdist movement by Martin Esslin.

Though Pinter's plays are characterized by a pervasive sense of threat and have hence jokingly been dubbed "comedies of menace," only one of his plays, *The Dumb Waiter*, qualifies as a crime play in the strict sense. The play's protagonists are two contract killers, Ben and Gus, who are sitting in a shabby basement flat waiting for instructions concerning their next job. Although their conversation deals mostly with trivialities, it gradually becomes clear that Gus, the second-in-command, has qualms about their last murder, because the victim was a woman. Instead of orders for their next murder, they receive orders for fancy foreign meals from the defunct upstairs restaurant via a "dumb waiter." The final twist is that Gus himself is the next intended victim, but before he is executed by Ben, the curtain falls. *The Dumb Waiter* was claimed by Esslin for the Theatre of the Absurd on the following grounds:

> *The Dumb Waiter* brilliantly fulfils Ionesco's postulate in completely fusing tragedy with the most hilarious farce. It also succeeds in turning the mysterious supernatural ingredient [...] into an additional element of comedy: the spectacle of the heavenly powers bombarding two solemn gunmen with demands for "*macaroni pastitsio, ormitha macarounada*, and *char siu* and bean sprouts" is wildly funny. Yet the main element of comedy is provided by the brilliant small talk behind which the two men hide their growing anxiety. These discussions of which football team is playing away on that particular Saturday, whether it is correct to say "light the kettle" or "light the gas", the desultory discussions of trivial news in the evening paper are utterly true, wildly comic, and terrifying in their absurdity.[2]

Esslin's reasons for claiming *The Dumb Waiter* for the Theatre of the Absurd are very persuasive, since Pinter makes use of the plot motifs of the crime play in much the same way Beckett had employed music hall routines. However, it is debatable whether Stoppard's plays really belong to the same category, as Esslin has argued in the revised version of *The Theatre of the Absurd* of 1978:

> The plays of Tom Stoppard also clearly show the impact of the Theatre of the Absurd, in spite of the obvious difference in other aspects of their approach and the tradition – that of English high comedy – which they represent. *Rosencrantz and Guildenstern Are Dead* (1966) uses structural elements of *Waiting for Godot*, while *Jumpers* (1972) concludes its brilliant and zany-absurdist exploration of the problem of good and evil in human existence with a direct paraphrase of famous lines from *Waiting for Godot*: "At the graveside the undertaker doffs his top hat and impregnates the prettiest mourner." To which the character of Archie, and surely the author of the play also, adds: "Wham, bam, thank you Sam." The play's debt to Samuel Beckett could not have been more clearly emphasized.[3]

Rosencrantz and Guildenstern Are Dead does not concern us here, but *Jumpers* (Old Vic, 1972) does, because right at the beginning of the play, a man is murdered. While participating in the creation of a "short, blunt human pyramid," Duncan McFee, professor of philosophy and member of the university gymnastic team, is shot by an unseen assassin. Suspects include beautiful Dorothy "Dotty" Moore, a retired musical star and wife to philosophy professor George Moore; her doctor (and presumably lover) Archie, the university's vice-chancellor; and finally McFee's mistress, who happens to be George's secretary. A police inspector named Bones arrives to investigate the case and perhaps to obtain an autograph of Dotty's, whom he has long admired. Archie adeptly and ruthlessly disposes of McFee's body in a large plastic bag, and tries to bribe Bones into discontinuing his investigation. Whether he intends to shield Dotty or whether he himself killed McFee is by no means clear – nor is it revealed at the end of the play. Thus, *Jumpers*, considered as a crime play, imitates a single earlier attempt by John van Druten to refuse to grant the audience closure by naming the culprit. The murder of McFee serves to unmask the ruthlessness behind Archie's efficiency: "It's not as if the alternative were immortality."[4] In *Jumpers*, Stoppard makes use of the crime play in order to investigate the philosophical question of whether moral absolutes can continue to exist in a totally secularized society. The play's protagonist, George Moore, is continually interrupted in his attempts to write a lecture on the great "whodunit" question: "Who created the universe?"; or as George puts it more economically: "Is God?" Though *Jumpers* thus poses an existential question, it does not present metaphysical uncertainty as an experience but as a form of intellectual gymnastics. It is therefore doubtful if *Jumpers* should be included in a list of plays of the absurd, for all of Stoppard's supposedly acknowledged debt. For the character of Archie, as the above plot summary makes abundantly clear, is not an author surrogate, as Esslin's commentary would seem to suggest.

In all the plays of his early phase, Stoppard clashed (at least) two literary or dramatic styles to achieve his effects: in *Rosencrantz and Guildenstern Are Dead*, he made Shakespeare collide with Beckett, in *Jumpers* he confronted George Moore of the *Principia Ethica* with the whodunit and with musical comedy, and in *The Real Inspector Hound* (Criterion, 1968) he combined the classic stage whodunit with the demolition of the fourth wall as suggested by epic theatre. *The Real Inspector Hound* is set in a theatre in which a traditional whodunit is being performed. Moon and Birdboot, two theatre critics, are watching a conventional *Mousetrap*-style murder mystery and let themselves be lured into the action on stage. Moon is obsessed with fantasies about the disposing of Higgs, his paper's main theatre critic, and Birdboot with sexual fantasies about the two leading ladies of the theatre company. After the first act, in which a character named Simon Gascoyne has been murdered – plus an unknown character whose body has been lying on stage since the curtain went up (albeit strangely ignored by the actors) – Birdboot enters the world of the play by answering a telephone ringing on stage. He is instantly caught up in the action, taking the part of Simon Gascoyne in the first act, while the plot seems to repeat itself. Moon, too, becomes involved in the play's action, when it is revealed that the unacknowledged body on stage is that of Higgs. It turns out that the murder mystery is a trap conceived by Puckeridge, Moon's stand-in, in order to get rid of both Higgs and Moon. *The Real Inspector Hound* is an intricately plotted crime play as well as a parody and mainly interesting for its satire on superannuated devices and mannerisms of the crime play.

The well-made play could be attacked in yet another way than by epic or absurd theatre. Instead of imitating the models of Brecht or Beckett, Joe Orton proceeded to demolish the well-made play "from within," by pushing farce to its limits and its logical conclusion. According to Bentley, the main feature of farce is "outrage to family piety,"[5] and this is a very apt formula to describe the plays of Orton. Orton aimed at provoking the audience with depictions of aberrant sexuality and gratuitous violence, and accordingly two of his three major stage plays revolve around crime. His first major stage success, *Entertaining Mr Sloane* (New Arts, 1964), begins with Mr Sloane, an attractive young man, renting a room in a house belonging to Kemp and his elderly daughter Kath, although Kemp believes he recognized Sloane as the murderer of his former employer. Feigning maternal affection at first, Kath soon seduces Sloane. Her brother Ed, equally attracted to the young man, offers him a job as a chauffeur. Their relationships become more strained when Kath gets pregnant and Sloane in a fit of rage kills Kemp. Kath and Ed agree that they will share Sloane's favours on a half-yearly basis and in return cover up the crime. The obvious criminal Sloane turns out to be no worse than the respectable middle-class family by whom he has been trapped. Instead of attacking the conventions of the well-made play, *Entertaining Mr Sloane* meticulously adhered to them; John Russell Taylor praised it as a "solid, well-managed commercial play"[6] and Terence Rattigan thought so highly of it that he invested £3,000 into its transfer to the West End.

Loot (Jeanette Cochrane, 1966, directed by Charles Marowitz) was an even greater success with the critics, winning the *Evening Standard* and *Plays and Players* awards for best play. "*Loot* is a farcical parody of the detective drama but its detective [...] is a bully who takes a bribe to protect the guilty and will arrange for McLeavy, the only innocent man in the play, to meet an 'accidental' death."[7] The play largely relies on shock tactics, demolishing the values of filial love, piety towards the dead, and law and order, depicting the dismemberment of corpses as well as homosexual practices. Traditional detective plays and stories are parodied in the character of Inspector Truscott, who, Holmes-like, draws exaggerated conclusions from insignificant clues, but, unlike Holmes, has no scruples about forging the evidence to obtain his ends. As Innes has pointed out, *Loot* also firmly belongs into the tradition of English mainstream drama, since it

> clearly echoed Ben Travers' 1928 classic, *Plunder*, in its title as well as basic elements of its plot: a nurse who cuts the family out of her patient's will; a double robbery that brings police investigation; the threat of arrest for murder, and the use of blackmail to provide a resolution. At the same time, where Travers satirically presented this frenetic criminal activity as normal, Orton exaggerated it to the point of absurdity.[8]

In 1967, the Royal Court staged a programme with the title *Crimes of Passion* consisting of two short plays by Orton, *The Ruffian on the Stair*, the revised version of a radio play broadcast by the BBC in 1964, and *The Erpingham Camp*, originally a TV play. *The Ruffian on the Stair* is summed up concisely by M. D. Allen as follows:

> Mike, a small-time crook, has killed someone in a road accident. The dead man's brother, Wilson, invades the home of Mike and his girl friend, Joyce, and eventually tricks Mike into shooting him. Mike has not only removed the obligation of living from the bereft Wilson but has also provided him with the revenge he desired, for Mike must now explain himself to the police.[9]

It must be added, however, that Mike relies on a mild verdict for a "crime of passion" and that Joyce is more distressed by the death of her goldfish than by Wilson's death. *The Erpingham Camp* exposes the latent humiliation and violence in so-called "entertainments" at a holiday camp. One evening's entertainment at the "Erpingham Camp," an institution run along the lines of a totalitarian state, gets out of hand and ends in looting, vandalism and the eventual killing of the manager Erpingham.

The two earliest crime plays by dramatist Simon Gray, *Wise Child* (Wyndham's, 1967) and *Dutch Uncle* (Old Vic, 1969) both in theme and in style closely resemble Orton's work. Gray had written some short fiction and TV plays before producing his first stage play, *Wise Child*, which was originally also intended for television, "but at the time [...] considered too bizarre for home viewing."[10] The play's protagonists are two crooks: Jerry is a youngish confidence man and "Mrs Artminster" a rather more masculine and violent criminal by force of circumstance obliged to pose as Jerry's mother. Jerry engages his "Mum" in elaborate role play, which is already in progress when the play starts. The two crooks are hiding in a hotel run by Mr Booker, a middle-aged homosexual strongly attracted to religion and to Jerry. "Mrs Artminster" alternately tries to persuade Jerry to slosh Booker over the head and rob his safe and to sell Jerry to Booker. There is a short interlude of Artminster trying to seduce Janice, the West Indian maid, and in the end he returns to his male personality, leaves Jerry and goes back on the run. Jerry blames Booker for the break-up between "Mum" and himself and in a sudden outburst of violence kills him. The play starts as a comedy but soon becomes thoroughly depressing in its highlighting of the characters' mutual dependence and their inevitable frustration, recalling Sartre's *Huis clos*. In the original production, the part of "Mrs Artminster" was taken by Alec Guinness, an unlikely casting choice.

Dutch Uncle is a black farce about an incompetent murderer inspired by the exploits of serial killer John Christie. Mr Godboy's two main ambitions in life have been to be a famous detective or else a famous criminal, preferably a notorious mass murderer.[11] Throughout the play he incompetently tries to gas his wife in a large walk-in cupboard and to subject his tenant Doris to the same treatment. The play repeatedly examines the relationship between sexuality and violence – though playfully rather than (psycho-) analytically – and includes police violence in its analysis. The recent abolition of stage censorship had made sexually motivated killing a possible topic for a stage play; however, audiences found it impossible to stomach a farce based on a true and recent murder case, and Gray recalls the play's opening night as being "the worst night in the British theatre."[12]

The instances of violence and sexual deviance depicted in *Entertaining Mr Sloane*, *Loot* and *Wise Child* are even more amazing considering that at the time of their first performance, censorship was still formally in operation. Until 1968 every play to be produced on the English stage had to be submitted to the Lord Chamberlain's office for approval. Plays could be banned for indecency, violence, blasphemous or abusive language or the unflattering portrayal of people alive or dead for less than 50 years. Successive Lords Chamberlain took increasingly liberal attitudes towards the portrayal of violence, to swearing and sexual content. On 28 September 1968, censorship was finally abolished, which paved the way not only for disciples of the continental Theatre of Cruelty but also for improvisation, which so far had been illegal since it was likely to render control and censorship impossible. *Dutch Uncle* was among the first crime plays to benefit from the abolition of censorship. As a rule, the genre of the crime play proved reluctant to make use of the new liberties, however, plays like John Hopkins' *This Story of Yours* (Royal Court, December 1968) or Barry England's *Conduct Unbecoming* (Queen's, 1969) probably would not have found a producer had censorship still been in operation.

This Story of Yours is one of the few attempts in the genre to deal with crime and punishment in a realistic manner. Police Sergeant Johnson beats up a suspect in the interrogation cell so viciously that the man dies. The play consists of three confession scenes: The first scene shows Johnson returning home after the incident. He is drunk, has a fight with his wife, and reveals that he is at the end of his tether, having been professionally exposed to sights of death and destruction for 20 years. Act II shows Johnson being questioned by his superior: this scene culminates in Johnson's confession of having discovered his own sadistic impulses in his encounter with suspect Baxter. Act III is a flashback showing the incident itself: the parallels between the rapist[13] and the interrogator become more and more obvious. Johnson's beating up of Baxter is accompanied by the sound of a woman reaching orgasm.

Conduct Unbecoming likewise examines how the pressures exerted on men serving in institutions such as the police or the army may eventually lead to violent outbursts of aggression. The play is set in the India of the 1880s. On the night of a ball at the officers' mess, a major's widow is attacked viciously in a manner recalling a local pastime in which soldiers attempt to penetrate the anus of a stuffed boar with their swords. Suspicion falls on young Millington, who molested the lady before; rather half-heartedly, however, since he merely tried to bring about his immediate dismissal from the regiment. An unofficial court martial is held to judge Millington, who is defended by Drake, a young officer who gradually has to revise his concepts of truth and honour. Drake matures in the course of the trial as he discovers the real culprit. He finds an Indian woman who was similarly attacked, but all the witnesses available swear that these atrocities were committed by one John Scarlett, a dead hero of the regiment, flayed alive during the time of the Mutiny. It turns out that Roach, a friend of Scarlett's, suffered from a guilt complex because he survived while Scarlett was killed, and thus brought him back to life by impersonating him – as the vicious and sadistic man Scarlett was in his lifetime. The cause for the sadistic excesses described in the play is revealed to be regimental discipline: the victim explicitly states: "It doesn't matter which one of these men it was. They are all the same. Stupid, cruel men, who treat pigs and women as though they were objects."[14] *Conduct Unbecoming* at the same time manages to fulfil generic expectations and undertake a serious investigation into the origins of violent behaviour.

Like Hopkins and England, Howard Brenton questions the previously clear-cut distinction between "the criminal" and "the policeman" in his two crime plays, *Revenge* (Royal Court, 1969) and the short play *Christie in Love*, first performed by the Portable Theatre, which also transferred to the Royal Court in 1970. Like *This Story of Yours* and *Conduct Unbecoming*, Brenton's plays profited from the recent abolition of censorship. Roger Cornish efficiently sums up *Revenge* as

> a two-act seriocomic vaudeville in which an ancient professional criminal, Adam Hepple, seeks to destroy his nemesis, MacLeish of Scotland Yard. Both men are driven by ideals, Hepple by a mobster's code, MacLeish by Calvinist fervor. But their motives are confusingly (and purposely) mixed: Hepple seeks to preserve a sense of traditional values; MacLeish models his toughness on American gangster Al Capone. At the finish, all characters die, narrating their deaths and biting blood bags, one of several theatricalisms which set the tone for Brenton's later work.[15]

It should be added that Hepple and MacLeish were played by the same actor.

Christie in Love, as the title suggests, is a play based on the murders of notorious serial killer John Christie. The play opens after Christie's arrest and shows him being questioned by the police, who have reconstructed the murders to a certain extent, but are at a loss as to motivation. The first murder is then reconstructed with one of the policemen playing a woman's part using a life-size puppet and Christie responding as he did to his first victim, an amateur prostitute. In the course of the encounter, Christie pretends to be a plainclothes policeman in order to obtain sexual favours from the prostitute. Schnierer comments on the role reversal inherent in this device: "Dramatically conclusive is the dislocation technique which turns the police constable into a whore [...] and Christie into a policeman."[16] The last scene shows Christie's execution. The utter helplessness caused by the confrontation with serial killing is expressed by the Inspector: "I dunno, it's disappointing. Why can't a mass murderer be just a bit diabolical? Why can't a pervert like you, already in the annals of nastiness, have fangs or something? [...] And Madame Tussauds has been onto us all day for a plaster cast of your head."[17] In order to manipulate audience sympathies, Brenton created Christie as a fully human, "three-dimensional" character, while the policemen interrogating him are deliberately designed as cardboard character types.

Apart from the impact of developments in theatre history, the crime play continued to be influenced by contemporary detective fiction, mainly because a number of detective novelists also produced works for the stage. Agatha Christie dramatized her novel *Towards Zero* together with Gerald Verner (St. James's, 1956) and *Five Little Pigs* as *Go Back for Murder* (Duchess, 1960). She also produced two original plays, *Verdict* (Strand, 1958) and *The Unexpected Guest* (Duchess, 1958). For *Verdict*, she departed from her successful whodunit formula and had the murder committed in full view of the audience. When it seemed, on the opening night, that she also intended to deny her spectators the customary happy ending – the curtain went down a few minutes early by mistake, cutting off the last-minute reconciliation – the play was booed. In spite of its lack of popular success, Christie thought highly of her play, ranking it second only to *Witness for the Prosecution*. *Verdict* was her attempt to write a problem play, demonstrating in the guise of a crime play that, in an imperfect world, uncompromising idealism will only lead to suffering. For *The Unexpected Guest*, Christie returned to her successful formula and achieved a run of over 600 performances. Michael Starkwedder, the "unexpected guest" of the title arrives at the house of the Warwick family in South Wales late one November evening and finds Laura Warwick holding a gun over the dead body of her wheelchair-bound husband. Starkwedder decides to save Laura by re-arranging the evidence so that it points to an outsider. *The Unexpected Guest* was Christie's last stage success; both *Go Back for Murder* and a trilogy of one-act plays that she produced in the following years failed to achieve the popularity of her earlier plays.

Besides Christie, a number of other detective novelists also produced works for the stage. Lawyer Henry Cecil adapted three of his detective novels, which usually feature some important problem of legal practice, for the stage with the aid of different collaborators. Michael Gilbert, who also combines the professions of solicitor and detective novelist, produced the crime plays *A Clean Kill* (Criterion, 1959), *The Bargain* (St. Martin's, 1961) and *The Shot in Question* (Duchess, 1963). *A Clean Kill* is a conventional crime play employing an amorous triangle and the somewhat hackneyed device of a stopped clock indicating the time of the murder. *The Shot in Question* likewise uses the adultery motif and is also primarily concerned with the reconstruction of the events leading to the death of the victim. While these two plays revolve around a central mystery that is elucidated in the course of the action, *The Bargain* is a play of intrigue and counter-intrigue about a lawyer who is tricked into compromising and criminal behaviour, closely related to Cecil's *Alibi for a Judge*. What is most striking about these plays today is the strict code of conduct applied to the members of the legal profession – even in the supposedly "swinging 60s," it was apparently thought compromising behaviour for a lawyer to spend the night under the same roof with a single woman.

Detective novelist and director William Fairchild wrote two original crime plays, *The Sound of Murder* (Aldwych, 1959) and *Breaking Point* (Lyric, 1962). *The Sound of Murder* is yet another play about adultery and murder, employing a sound-recording device and introducing a villain who is cold and egoistic in private life and at the same time an acclaimed author of warm-hearted children's books. This character was played by Peter Cushing, the Frankenstein of the Hammer horror films, and described by J. C. Trewin as "quite the most damnable personage I have met in the theatre."[18] *Breaking Point* is mainly remarkable for its innovative setting: in order to escape from the drawing-room, Fairchild moved the action to a hut in the Canadian Arctic. It should be added that in Fairchild's plays, the main villain is not the murderer, which somewhat undermines the simplistic moral scheme of traditional crime fiction and drama.

Finally, Roy Vickers, an author of short detective fiction, made a contribution to the genre of the crime play by publishing his short story "Double Image," which was turned into a successful crime play by the team of Roger MacDougall and Ted Allan and produced at the Savoy in 1956.[19]

As the above survey shows, the crime play and detective fiction were still strongly interrelated. There is, however, one important exception: in the field of spy fiction, drama increasingly moved away from the model of the novel. At the height of the Cold War, popular interest in spy novels increased: the 1950s saw the emergence of the new secret agent hero James Bond in fiction, and the 1960s his first appearance on screen. In the 1960s, John Le Carré began writing spy fiction less inclined towards hero worship, most notably his novel *The Spy Who Came in from the Cold* (1963). The crime play, however, produced only two stage plays revolving round espionage, John Chapman's *Simple Spymen* (Whitehall, 1958) and Arthur Watkyn's *Out of Bounds* (Wyndham's, 1962), and these two plays treat the theme of espionage in a distinctly comic manner. The immensely successful *Simple Spymen,* which saw over 1,400 performances, is a roaring farce revolving around two buskers mistakenly engaged by the War Office to get hold of an important invention by a foreign scientist called Grobchick. While the War Office believe that the "atomic pile restorer" Grobchick has invented may be used for the fabrication of atomic weapons, it finally turns out to be nothing but a carpet restorer. In the course of the action, everybody impersonates somebody else, which shows that the *theatrical* interest of espionage lies in its role-playing. While the James Bond saga had glorified the virtues of patriotism and courage, Chapman's play deliberately undermined them. For instance, when one of the buskers has just given the other a pep talk about courage in the face of danger and notes to his dismay that the other's knees are still trembling, he asks: "What's the matter with your knees?," getting the reply "They're not so easily fooled."[20] "They're not so easily fooled" neatly sums up the audience response Chapman hopes for; theatre audiences are apparently expected to be more sceptical towards Cold War propaganda than cinema audiences or the readers of spy novels.

Watkyn's play is clearly inspired by *Simple Spymen,* as well as by Graham Greene's spy novel *Our Man in Havana*. In *Out of Bounds,* once again an ordinary individual – in this case the headmaster of a Somerset school – is conscripted as a secret agent and manages the situation admirably. *Out of Bounds* was more sober in mood than *Simple Spymen* and largely served as a vehicle for Michael Redgrave, who played the headmaster.

The period from the mid-1950s to the mid-1960s also saw the production of a number of conventional crime plays that either imitated specific successful earlier models or in general conformed to the established formulas of amorous triangle and murder of spouse. *The House by the Lake* (Duke of York's, 1956) by Hugh Mills for instance was a reworking of the central theme of *Gas Light*: a childishly dependent woman breaks free from her domineering husband and proves him to be a murderer. David Ellis' *Make Me a Widow* (Comedy, 1964) recycled the plot of *Someone Waiting,* using changed genders: here it is a woman who gets revenge on the man responsible for her daughter's death, sacrificing her own life in the process. The amorous triangle leading to murder featured prominently in Leslie Sands' *Something to Hide* (St. Martin's, 1958), Philip Mackie's *The Big Killing* (Princes, 1962), Jack Popplewell's *Policy for Murder* (Duke of York's, 1962) and Robert Bloomfield's *Portrait of Murder* (Savoy, 1963). Murder of spouses occurred in all of the above, and in Mackie's *The Key of the Door* (Lyric Hammersmith, 1958). Arthur Watkyn's *Amber for Anna* (Vaudeville, 1964) was a traditional whodunit with suspicion resting on everyone in turn. In spite of their conventional plot structure, several of these plays tried to introduce innovative features: *The Key of the Door*, for instance, made extensive use of flashback, and *Portrait of Murder* used a *doppelgänger* motif and cleverly employed the device of doubling. *Amber for Anna*, which was summed up by Hagen as "[a] group of people returning from a thriller relive another at home,"[21] is made more interesting by an extensive metadramatic debate on the improbability of crime plays, which is then interrupted by real crime intruding. The discussion of the crime play in a crime play points to an increasing self-consciousness and an undeniable exhaustion of the genre.

Besides, the type of crime play based on the homicidal triangle and the murder of spouse set in the middle-class drawing-room was the variety most threatened by the increasing dissemination of television, since this kind of play could be easily and cheaply produced for TV and much more comfortably and cheaply consumed. According to BBC statistics, in 1955, about 40 per cent of the population had access to a TV set, while in 1966, the percentage had gone up to 93. In 1955, commercial television was introduced in Great Britain, and the BBC tried to compensate for its lost monopoly by increasing its output of television drama. The crime play on stage had to compete not merely with televised crime plays[22] but also with police serials specifically created for TV, such as *Dixon of Dock Green* (1955), *Scotland Yard* (1960), *Who Me?* (1961), *Jacks and Knaves* (1961) and *Z Cars* (1962). Accordingly, once the country had become saturated with TV sets, the genre of the stage crime play experienced a marked decline in absolute numbers of new plays produced, but some of the plays that were performed were decidedly more original, because they had to offer their audiences something they could not have at home.

Parallel to the conventional crime play discussed above, crime comedy continued to flourish – and indeed turned out to be the only subgenre of the crime play to spawn sequels to successful plays. The crime comedies of the period are based on the principle of inadequacy, casting unlikely characters as criminals or detectives. Following the models of *Arsenic and Old Lace* and George Pollock's films about Miss Marple, playwrights frequently chose elderly women either as criminals or detectives. In *Tabitha* (Duchess, 1956) by Arnold Ridley and Mary Cathcart Borer, a group of old women conspire to poison their landlady but soon get qualms of conscience. The title of Peter Coke's *Breath of Spring* (Cambridge, 1958) ostensibly refers to an expensive variety of mink coat, but in the course of the action assumes the symbolic significance of the rejuvenating experience of crime. Dame Beatrice, an elderly impoverished gentlewoman, and her lodgers find a new interest in life by committing a number of successful thefts. This play was so popular – it saw 431 performances – that it spawned a sequel called *Midsummer Mink* (1965), which, however, did not attain the same amount of popularity.

The elderly woman as a detective was embodied by Irene Handl in Jack Popplewell's *Busybody* (Duke of York's, 1964) and its sequel *Dead Easy*

(St. Martin's, 1973) as well as in *My Giddy Aunt* (Savoy, 1968), a broad farce written by John Chapman and Ray Cooney. In *Busybody*, Handl played the part of Lily Piper, office cleaner at Marshall Developments Ltd. The play starts off with Mrs Piper discovering a corpse in the office and ringing the police. Strangely, the corpse seems to have moved – a "busy body" – and, less strangely, to be that of Mr Marshall, who is a Most Likely Victim, since everybody in the office, plus his wife, had a motive for murdering him. Confusion reaches its height when Marshall turns out to be alive, but another corpse is soon discovered some way off in Notting Hill and turns out to be that of James Cameron, Mrs Marshall's lover. Marshall escapes, and in the third act is revealed to be an innocent suspect framed by his partner Robert Westerby, who would have profited financially as well as romantically – he, too, is in love with the boss's wife – had Marshall been convicted for murder. Thus, each of the three acts conforms to another sub-genre of the crime play: the ordinary whodunit (I), the love triangle murder (II) and the innocent suspect hunted for a murder he did not commit (III). The crime is solved by Mrs Piper, the amateur detective of the play, who provides most of the humour, particularly in her exchanges with the policeman in charge, Police Superintendent "Arry" Baxter, an old acquaintance of Mrs Piper's, who ought to be in bed nursing his cold. *Busybody* is a very self-conscious crime play, using the typical guiding questions that Suerbaum identified as the main characteristic of the true detective novel. In I.iii, for instance, Baxter asks the assembled suspects: "Did you rid yourself of Richard Marshall to clear the path for your real love? Or you, Mr Westerby? Is your story true? Or are you indeed the lover of Claire Marshall?"[23] Lily Piper reappeared in *Dead Easy,* which makes her the only serial detective in an original crime play, in spite of – or perhaps because of – the fact that she is an unlikely detective being middle-aged, female, and working-class. In *My Giddy Aunt*, Handl played both the elderly, mentally confused Lady Eppingham, who is murdered in the course of the action, and her practical energetic sister Beatrice who investigates the crime.[24]

An inadequate elderly male criminal figured as the protagonist of Arthur Watkyn's *Not in the Book* (Criterion, 1958). Wilfrid Hyde-White played Andrew Bennett, a respectable if somewhat boring businessman who must dispose of a blackmailer and relies on a meticulously worked out murder plan lifted from a detective novel manuscript. He hopelessly messes up the projected murder, almost accidentally poisoning the local police inspector in the process. However, the decisive factor is that he just cannot bring himself to go through with the murder and warns his victim just as he is about to swallow his poisoned coffee. The use of a novel as a blueprint for a real murder raises the question of the relationship between fictional and actual crime and allows the author to accommodate a lengthy discussion of narrative conventions in detective fiction, similar to the one occurring in his later play *Amber for Anna*.

The theme of the battle of the sexes in evidence in a lot of the traditional crime plays of the period also produced a new plot formula which may be summed up as: "female hostage turns the table on her kidnapper(s)." The first example of this type of play is Monte Doyle's *Signpost to Murder* (Cambridge, 1962). The play's plot revolves around the rule that once an inmate of a psychiatric asylum has been at large for four weeks, a new order must be issued to have him recommitted. Overpowering a male nurse, Roy Collier escapes from a lunatic asylum he has been confined to after supposedly murdering his wife in her bath. He enters the cottage of Sally Thomas in the vicinity of the asylum and takes her hostage for 28 days, during which a curiously intimate love-hate relationship develops between the two.

The new formula was brought to perfection by Frederick Knott in *Wait until Dark* (Strand, 1966), which saw over 1,600 performances in London, successfully transferred to New York and was later filmed by Stanley Donen. After being approached by the beautiful criminal Lisa, young London photographer Sam Henderson has unwittingly smuggled a doll containing dope from Holland to London. Lisa's two accomplices, con men Mike and Croker, and the somewhat more dangerous psychopathic criminal Roat are now trying to make Henderson's blind wife Susy hand over the doll to them. Roat has killed Lisa, and the criminals manage to persuade Susy that the doll may implicate Sam in the murder in order to make her search the flat. Susy, however, is not without resources of her own: with the help of the neighbours' daughter Gloria, she sees through Mike's fictions, and she finally defends her territory against the dangerous Roat by blacking it out completely. The enormous impact of the play is hardly to be explained by the mere summary of its plot, which even contains a major logical flaw.[25] The intense excitement is mainly caused by formal devices, by the constellation of characters and the use of recurrent symbols. The play's list of characters is beautifully symmetrical. There are three male criminals, a good one (Mike), a bad one – Roat, who is conceived as an image of perfect, that is, protean, evil: he appears first as an old man, then as a young man, and finally as definitely subhuman, "like a reptile"[26] – and an indifferent one (Croker), three female victims, Susy (good), Lisa (bad) and Gloria (indifferent), and three detectives, Sam and two policemen, who, however, come to the rescue belatedly. Susy's blindness is particularly interesting with respect to changing gender roles. In 1966, the "damsel in distress" is no longer conceived as weak simply because she is female – Knott adds an extra handicap to manipulate the odds sufficiently against Susy's survival.

Two individual crime plays produced in the period before 1970 still deserve closer attention – not because they established a new type, but because they took up an already existing formula and realized it in a particularly convincing manner. Rodney Ackland's *A Dead Secret* (Piccadilly, 1957) is a period piece set in 1911, based – as for instance *Rope* and *The Winslow Boy* – on a true criminal case. Frederick Dyson, a self-confident and self-satisfied insurance agent and a representative of the rising capitalist business ethics, persuades one of his lodgers, elderly Miss Lummus, to make over her property to him in return for a small annuity. When she dies of arsenic poisoning, Dyson's greedy and miserly behaviour casts suspicion on him, and the celebrated K. C. Arthur Lovecraft summoned to his defence declines his brief, entertaining grave doubts concerning his client's innocence. The last scene of the play foreshadows the death sentence against Dyson, but it adds a conciliatory note by granting Dyson one last conversation with his wife, during which they declare their mutual love. *A Dead Secret* was nominated one of the "best plays of the year" by *Plays and Players*, normally reticent in the praise of crime plays. This critical approval may be due to the fact that Dyson's innocence or guilt is not the central issue of the play: the main interest is in the portrayal of human greed, individualism and independence as qualities engendered by the rise of capitalism.

While Ackland had based his plot on a true criminal case, Jack Roffey in *Hostile Witness* (Haymarket, 1964) employed the format of courtroom drama that had previously been successfully employed for instance by Edward Wooll and Agatha Christie. The play's central character is Simon Crawford Q.C., a self-reliant and independent man, who suddenly finds he must rely on others for help when he is arrested for murder. The evidence against Crawford is fairly damning – for instance, his fingerprints are found on the murder weapon, and the victim's blood on his coat. Of course Crawford is the victim of an elaborate frame-up, engineered by a man he once sent to prison. The convict took up employment in Crawford's chambers under an assumed name and waited for an opportunity for revenge. The true culprit is finally identified due to a fatal weakness: he is colour blind. The colour blindness of the criminal mirrors another type of "moral blindness" in Crawford, who fails to recognize his clerk as the man he sent to prison, and who bullies Sheila, a junior counsel in his chambers, almost beyond endurance. The trial has a humanizing effect on Crawford, who learns to rely on Sheila and to be more aware of the people who daily surround him. *Hostile Witness* turned out to be the last courtroom play on the English stage to use the motif of the innocent accused. In the courtroom plays of the following decades, the issue usually was not the guilt or innocence of the accused but the fairness of the judicial system – as in Hare's *Murmuring Judges* or Norton-Taylor's *The Colour of Justice* – or a debatable moral standard, as in Terence Rattigan's *Cause Célèbre* and Terence Frisby's *Rough Justice*.

6
From *Sleuth* to In-Yer-Face Theatre

The 1970s saw the emergence of a new type of crime play, the type christened the "postmodern comedy thriller" by Marvin Carlson. What makes these comedy thrillers postmodern is that they use the parody of the established genre of the thriller as their starting point, in the manner of Stoppard's *The Real Inspector Hound*. The traditional crime play turned out to be a particularly suitable genre for parody because of its rigid conventions, and because the necessity for innovation – the constant need to surprise audiences – is inscribed in the rules of the genre itself. As Carlson explains,

> a genre dedicated to surprising audiences with unexpected turns inevitably finds audiences becoming more and more difficult to take off guard. On the other hand, they also become open to more and more radical and complex subversion of expectations. The result is a machine for increasing destabilization of generic givens and creating their replacement by ludic experimentation.[1]

Carlson isolates the following main characteristic features of the postmodern comedy thriller: generic self-referentiality, parody, a scepticism towards the positivistic optimism implied by classic detective fiction and "the simultaneous awareness and playful deferral of death."[2]

The first play of this type and the one to conform most exactly to Carlson's specifications was Anthony Shaffer's *Sleuth*, which opened at the St. Martin's in 1970. The play's protagonist is detective story writer Andrew Wyke, who has for some time been betrayed by his wife Marguerite. Her lower-class lover Milo comes to see Andrew, but instead of making a jealous scene, Andrew invites Milo to steal Marguerite's jewellery, which is heavily insured, so Milo will be able to offer Marguerite the life she is accustomed to. Andrew persuades Milo to disguise himself as a clown in order to commit the theft, but while the burglary is in progress, Milo learns that Andrew had an ulterior motive: he intends to shoot him as a presumed intruder. Act I ends with a shot and Milo dropping down the stairs. In Act II, Andrew is visited by Police Inspector Doppler who suspects him of having murdered Milo. When Andrew breaks down, Doppler drops his disguise and reveals himself as Milo. Andrew wants to make friends with Milo who has just qualified as a suitable partner for his elaborate role plays, but Milo insists that putting people in mortal fear is not a game. He tells Andrew that he has laid another trap for him. After strangling Téa, Andrew's Finnish mistress, Milo has hidden three vital clues all over the room which incriminate Andrew. Andrew has to find them before the police arrive. He manages to do so in the nick of time, only to learn that this was another hoax in which Téa willingly participated. Andrew invites Milo to stay and live with him, but Milo refuses, exposing Andrew as impotent, sexually as well as emotionally. When Milo goes to collect Marguerite's fur coat, Andrew resorts to his original plan and shoots Milo. The play ends with Milo's death and the arrival of the police.

The play was extremely popular with audiences – it saw 2,359 London performances and over 1,000 in New York – and critics alike. The only discontented reviewer, Martin Gottfried, complained that *Sleuth* was "both a mystery and a parody of one, mocking the clichés and mechanics of detective stories while capitalizing on them."[3] Shaffer himself expressed the same idea in slightly different terms, remarking that "his purpose in *Sleuth* was to spoof Agatha Christie and the 'cozy crime' genre and 'at the same time to use it – to have my cake and eat it too.'"[4] This is precisely what, according to Carlson, makes the play a postmodern comedy thriller.

> A murder plot, a victim, a police inspector, and a murder investigation – all the standard features of the conventional detective story are present, and yet none of them is what it seems. Each of the conventional generic codings the audience can be expected to make is subverted and the subversion itself utilized, not to make fun of the genre, as in Stoppard, but to fulfill its conventional goal of shocking and surprising the audience by unconventional means that, as Hutcheon claims for postmodern practice, "use and abuse, install and then subvert" the machinery of the genre itself.[5]

Like *Sleuth*, Shaffer's next play, *Murderer* (Garrick, 1975), also revolves around a man who enjoys playing murder games: Norman Bartholomew delights in re-enacting notorious historical murder cases. While demonstrating "murder as a fine art" the play at the same time questions the validity of this concept. The initial scene of Norman's apparent murder and dismemberment of a helpless female victim is particularly gruesome and instantly became famous – it was intended by the author that it should last at least 30 minutes. After the extended shock effects of *Murderer*, Shaffer returned to a more playful mood in *The Case of the Oily Levantine* (Her Majesty's, 1979), which transferred to New York in 1982 under the title of *Whodunnit*. This is a deliberate and obvious parody of 1930s' murder mysteries with the main target of satire being *Ten Little Niggers*. Besides, the play contains overt references to other authors and works, most notably to Chesterton and John Dickson Carr. The "oily Levantine" of the original title blackmails all the house guests of Champflower Manor, plus the butler, and in return is beheaded with an ancient sword. Everybody present had both motive and opportunity for committing the murder. Inspector Bowden's investigations soon reveal that the guests are not what they seem – in fact, they are all actors, hired for a murder charade by their mysterious absent hostess Angela Bazeby. The parody is very clever and accomplished and genuinely funny, as is the constant suspension of the play between different levels of reality.

The elaborate "murder games" of Shaffer's plays and their self-conscious genre parody makes them exemplary specimens of the postmodern comedy thriller. Other examples of the type on the English stage are Simon Gray's *Stage Struck* (Vaudeville, 1979), *Murder by the Book* by Greenwood/King[6] and Gerald Moon's *Corpse!* (Apollo, 1984). These plays also revolve around murder games and continuously challenge the audience to reassess the ontological status of the events presented on stage – are they meant to be "real" or are they "play"? The "simultaneous awareness and playful deferral of death" is in evidence in the shape of an abundance of false deaths – in *Murder by the Book*, for instance, husband and wife shoot each other with blanks and poison each other with placebos, whereupon the "victim" enacts a convincing "death scene." And "*Corpse!* piles up false deaths with the kind of maniacal repetitiveness suggestive of traditional farce."[7] To the basic confusion between what is real and what is play, *Corpse!* adds a confusion of identities: the play's plot revolves around an unsuccessful actor plotting the murder of his twin brother. The *doppelgänger* motif and the accompanying theatrical device of doubling were becoming increasingly popular in the 1980s and appear for instance in Charles Ross' *Dead Ringer* (Duke of York's, 1983), Rees/Elice, *Double, Double* (Fortune, 1986) and Stoppard's *Hapgood* (Aldwych, 1988).

A play Carlson does not discuss although it fits his definition of the postmodern comedy thriller admirably is Tudor Gates' *Who Killed "Agatha" Christie?* (Ambassadors, 1978). The play's title not merely expresses a pronounced hostility towards the "Queen of Crime" (not uncommon among mystery writers) but also defines the basic intention of the play, "mocking the clichés and mechanics of detective stories while capitalizing on them:" An unsuccessful actor-playwright commits the perfect murder killing an elderly homosexual drama critic jokingly referred to as "Agatha" Christie. The murder weapon is a crime play. John Terry invites Arthur Christie to his house, pretending to have knowledge of a clandestine affair between Terry's wife Jo and Christie's lover Brian. In the course of their conversation it transpires that Jo and Brian are having a rendezvous downstairs and can be overheard in their lovemaking via Terry's advanced hi-fi equipment, and that Terry plans to kill them *in flagrante delicto*. It then turns out, however, that the real target of his malice is Christie, who wrote a few scathing reviews of Terry's work, dismissing it as melodramatic and unconvincing. The climax of the play has the sound of the lovemaking between Brian and Jo turned on full, with Terry storming downstairs and heard to cut the throats of Brian and Jo, while Christie is helplessly tied to his chair. Terry rushes upstairs again with Brian's head in his hand and declares that he will kill Christie as well and frame him for murdering Brian, whereupon Christie dies of a heart attack. In his final soliloquy Terry reveals that the sounds they were listening to came from a radio play in which Brian was starring, the head was a stage prop for which Brian obligingly sat, and that Brian has indeed a lover called Jo – John Terry. *Who Killed "Agatha" Christie?* is a highly metatheatrical two-hander which drops names of famous writers and critics by the dozen.

Another almost perfect example of the postmodern comedy thriller had not yet appeared when Carlson's study was published: *Mindgame* by Anthony Horowitz (Vaudeville, 2000), which constantly undermines the audience's assumptions about what is "real": Styler, an author of true crime paperbacks, comes to see Dr Farquhar, head of Fairfields, an asylum for the criminal insane, in order to obtain his permission to interview one of the inmates, serial killer Eastermann. Nurse Plimpton tries to warn Styler to leave as quickly as possible, but he persists, discovering too late that, as in Poe's "System of Doctor Tarr and Professor Fether," the patients have taken over the hospital and killed the staff, apart from a few doctors, among them "Nurse Plimpton," whom they have kept alive for sadistic role-play. Farquhar really is Eastermann, and he gradually manages to persuade Styler to cooperate in the torture and murder of Plimpton. When Styler finally owns up to his own aggressive impulses, another reversal takes place, this time borrowed from the classic German horror film *Das Kabinett des Dr. Caligari*. It is revealed that Plimpton and Farquhar are indeed therapists, and that "Styler" is Eastermann who has just taken part in an extraordinarily lively and challenging psychodrama session.

While *Who Killed "Agatha" Christie?* and *Mindgame* radically undermine the distinction between play and reality, a number of plays Carlson does discuss are not really sufficiently subversive to be counted as postmodern comedy thrillers. This applies in particular to Francis Durbridge's plays *Suddenly at Home* and *Nightcap*, which are more appropriately discussed in a different context.

As pointed out in the previous chapter, by 1970 virtually every inhabitant of the UK had access to a TV set. During the last decades of the century, the impact of TV on the genre of the crime play was accordingly becoming more and more noticeable. The career of Francis Durbridge is a case in point. Durbridge had started his career in the early 1930s with a number of BBC radio serials featuring the novelist-detective Paul Temple. He adapted two of these serials into novels and from the 1940s onwards produced both radio serials and original novels. From the early 1950s, he became an extremely successful writer of TV serials through which he shaped the viewing expectations of millions of viewers. In 1971, his first stage play, *Suddenly at Home*, opened at the Fortune. The plot revolves around Glenn Howard's murder of his wife Maggie, which he carries out with the help of his mistress Sheila. His elaborate plan also includes the framing of Sam Blaine, Maggie's former lover, as the culprit. Sam, an author of crime novels, strikes back: he enlists the help of an unemployed actor to perform the part of "Police Inspector Remick," whose intervention gradually brings Sheila to the end of her tether. This crime play has all the usual ingredients: not one or two but three love triangles, the elaborate framing of an innocent person, the murderous husband, the return of a person presumed dead, here repeated twice, and the bogus policeman. As in Durbridge's other plays, the correct solution largely depends on realizing who is cooperating with whom. The murders occur on the open stage, and much of the play's actual puzzle aspect is connected to Sam's attempts to clear himself. This play works quite as well when presented on TV as it does on stage – in fact, it *was* presented on TV in Germany – and the same applies to Durbridge's other stage crime plays. What makes his plays so suitable for the medium of TV is their "well-madeness:" Intended for the picture-frame stage, these plays make no attempt to "smash the fourth wall" and acknowledge the presence of the audience, hence they may easily be transferred to the "box." Not only is there no direct audience address, a declamatory style of delivery is also out of place; the dialogue aims at easy conversational realism, and the acting is naturalistic. The plot is suspenseful and contains several "surprise reversals," which, instead of destabilizing the audience's expectations and evoking a feeling of unease, gives the spectators the comfortable feeling of watching a well-crafted piece of work. The cast is limited to a small number of characters belonging to a social class slightly above that of the general viewer:

> The cast is normally drawn from what is popularly termed the gin-and-tonic set – upper-middle-class estate agents, photographers, and antique dealers abound, as do others who appear to have ample time to indulge in deception and criminal intrigue. [...] Then there are his settings – switching from London to Thames-side commuter country, and back again. When the hero is lured to a cottage at Marlow or a houseboat on the river, we can be sure he will find a door ajar and a corpse just beyond.[8]

Of Durbridge's stage crime plays, two – *Suddenly at Home* and *Deadly Nightcap* (Westminster, 1986) – are based on the homicidal amorous triangle, and two – *House Guest* (Savoy, 1981) and *A Touch of Danger* (Whitehall, 1988) – revolve around the motif of professional crime intruding into middle-class family life. His second West End play, *The Gentle Hook* (Piccadilly, 1974) combined the themes of professional crime and the disintegration of a marriage. In spite of Durbridge's immense popularity as a writer for TV, his stage sets fail to acknowledge the existence of television. There are no TV sets in his living-rooms – unless we conclude from the position of the sofa that the TV is meant to be placed where the audience is sitting.

A handful of other playwrights, however, not merely acknowledged the existence of television but actually wrote crime plays centrally concerned with TV's exploitation of crime. This applies for instance to Richard Harris' *The Business of Murder* (Duchess, 1981), Alan Ayckbourn's *Man of the Moment* (Globe, 1990) and Ben Elton's *Popcorn* (Apollo, 1997). *The Business of Murder* is a play about "the worm that turns" and gets revenged on two people who destroyed his life, strongly reminiscent of Williams' *Someone Waiting*. Six years ago, Stoner was suspected of having murdered his divorced wife and nine-year-old son. Police Inspector Hallett arrested him and tried to enforce a confession, and a budding young playwright, Dee Redmond, had her first big success on TV with a semi-documentary about the case. Since then, Dee and Hallett have been conducting an adulterous affair. Stoner lures Hallett and Redmond into his flat and gradually reveals that he has killed Mrs Hallett earlier that evening and carefully framed Hallett and Dee for the murder. Dee believes him, but Hallett realizes that this is only a cleverly fabricated fiction. In the showdown, Stoner manages to provoke Hallett to enter into a fight with him and forces him to stab him. In a deliberate challenge of audience expectations, actor Francis Matthews, Durbridge's television hero Paul Temple, was cast for the part of Stoner.

Man of the Moment is more cynical in outlook. Fourteen years ago, timid and friendly bank manager Douglas Beechey in a sudden burst of heroism attacked a robber who had entered his bank. Beechey was celebrated as a national hero, while Vic Parks, the bank robber, served his prison sentence. Since then, a reversal of fortunes has taken place: Parks made a brilliant career in the media while Beechey, who got married to a hostage wounded in the course of the robbery, has to struggle to make a living. The main target of this satirical play is the way the media industry distorts actual events beyond recognition to make them fit preconceived ideas about good and evil, heroism and cowardice.

Ayckbourn's satire seems mild, however, when compared to the attack on the media industry undertaken by Elton's *Popcorn*, which discusses the interrelation between "true crime" and representations of crime in the mass media. The edge of satire is somewhat blunted, however, by the fact that *Popcorn* is set in the United States, in the Beverly Hills home of film director Bruce Delamitri, a character obviously modelled on Quentin Tarantino. On returning home after winning his first Oscar for *Ordinary Americans* (a travesty of *Natural Born Killers*) with *Playboy*-centrefold Brooke, Bruce is taken hostage by the notorious "Mall Murderers" Wayne and Scout. Wayne intends to escape capital punishment for his crimes by making Bruce declare to running TV cameras that he himself is morally responsible for the killings committed by the Mall Murderers, since his films inspired their crimes. As the evening progresses, TV ratings fall and only pick up again once Wayne has shot Farrah, Bruce's estranged wife. He announces that the killing will continue as long as TV ratings keep rising. In the epilogue, it is revealed that Bruce and Scout were the only people to survive the massacre and that viewers and victims' relatives have already begun to sue the police and TV companies all over the country.

Since the play is predominantly concerned with representations of crime in the different media, it is worth noting that *Popcorn* itself exists in various incarnations; it was produced on stage and issued as a novel almost simultaneously, and there has been talk of turning it into a film to be directed by Joel Schumacher. Elton himself is familiar with a variety of different media, having written not merely novels and stage plays, but also screenplays and TV plays – he is probably best known for his contribution to the TV series *Blackadder*. Due to its extensive use of graphic violence, *Popcorn* has frequently been discussed in the context of in-yer-face theatre, although critic Aleks Sierz has complained about "Elton's frequent use of the same sick humour he claims to be satirizing."[9] Elton obviously makes use of the postmodern practice to "use and abuse, install and subvert" at the same time, and is thus more closely related to the postmodern comedy thriller than to in-yer-face theatre.

The medium of television has affected the stage crime play not merely in shaping audience expectations and writing practices (as in the case of Durbridge's plays) or by presenting a topic for the crime play (as in Harris or Ayckbourn) but first and foremost as a formidable competitor. This becomes most obvious when considering the adaptation of detective novels. A number of the current Queens of Crime – for instance Minette Walters, Ruth Rendell, Elizabeth George and Val McDermid – have had their work broadcast on TV but not adapted for the stage. Whenever a new female English writer of detective fiction emerges, she is hailed by the press as Agatha Christie's successor, but so far none of them has shown Christie's interest in work for the stage. The lack of interest of the younger generation of authors of detective fiction in work for the theatre has caused the crime play to disengage itself almost entirely from contemporary detective fiction. Instead of transferring new detective fiction to the stage, the crime play nostalgically resurrected the heroes of Golden Age detective fiction. Leslie Darbon, for instance, adapted Agatha Christie's novels *A Murder Is Announced* and *Cards on the Table* for the Vaudeville, and Graham Greene revived E. W. Hornung's gentleman cracksman Raffles in his "Edwardian comedy" *The Return of A. J. Raffles* (Aldwych, 1975). In contrast to Darbon's plays, Greene's comedy is not a dramatization of any particular Raffles novel but merely "based somewhat loosely on E. W. Hornung's characters in *The Amateur Cracksman*."[10] In Greene's play, Raffles crosses the path of another (in)famous fin-de-siècle character, Wilde's "Bosie" Lord Alfred Douglas. Raffles has survived his reported death in the Boer War and returns to his close friend Bunny, who has struck up an acquaintance with Lord Alfred Douglas, lonely since Wilde's departure to Paris, and impecunious since the Marquess of Queensberry stopped his allowance. Bosie persuades Raffles and Bunny to burgle the Marquess's home where they encounter the Prince of Wales, who is staying there incognito to meet his mistress. Raffles ensures that some compromising letters are returned to the Prince which were pinched earlier in the evening by a German agent (who, like Raffles himself, was disguised as a waiter), and the gratitude of Royalty protects Raffles from the more severe consequences of his crime. At the end, Raffles and Bunny go on a burgling spree to the Loire while in England they will from now on play nothing but cricket.

Jeremy Paul's *The Secret of Sherlock Holmes* (Wyndham's, 1988) is another creative rewriting of a detective fiction classic rather than a stage adaptation of Doyle's novels. Taking his cue from contemporary literary criticism that considers the detective the criminal's "double," Paul produced a play largely epic in character: In turns, Watson and Holmes tell and re-enact the story of their extraordinary friendship, culminating in a hypothetical explanation for what author Jeremy Paul thought to be the central problem of their relationship: how could Holmes let his best friend believe in his death for three years after the incident at the Reichenbach Falls? The answer, presented as a duel of wits, is that Holmes himself is Moriarty, creating a master criminal as one of his many disguises because he needed the stimulus of a dangerous adversary. Jeremy Brett and Edward Hardwicke starred, recreating their TV interpretations of Holmes and Watson on stage.

Playwright David Hare, by contrast, was more inspired by the classic American detective fiction of the hardboiled variety. His crime play *Knuckle* (Comedy, 1974) bears a motto from Ross MacDonald and jokingly juxtaposes George Moore's *Principia Ethica*[11] and the writings of Mickey Spillane. The play's attack on immoral capitalist business practices and its morally ambiguous hero place it firmly in the tradition of the writings of Dashiell Hammett. Curly Delafield left his home town of Guildford decades ago and has now returned to investigate the disappearance of his sister Sarah. Sarah vanished from Eastbourne Beach, leaving behind her coat and a handbag containing two first-class return tickets. She may either have committed suicide or met a man, presumably her lover, who subsequently killed her. Even in adulthood, both Curly's and Sarah's lives are still dominated by their father Patrick, a smooth city banker. While Sarah hysterically voiced her dissent with her father's business practices, Curly expressed his disapproval by becoming an arms dealer. It finally turns out that Sarah is alive and (reasonably) well, and that she only staged her disappearance in order to implicate her father, who had just been involved in a rather shady property speculation including the brutal eviction of tenants. Curly decides not to expose his father's business practices and thus loses the affections of Jenny, a girl he has admired since the beginning of his investigation. In Schnierer's analysis, *Knuckle* becomes the "ultimate" crime play, in the sense of being the crime play to end all crime plays:

> Hare employs the simple investigative structure of the Mickey Spillane thriller to show the total inadequacy of such an approach in the face of a world which cannot be changed by it, because it is rigid and circular. [...] The hermetical nature of the play, its circular structure, does away with the need for the champion of truth, the redeemer figure, the *deus ex machina*, the catalyst, the evil antagonist and the funny copper – in short, all the tasks a policeman or a detective, literary or otherwise, could fulfil on stage.[12]

Knuckle received mixed reviews but was praised by the *Times Literary Supplement* and won the John Llewellyn Rhys Prize. Hare expressly declared his intention to write a "play *noir,*" and indeed *Knuckle* is more strongly reminiscent of a *film noir* than of hardboiled detective fiction. "*Knuckle* has the flow as well as the plot of a film: its 17 scenes move rapidly through time and space like the major sequences of a film while retaining the surface patina of tough-guy realism."[13] In his crime play about revenge *Wild Justice* (Theatre Royal Stratford East, 1990), Barrie Keeffe also self-confessedly aimed at writing a "*film noir* for the stage" and in his production note demanded "*Citizen Kane*-style" light and shadow effects. The classics of *film noir* obviously proved more inspirational for playwrights than the classics of their own genre. Appropriately, David Hare next tried his hand at the genre of the crime film with *Paris by Night* and *Wetherby*.

Due to the growing influence of the mass media, the imminent demise of the stage play was repeatedly predicted. In the 1990s, however, British theatre experienced a sudden increase in popularity. A new generation of playwrights produced exciting new plays with tremendous popular and critical success that were soon exported to the Continent. These new British plays had a number of features in common and were collectively christened "In-yer-face theatre" by *Tribune* drama critic Aleks Sierz. In his book entitled *In-Yer-Face Theatre: British Drama Today*, Sierz gives the following definition:

> The widest definition of in-yer-face theatre is any drama that takes the audience by the scruff of the neck and shakes it until it gets the message. It is a theatre of sensation: it jolts both actors and spectators out of conventional responses, touching nerves and provoking alarm. Often such drama employs shock tactics, or is shocking because it is new in tone or structure, or because it is bolder or more experimental than what audiences are used to. Questioning moral norms, it affronts the ruling ideas of what can or should be shown onstage; it also taps into more primitive feelings, smashing taboos, mentioning the forbidden, creating discomfort. [...] How can you tell if a play is in-yer-face? It really isn't difficult:

> the language is usually filthy, the characters talk about unmentionable subjects, take their clothes off, have sex, humiliate each other, experience unpleasant emotions, become suddenly violent.[14]

This dominant new theatrical style was represented by the works of writers like Sarah Kane, Mark Ravenhill and Anthony Neilson, which were staged, at least initially, at venues dedicated to the fostering of new writing such as the Royal Court, the Almeida or the Hampstead Theatre or pub theatres like the Bush and the Finborough.

Since in-yer-face theatre was predominantly concerned with taboo-breaking and violence, quite a number of plays turned out to be crime plays. Anthony Neilson's *Normal* (Finborough, 1991), for instance, was based on the homicidal career of the "Düsseldorf Ripper" Peter Kürten. Most of the action is set in 1930s' Germany, and the play begins with a flashback: visiting a chamber of horrors in an amusement arcade, lawyer Justus Wehner finds himself reminded of his first brief as a young lawyer, which consisted of defending Kürten. The inexperienced middle-class young man gradually came under the spell of the monstrous Kürten and was forced to acknowledge his kinship with the murderer, who was eventually found to be "normal" in the legal sense of being accountable for his actions and, thus, executable. In the epilogue, a parallel between Kürten's murders and the crimes of the Third Reich is drawn, raising the question of whether evil is inherent in certain individuals, particular societies, or even in human nature in general. The play's choice of title is of course highly ironic, since it is precisely what we understand by being "normal" that is the issue. As Neilson himself put it: "You name the perversion and he embraced it. When he's declared sane, you're meant to think: 'What do I have to do to be judged insane?'"[15] Sierz relates Neilson's play to the public's increased interest in serial killers in the early 1990s, which is also in evidence in several Hollywood films of the period, most notably *The Silence of the Lambs* (1991) and *Natural Born Killers* (1994). Among British playwrights, Karen Hope also made use of the theme in *Foreign Lands* (Finborough, 1993), which is loosely based on the exploits of Myra Hindley and poses the question of what comes after crime and punishment.

Simon Donald in *The Life of Stuff* (Donmar Warehouse, 1993), by contrast, chose organized crime as his topic. Sierz calls it "an urban fairy tale fuelled by drugs and violence,"[16] but it is perhaps more aptly described as a gangster play centring on rivalry in a drug racket. In its excessive use of violence – although most of it suggested rather than shown – the play recalls the hardboiled tradition. The verbal style also seems to echo the laconic language used for instance by Hammett in *Red Harvest*, but Donald employs the characteristic hardboiled language in order to reveal his characters' lack of ability to reflect and their complete dependence on animal impulse – sex, fear, pain – and drugs. The play makes extensive use of disgust: people drink urine and nibble a severed toe by mistake.

Like Donald's play, Jez Butterworth's *Mojo* is set in a nightclub. The play gained immediate critical acclaim, winning several Most Promising Playwright Awards and being compared to the plays of Pinter and Beckett as well as to the films of Quentin Tarantino. The play, however, more closely resembles American plays like those of Sam Shepard and, particularly, David Mamet. Ezra, the owner of a Soho night club, changes from juke boxes to live acts after discovering "Silver Johnny," a teenage boy who has an apparently irresistible attraction for young girls. The promise of increased profits makes the club interesting to East End gangster Sam Ross, who prepares a takeover: he takes Johnny with him and kills Ezra, who is found in two dustbins by his business associates Mickey, Sweets, Potts and Skinny and "Baby," Ezra's son. They are mortally frightened that Ross will return to complete the takeover; at the same time, rivalry breaks out within the group as to who is to be Ezra's successor. At first it looks as if Mickey, the oldest, will remain in charge, but in a dangerous spontaneous action, Baby goes and kills Ross with a cutlass, thus establishing himself as the rightful heir. Ezra and Ross never appear on stage; the whole action is seen through the bewildered eyes of Mickey, Baby, Sweets, Potts and Skinny, who are ridiculously underequipped for gang warfare. Sierz praises Butterworth's elegance of writing and discovers problematic male-male relationships as the play's underlying central theme:

> With its all-man cast, *Mojo* is clearly a study of blokedom. While much of the plot, which tiptoes around the power struggle between two gangsters, happens offstage, what you actually see is male relationships: Sweets and Potts, Baby and Skinny, Mickey and Skinny, and Baby and Mickey. So *Mojo* starts as a gangster thriller and ends as a study of absent fathers.[17]

Similarly, Simon Bennett's *Drummers* (Royal Court, 1999) at first glance seems to be about professional crime – "drummer" is a slang term for a housebreaker – but actually is concerned predominantly with family life, as Bennett maintains: "*Drummers* isn't really about crime. It's about love. It's about how love's not smooth. Especially in families. The characters just happen to be criminals."[18] The play accordingly shows how conflict and mistrust develop in families that live by crime – there is a father-son conflict, a mother-son conflict and a conflict between two brothers, the two "drummers" Ray and Barry, with whom the concept of brotherly love takes on the curious form of anal rape.[19] The conflict between Ray and his mother Ella, however, ends in a reconciliation that strangely echoes the ending of *The Mousetrap*: Ella has made Ray's bed, and he finally goes to sleep.

The playwright to go furthest in his use of violence and shock tactics was Irvine Welsh (the author of *Trainspotting*) in his play *You'll Have Had Your Hole*, which premiered at the West Yorkshire Playhouse in 1998 and was shown at the London Astoria 2 in 1999. Even Sierz (who is surely used to some amount of stage violence) called the play "excruciatingly cruel"[20] and noted "a delight in torture that was positively Jacobean."[21] The plot revolves around two small-time gangsters who have kidnapped another petty criminal and taken him to a sound-proof recording studio in order to torture him at their leisure. Significantly, *You'll Have Had Your Hole* was less successful than the previous offerings of in-yer-face theatre, which may indicate that the trend was beginning to abate; sooner or later, there will be no taboos left to be broken. Apart from the ubiquitous anal rape, in-yer-face crime plays featured cannibalism (Debbie Isitt, *The Woman Who Cooked Her Husband*, Royal Court, 1991) and matricide (Phyllis Nagy, *Butterfly Kiss*, Almeida, 1994). Doug Lucie's *Gaucho* (Hampstead, 1994), a play based loosely on the career of Oxford drug baron Howard Marks, contains a sequence in which a traitor is tortured by a woman who performs aggressive oral sex on him. It may be concluded that in-yer-face theatre largely achieves its effects of shock and disgust by combining basic physical activities – such as eating and defecation, violence and copulation – which our culture requires to be kept separate.

Largely unaffected by major trends like the postmodern comedy thriller and in-yer-face theatre, a few established subgenres of the crime play continued to evolve, most notably the spy play, courtroom drama and crime comedy. As the end of the Cold War approached, spy plays interestingly were becoming markedly less humorous. In *Hapgood* (Aldwych, 1988), Tom Stoppard used the format of the spy play to continue his discussion on the validity of moral absolutes that had already been the theme of his previous crime play *Jumpers*. But while *Jumpers* had been inclined towards the affirmation of moral absolutes, *Hapgood* questions the existence of absolute moral values, at least in political, though possibly not in personal life. The play revolves around a mole hunt in the British Secret Service. Elizabeth Hapgood, head of the British Intelligence Network, suspects her assistant Ridley of being a double agent and, incidentally, one of a pair of twins working for the KGB – the *doppelgänger* motif here serves as an image of dual loyalties. She traps him in order to clear the prime suspect, Joseph Kerner, a Russian defector and former lover of hers, who is also the father of her only son Joe. Kerner, a physicist, supplies the scientific analogy to the moral problem of oscillating political loyalties by stating that a double agent is like a trick of the light. Hapgood finally catches Ridley by pretending to be her own vulgar twin sister Celia Newton, who has to cooperate with him in carrying out an exchange of secret information for Hapgood's son Joe, supposedly kidnapped by the Russians. When Ridley discovers that he has been fooled, he tries to kill Hapgood who shoots him first. The last scene shows Hapgood saying good-bye to Kerner, who is returning to Russia. The play deals with loyalty in political as well as in personal life and ventures the suggestion that, as modern physics did away with the concept of certainty, human behaviour also is no longer governed by absolutes. As this plot summary reveals, *Hapgood* shows a number of characteristic features of the postmodern comedy thriller in its use of the *doppelgänger* motif and its scepticism towards moral certainties. However, Carlson does not include *Hapgood* in his analysis, although he discusses Heisenberg's contribution to the development of postmodern uncertainty: "Werner Heisenberg, in his highly influential *Physics and Philosophy* (1962), argued that the epistemological and ontological uncertainty introduced to the world of physics by quantum theory and relativity theory must inevitably affect the world of philosophy as well."[22] Ten years after *Hapgood*, Michael Frayn reused the uncertainty principle as a metaphor for moral ambiguity and political betrayal in his play *Copenhagen* (National, 1998).[23] However, the dates of the publication of *Physics and Philosophy* and the first performances of *Hapgood* and *Copenhagen* suggest that Heisenberg's uncertainty principle had a lesser effect on the abolition of moral certainties than the collapse of Communism.

Like *Hapgood*, Alan Bennett's double bill programme *Single Spies* (National, 1988) raises the question of what sense the concept of "treason" makes in modern politics. The programme consists of two one-act plays, *An Englishman Abroad* and *A Question of Attribution*, inspired respectively by the cases of Guy Burgess and Anthony Blunt. *An Englishman Abroad* depicts Burgess' loneliness and homesickness in Russian exile, while *A Question of Attribution* is a conversation piece concerned with art and history, forgery and false attribution, leading to the conclusion that fake is in the eye of the beholder.

For his spy play *Cell Mates* (Albery, 1995) Simon Gray also chose a celebrated historical case, that of George Blake, who was famously sentenced to a 40-years' imprisonment at Wormwood Scrubs. In prison he meets the Irish petty criminal Sean Bourke. The two men establish a friendship, and after his release, Bourke manages to organize Blake's escape single-handedly. Blake flees to Moscow where Bourke comes to visit him only to learn to his dismay that he has attracted the suspicions of the KGB and will have to remain in the country. In the play's last scene, however, Bourke discovers that the threat of the KGB was nothing but a clever invention by Blake who wanted his friend to stay with him. While the play's concern with the fate of English defectors in Russia echoes *An Englishman Abroad*, the theme of two men bound together by mutual dependence recalls Simon Gray's very first play, *Wise Child*. By 1995, Gray had obviously abandoned the postmodern aesthetics to which he had aspired in his comedy thriller *Stage Struck*. Reviewing Gray's contribution to the genre of the crime play, one notes that his plays were generally modelled on earlier successes: *Wise Child* and *Dutch Uncle* imitated the plays of Orton, *Stage Struck* was inspired by Shaffer's *Murderer*, *Cell Mates* by *An Englishman Abroad*, and *Molly* (Comedy, 1978) by Terence Rattigan's *Cause Célèbre* (Her Majesty's, 1977).

Cause Célèbre was the earliest of a handful of courtroom plays produced during the period presently under survey that discussed some kind of social or moral issue. Although *Cause Célèbre* ostensibly is a courtroom drama based on the notorious Rattenbury trial of the 1930s, Rattigan keeps his action largely out of the courtroom (as in his previous *The Winslow Boy*), and the play's central theme is not murder but sexual hypocrisy. The moral problem at issue in Jeffrey Archer's *Beyond Reasonable Doubt* (Queen's, 1987) and Terence Frisby's *Rough Justice* (Apollo, 1994) is euthanasia – in the former play, a husband kills his terminally ill wife; in the latter, parents kill their handicapped child. The protagonists of both plays stand trial for murder, defend themselves and leave the courtroom as free men. From the point of view of the ideology transported in popular crime plays, these two plays indicate a wide-spread acceptance of euthanasia. From the point of view of plot construction, it is worth noting that in contrast to earlier courtroom drama, the accused in late twentieth-century courtroom plays tend to be guilty of the crimes they are accused of.

This also applies to the most interesting courtroom drama of the period, David Hare's *Murmuring Judges* (Olivier, 1991). *Murmuring Judges* was the second part of Hare's trilogy on British institutions, appearing after *Racing Demon* (on the Church of England) and before *The Absence of War* (on the political parties). The play follows three young people through the British penal institutions of the police, the court of law, and the prison. Although a first time offender, young Irishman Gerard McKinnon is sentenced to five years' imprisonment after collaborating in a robbery as the gang's driver. It turns out that two policemen eager for promotion – Barry Hopper and Jimmy Khan – planted drugs on McKinnon's accomplices to enforce a confession. Irina Platt, a junior lawyer in the chambers of conservative barrister Sir Peter Edgecombe, tries to make sure that McKinnon receives a fair trial in his appeal but finds that she is up against numerous time-honoured rules of behaviour and institutional loyalties. As a result of her efforts, McKinnon's sentence is reduced to four years and a half, but the whole affair has a number of more far-reaching consequences: Barry is promoted to the Flying Squad, but his girlfriend Sandra, a young police constable, decides to report him to the authorities; Irina quits her post with Sir Peter, and Gerard for the first time discovers his affinities with the Irish Republican Army. A problem play in the tradition of Galsworthy's *Justice*, this play is even more scrupulously fair than Galsworthy's plays on the legal system, in spite of the fact that the older ranks of the Bench and Bar are mildly caricatured and portrayed as lacking both a sense of reality and of responsibility.[24]

While the main target of criticism in *Murmuring Judges* had been the judiciary, Richard Norton-Taylor's *The Colour of Justice* (1999) examined racism within the police force. *The Colour of Justice* is a special case among the plays surveyed in this study, since it was based entirely on the transcripts of the Stephen Lawrence inquiry that were merely edited by *Guardian* journalist Norton-Taylor. In 1993, 18-year-old Stephen Lawrence died as the result of a racist attack. The police failed to arrest five likely suspects – whether out of sheer incompetence or because of their own racist prejudice was what the inquiry attempted to establish. The amazing amount of records that got lost or mislaid in this case clearly suggests that these losses occurred intentionally.

However, crime plays intent on social criticism did not necessarily choose the format of courtroom drama. Barrie Keeffe's *Sus* (Soho Poly, 1979) used the framework of a police interrogation in order to criticize racism in the police. The blurb of the published version sums up the plot as follows: "Two detectives keenly await the Tory landslide while interrogating a black suspect – not for the 'sus' charge which he so cheerily expects but for the brutal murder of his heavily pregnant wife." "Sus" is a technical term allowing the police to detain a suspect for a certain period of time without any evidence. Here it refers to the automatic suspicion against the husband of having performed an illegal abortion. It is finally revealed that the woman died of a haemorrhage brought about without any outside interference. The political statement rests on the depiction of the cruel interrogation methods used by the two racist policemen, but, regrettably, Keeffe needs to resort to the convention of the innocent suspect in order to make his point and to avoid compromising the impact of his play.[25]

Unimpressed by popular trends, Alan Ayckbourn throughout the 1990s continued to write and perform his farces at his Scarborough laboratory, the Stephen Joseph. *Man of the Moment* has already been discussed above as an example of media critique. While the title of Ayckbourn's next crime play, *The Revengers' Comedies* (Strand, 1991), jokingly alludes to Jacobean drama, the play itself is actually based on the principle of Patricia Highsmith's (and Alfred Hitchcock's) *Strangers on a Train*. Two suicidal people meet on the Albert Bridge at midnight and decide to swap revenges: Karen Knightley will get revenged on Lembridge Tennit, the company that has just made Henry Bell redundant, and Henry in return will endeavour to hurt Imogen Staxton-Billing, the wife of Karen's lover. The shifts between the two parallel strands of the action present a considerable difficulty for the dramatist – in all likelihood, this is what attracted Ayckbourn to the basic plot in the first place. With its numerous shifts, *The Revengers' Comedies* arguably is better suited for cinema and accordingly was made into a film in 1997.

Communicating Doors (1995) deals with time travel and hence contains a strong fantasy element. Three women of different ages and backgrounds get involved with shady businessman Reece Welles over a period of 40 years. At one point in time they are each in danger of being murdered by Reece's business associate and best friend Julian Goodman. A time-warp device installed in the "communicating doors" of their hotel suite enables them to move back in time in order to warn their predecessors about Julian's designs and thus change their individual life histories for the better. The play turns out to be a surprising celebration of female solidarity and is very funny in parts. The intellectual appeal of the play, however, does not derive from the murder mystery element but from the involved arguing out of the logical paradoxes created by the possibility of time travel.

Considering the development of the English crime play in the course of the twentieth century as a whole, one may discern several general tendencies. Throughout the century, the crime play quickly adopted – and was adopted by – new trends in drama: after crime melodramas, there appeared well-made crime plays, absurdist crime plays, postmodern comedy thrillers and "in-yer-face" crime plays. In this respect, the crime play resembles the detective novel, which was also frequently used for (mainly postmodern) literary experiments. The crime play also showed a remarkable sensitivity to the cultural preoccupations of its time, taking up the craze for spiritualism in the early twentieth century as well as the theme of changing gender relations and the ensuing marital discord in mid-century and discovering the "angry young man" of the postwar period as a protagonist in drama before that term was ever coined.

Major shifts with respect to formal aspects of the plays' construction, such as the management of plot, time and space, coincide with larger cultural shifts. Crime melodrama disappeared when the emerging medium of cinema adopted the dramatic techniques of the type of play dominant at the period of its inception – Victorian melodrama – and thereby hastened the demise of melodrama. Similarly, TV at the beginning of its mass dissemination adopted the dominant dramatic techniques of its period, the conventions of stage naturalism and the well-made play. This forced producers of stage drama to experiment with theatrical formats that could not easily be adopted by TV, mainly by exploiting the fact that audience and actors in the theatre share the same space. Presumably TV contributed more to the demolition of stage naturalism than the combined examples of Brecht, Thornton Wilder, and Beckett.

These observations apply to commercial drama in general; a historical development that affects the crime play more specifically is its divorce from contemporary detective fiction which occurred in the 1960s. During the last decades of the twentieth century, contemporary detective novelists have had their work adapted for TV but not for the stage, while playwrights have proved to be less inspired by current trends in detective fiction than by the formats of the Golden Age "cozy" or the hardboiled thriller and its cinematic relative, *film noir*.

Throughout the century, new trends in the crime play first and foremost evolved in the realm of comedy. This may be explained by crediting comedy in general with a particular innovative potential, but it may also be argued that comedy (and humour in general) is of special importance for the treatment of the gruesome theme of crime. After all, George Grella has persuasively argued that the Golden Age detective novel is but a new version of the comedy of manners.

Part II
A Poetics of the Crime Play

7
Settings and Stage Sets

In his discussion of the Victorian heritage in twentieth-century English drama, Heinz Kosok explains how the requirements of stage realism and the necessities of theatre architecture and theatrical production bring forth an interminable succession of plays confined to a single living-room set.

> The technical equipment of the theatre frequently is still in the state of the 19th century; that is, there are rarely any revolving or sliding stages, high flies or complicated lighting equipment that would allow quick set changes. Accordingly, English theatre tends to confine itself to a single set that – due to a lack of sufficient depth of the stage – almost always has to be an interior. Since the programme does not change nightly, the set may be built solidly and furnished for a long run. A dramatist who wants to see his work performed has to adapt to these circumstances. The result is an endless succession of drawing-rooms as the settings of English plays, where a complex plot may unfold most easily.[1]

The crime play conforms to this general rule, and a number of crime plays that have attained classic status made a virtue of necessity and used the confinement to a single set as a means of increasing suspense, the type christened the "murder house mystery" by LaBorde. The trend towards the single set continues into the 1990s, but from the late 1980s onwards, the opposite tendency is also to be observed. A number of productions employ non-realistic multifunctional sets that allow the action to move freely in space and in time, a decision that is perhaps informed by the influence of the aesthetics of cinema.

The solidity of set and furniture helps to reinforce the illusion of stage realism and thus conforms to the doctrine of verisimilitude governing both the well-made play and the genre of detective fiction. In detective fiction, the "surface texture of the real world"[2] is meticulously imitated in order to distract the reader from the lack of realism in characterization and plot construction. A similar attempt to make an improbable plot more realistic and hence more acceptable to audiences had already been made in Victorian domestic melodrama, as Booth has noted: "It is with domestic melodrama that the curious paradox begins of a mostly unreal content combining with increasingly realistic settings, a dream world disguised as a true one."[3] Percy Fitzgerald explains that the pleasure of seeing everyday objects on stage resides in their very familiarity:

> [T]he most complicated and familiar objects about us are fearlessly laid hold of by the property man and dragged upon the stage. Thus, when we take out dramatic pleasure, we have the satisfaction of not being separated from the objects of our daily life ...[4]

The key term in this description is "familiar": audiences are expected to take pleasure in seeing objects on stage with which they are thoroughly acquainted in real life, and perhaps seeing them put to new and surprising uses.

In his Chapter 2, "The Scene of the Crime," Carlson describes the typical crime play set as follows: "The setting for such works is most commonly a well-to-do British country home, preferably one, as Lehman notes, isolated by both geography and inclement weather to intensify the aura of entrapment"[5] and La Borde adds that "gloomy, forbidding rooms are most desirable,"[6] particularly if they contain secret rooms or passageways. This convention is deftly parodied in Tom Stoppard's *The Real Inspector Hound*, when the housekeeper answers the phone with the words "Hello, the drawing-room of Lady Muldoon's country residence one morning in early spring?"[7] and later describes the location more specifically thus: "I've known whole week-ends when Muldoon Manor, as this lovely old Queen Anne House is called, might as well have been floating on the pack ice for all the good it would have done phoning the police."[8] The country-house setting in a remote location, though so prominent in the minds of theatre critics, turns out to be not quite as frequent as might be expected. Country-house sets proper occur in early twentieth-century crime plays, but become considerably less frequent as the century progresses – just as the country-houses themselves. After 1970, country-house sets occur only in period crime plays like Graham Greene's *The Return of A. J. Raffles*, or in parodic crime plays such as Anthony Shaffer's *Sleuth*. Yet the term "country-house" is a helpful concept since it provides a number of useful categories for distinguishing different types of set. Since it is a *country*-house, it suggests the opposition between rural and urban location, since it is a country-*house* it moreover points to the distinction between public and private spaces. The fact that a country-house belongs to an upper-class extended family, housing numerous poor relatives, servants and other dependants, distinguishes it on the one hand from middle-class and working-class settings and on the other hand from homes inhabited by the typical nuclear family of the twentieth century. In the course of this chapter all the four possible distinctions will be followed up.

To begin with the opposition between rural and urban settings: W. H. Auden has stated about the detective story that a rural setting is preferable to the town. The setting "should be the Great Good Place; for the more Eden-like it is, the greater the contradiction of murder."[9] In the tradition of Auden's essay, detective fiction criticism has tended to locate British crime fiction in small villages – hence the derogatory term of the "little world of Mayhem Parva" coined by Colin Watson – and American crime fiction in large cities, frequently Los Angeles or San Francisco, towns that – in Auden's terminology – represent the "Great Wrong Place." In American crime fiction the choice of an urban setting often coincides with a pronounced interest in organized crime. The English crime play, however, prefers an urban setting but still concentrates on domestic crime. London is by far the most frequent location for a crime play – more than 50 per cent of the plays considered are set in London. This may be yet another instance of the "familiarity" to which the crime play aspires: London audiences apparently prefer to watch London plays. This tendency was already noted by Dion Boucicault as early as 1864, when he called the same play *The Streets of London* in London, *The Poor of New York* in New York, *The Poor of Liverpool* in Liverpool – and in Leeds, where it had try-out runs, the play was named *The Poor of Leeds*. Boucicault himself commented: "I localize it for each town, and hit the public between the eyes; so they see nothing but fire."[10] A twentieth-century example of the practice of re-localizing a crime play depending on the city of performance is Arnold Ridley's *The Ghost Train*, which was given an American setting for its production in the US.

Among rural locations, the counties adjoining the Greater London Area are chosen most frequently, particularly Kent, Surrey and Sussex. A setting in the home counties first of all denotes wealth: Constance Cox in *The Murder Game*[11] dubs this area the "stockbroker belt." Moreover, this choice of location may have to do with the mechanics of plot, since characters may travel down to London to gather information at Scotland Yard or Somerset House within one day. Accordingly, the distribution of crime play locations across Britain may be visualized as a magnetic field with London at its centre.

The only rural area remote from London to occur in crime plays with some frequency is the South West of England, particularly Cornwall and the Devon coast. The majority of these plays, however, are adaptations of novels, which capitalize on the picturesque and sublime character of the landscape to a far greater extent than the drama versions. This difference becomes most evident when comparing the stage adaptations of *A Man with Red Hair* and *Rebecca* to their respective novel versions, in which the scenery assumes the same degree of importance as the characters of the action themselves.

In the context of the opposition between the rural "Great Good Place" and the urban "Great Wrong Place," it is noteworthy that in crime plays, moving from the city to the country often connotes danger. In *Love from a Stranger*, a young woman follows her serial killer husband to "a very charming country cottage" in "an enchanting garden"[12] after a whirlwind courtship in London, and it is the charming country cottage that sees the showdown between the murderer and his intended victim. In *Trunk Crime* the mentally disturbed protagonist drugs a fellow student in his Oxford college and puts him into a trunk, intending to let him suffocate in the swamp behind his cottage in the Oxford Marshes. To urban theatre audiences, the city seems to appear not only more pleasant but also safer than the country. A parallel case in crime fiction may be found in the tales of Sherlock Holmes, for although Holmes famously has his practice in 221b, Baker Street, London, the crimes he investigates are frequently committed in the country or the suburbs.

Considering the opposition between private and public, we find that the majority of crime plays throughout the century is set in private spaces, in a family home. This is important for the mechanics of plot since only a limited number of people have access to the scene of the crime; and it also corresponds to the genre's preference for domestic crime. There are, however, public types of set as well, which may be grouped in the following categories: 1) Pub/hotel; 2) professional; 3) legal institutions.

As has been noted by Ulrich Suerbaum, the hotel or boarding-house setting is an adaptation of the country-house set for less affluent times:

> The decline of the stately homes and their owners is reflected, albeit somewhat belatedly, in the scene of the crime. The manor houses are becoming more modest. [...] In late Christie, the country-house that sets the scene for the crime has been converted to a boarding-house (as in her successful stage play *The Mousetrap* of 1954) or marked for demolition (*Endless Night*, 1967).[13]

In itself, this type of set therefore marks the end of an era that was still persistently celebrated in detective fiction long after it had ended in real life. Hotel and boarding-house sets occur throughout the century. *Somebody Knows* is an interesting study of a boarding-house community as a surrogate family. A bitter travesty of this motif occurs in *Wise Child*, where the illusion of the surrogate family is dispelled almost immediately. The two earliest crime plays set in hotels – *The Man Who Stayed at Home* and *Murder on the Second Floor* – present the boarding-house as an adaptation of the "robbers' den" of fairy tale. The same applies to most of the pub settings, for instance the (roughly contemporary) *The Crooked Billet* and *A Spot of Bother*. The motif of the robbers' den was revived in late twentieth-century gangster plays, for instance *Mojo* and *The Life of Stuff*, which were set in nightclubs.

Professional sets mostly belong to the medical professions as for instance in *The Amazing Dr. Clitterhouse*. A number of plays combine public and private settings by having the action unfold in a parlour behind a shop (*Blackmail*) or by blending sitting-room and consulting-room (in *The Shot in Question*) or living-room and artist's studio (in *Murderer*). Another popular professional setting is a barrister's chambers, but this type of set had perhaps best be discussed in the context of plays set in legal institutions, like police headquarters, courtrooms and prisons.

Taken together, the latter settings occur in about one eighth of the plays discussed and thus are less frequent than might be expected in a genre predominantly concerned with crime and punishment. This suggests that the crime play does not mainly strive to celebrate – or even defend – the workings of the organs of institutionalized detection and punishment. In his chapter on "Police Procedurals," LaBorde sums up the crime play's attitude to the organs of justice, more specifically towards the police, thus: "the lair of the police serves as a stronger suggestive force than as something actually seen [...] Headquarters is spoken of as being a place where unknown peril awaits guilty and innocent alike."[14] Few of the plays under discussion are mainly concerned with the methods of the police, and those that are, are mostly hostile towards the police. The hostility towards the official agents of justice is a characteristic the crime play shares with the detective novel, which generally favours an amateur investigator. Accordingly, many of the plays set entirely or in part in prison aim at a reform of penal justice. It is extremely rare for a prison set to be presented as the fitting end and logical outcome of a criminal career. In the whole of the twentieth century I have found only two examples, Frank Vosper's *People Like Us* and Phyllis Nagy's *Butterfly Kiss*. Both plays are mainly concerned with the psychology of the murderess, and the portrayal of the heroine in prison only serves to complete the picture.

Courtroom plays are also less frequent than might be expected. Plays in which the essential action takes place during a court trial are *Libel!*, *Witness for the Prosecution*, *Hostile Witness*, *Beyond Reasonable Doubt*, *Rough Justice* and *The Colour of Justice*. It is perhaps noteworthy that in two of these plays the trial ends with a miscarriage of justice. However, courtroom drama may sometimes unfold in settings other than the actual criminal court, as for instance in *Guilty Party* and *Settled Out of Court*, where the procedure of a court trial is transferred to an ordinary living-room, or in plays set in a court martial, for example *Carrington, V.C.* and *Conduct Unbecoming*. Again, two of these plays end with a miscarriage of justice. What a closer look at the plays set in legal institutions reveals above all is a profound scepticism towards these very institutions.

The family home in which the majority of crime plays are set typically belongs not to an upper-class family but to members of the professional upper- to lower-middle classes. Middle-class settings appear in about one half of the plays discussed and get somewhat more frequent from 1930 onwards. Upper-class environments, which are popular at the beginning of the century, decline from nearly 50 per cent to about one sixth of the plays performed. Working-class sets, which initially form the backdrop to no more than 5 per cent of the plays produced, steadily increase in frequency up to about 20 per cent.

Among the plays set in upper-class environments, some celebrate upper-class gentlemanly virtues, while others make use of the stereotypical Victorian villain, the wicked squire, who is confronted by a middle-class detective. While in plays like *The Fourth Wall* and *Someone at the Door* the landed gentry is represented by the "bright young things" popularized by various 1930s entertainments, *Rope* and *Trunk Crime* show the bright young things degenerated into psychopathic killers. Plays like *Breath of Spring* and *Write Me a Murder* likewise show the upper classes in decline, while in the 1970s plays *Conduct Unbecoming* and *Coat of Varnish* upper-class codes of conduct are retrospectively reconsidered.

Important early plays set in middle-class settings are *Alibi* and *People Like Us*. As in *People Like Us*, a middle-class environment is often presented as depressing and sordid, for instance in *Payment Deferred* and *Night Must Fall*. In Priestley's *Laburnum Grove*, by contrast, the suburban middle-class family home is presented as idyllic. The respectable professional middle classes coming into contact with crime are portrayed in the plays of R. C. Sherriff and Barré Lyndon. Later in the course of the century middle-class settings appear in plays on marital conflict like *Murder Mistaken* or virtually anything by Francis Durbridge. In spite of their diversity, all these plays rely on the audience's preconceived notion of the incompatibility of criminal activity and middle-class morality.

The crime play portrays urban working-class families (*Power without Glory*) as well as rural ones (*Gathering Storm*), although the latter are considerably less frequent. The working-class detective as hero(ine) appears in *Saloon Bar* and *Busybody* and *Dead Easy* – notably these three plays are set in public "open" environments rather than family homes. In the late twentieth century, plays about underprivileged families are increasingly staged (*Drummers*) along with plays set in a sordid criminal milieu (*The Life of Stuff*) while so far criminals had been either dangerously glamorous or comic.

The proportion between plays set within extended families and nuclear families remains yet to be discussed. It turns out that both types are approximately equally frequent. There is no obvious relation between the period of writing and family size, although the extended family is a little more frequent before 1930, when it appears even in plays that do not conform to the whodunit formula. After 1930, extended families continue to exist, albeit mainly in whodunits, where they are necessary to provide a sufficient number of suspects. Typical examples of plays presenting large households with numerous dependants are the Christie plays *Black Coffee*, *The Hollow* and *Towards Zero*. Other plays conforming to the whodunit formula and therefore employing the extended family are *To Kill a Cat*, *The Key of the Door* and *Amber for Anna*. This proves fairly conclusively that family size is a function of plot structure and not intended to reflect actual living conditions of families in the period, although in a concession to verisimilitude *Amber for Anna* has *au pair* girls replacing the servants of earlier crime plays and novels.

Since nuclear families do not offer a sufficient number of suspects for the whodunit formula, they tend to appear in plays in which the identity of the culprit is revealed from the outset. The frequency of this type of play makes it seem appropriate to claim the existence of another crime play formula, co-existing side by side with the country-house whodunit: the crime play set in the urban middle-class nuclear family, in which the puzzle element has been largely abandoned.

The specific shape the family home takes in the crime play is the living-room set. The living-room set may undergo various permutations and appear as the library, the conservatory or as the observation saloon of a Nile steamer, but as a look at the floorplan (thoughtfully provided at the end of each play by the publisher French) shows, it is basically the same set. The living-room is the place in which the members of the family meet, the place in which guests are received and through which outsiders intrude upon the privacy of the family home. It is also the place to which the police will be admitted when investigating anything suspicious going on within the family. The living-room set is therefore popular not only because all the members of the family have equal access to it but also because it represents the interface between the private and the public realm. Negotiating the boundary between public and private is a primary concern of the crime play, as is also indicated by such frequent plot motifs as blackmail or family loyalty. An advertisement of the publisher French testifies to the popularity of living-room sets. On the inside of the back cover of several plays they inform aspiring amateur theatre companies:

> Our stock of scenery consists of: – Two Oak Chamber Scenes/ A Grey Panel Scene/ A Drawing Room Scene/ A Woodland Scene/ An Exterior of a House/ A Cottage Interior Scene/ A Proscenium/ A Landscape Backing/ Interior and Exterior Doors/ Modern and Old-fashioned Fireplaces/ Interior Sash Window/ Interior Casement Window/ Stone Balustrade/ Wood Planking/ Bookcase Backing/ and all Necessary Accessories for the Complete Equipment of our Scenery.

Accordingly, a producer who needs a living-room set will have five different sets to choose from, while a play set outdoors will necessarily have to make do with a single "Woodland Scene," which strongly supports Kosok's statement quoted at the outset of this chapter.

Plausible drawing-room sets had been introduced onto the stage during the heyday of Victorian melodrama. Booth comments that when Marie Wilton took over the Tottenham Street Theatre, "she provided furniture for the stage that would make interiors look like the interiors her patrons, rich or poor, had come from."[15] Thus, when the curtain rose the living-room set alone already provided an abundance of signs that the audience were competent to read from their personal experience of visiting other people's homes, and the audience could be relied on to quickly decode signs denoting prosperity and taste, but also social decline and neglect. The part of a crime play to consult in order to establish the signifying character of the set is the initial stage direction. Stage directions declare that the set itself can be "read" by spectators as signifying not only wealth and social status, but also upward social mobility or decline, taste and education, and even the owner's character. The initial stage directions of crime plays (and, presumably, other well-made plays, too) might even be used for writing a semiotics of interior decorating.

The set description of *Something to Hide* seeks to convey an atmosphere of wealth, taste and ease.

> The Toll House at Kingsmead in Essex is a charming old property which has been transformed with care and taste into a luxurious week-end retreat. The large living-room is its main feature. [...] The furnishings are costly and well-chosen.[16]

Subtle differences in wealth and social status may be signalled by two contrasting living-room sets within the same play. In *Mile-Away Murder*, the study of Sir Robert's house in Sussex is described as follows: "The room is pleasantly furnished and obviously belongs to a gentleman of culture, taste and plenty of money,"[17] "culture" being documented by built-in bookshelves and the presence of an *Encyclopaedia Britannica*. The main room in the house of Sir Robert's envious younger brother, by contrast, is a "mixture of study and dining room, furnished in country fashion. The keynote is comfort and solidity rather than luxury, everything being fairly good but definitely not expensive."[18] A stage direction from *The Man in Half Moon Street* specifies what is understood to be "good furniture": Hepplewhite chairs, a Chippendale side-table, a Sheraton cabinet, an Adam table.[19] Social decline is exemplified in the derelict country-house as described in *Write Me a Murder*: "The Manor has been the country seat of the Rodinghams for 500 years and is now in a most dilapidated condition. The two rooms are the den of an English country squire. [...] In both rooms the darkened wallpaper is peeling off in places and the ceiling is cracked."[20]

Proceeding down the social scale we find a variety of distinct shades of prosperity, education and taste in firmly middle-class settings. The educated middle-class living-room belonging to the innocent suspect of *The Day's Mischief* is described thus:

> The room reflects the taste of a man of considerable scholarship on a salary of about one thousand pounds a year. There are rows of books on either side of the door, which is in the back wall, and some good pieces of antique furniture, well cared for [...]. Mrs Barlow collects Bristol glass and china, and there is evidence of this among the ornaments. One or two good pictures and etchings of Oxford hang on the walls.[21]

The policeman's home in *This Story of Yours* represents a considerably lower degree of income, taste and education:

> The house is immediately pre-war in design, arid and featureless. The room is filled to capacity with furniture; chairs and a sofa, a suite and an extra armchair, tables of various sizes, a large modern radiogram, a television set, the traditional department store reproductions on the walls and a row of flying ducks above the fireplace.[22]

The tightly packed furniture of this set is likely to induce claustrophobia in the viewers as soon as the curtain goes up. It is also noteworthy that the etchings of the teacher's home in *The Day's Mischief* are replaced by department store reproductions, while the flying ducks have become proverbial for poor taste.

Upward social mobility is illustrated by the two contrasting sets in *People Like Us*. The Underwoods' parlour at Norbury "is a lofty and rather gloomy room of mahogany furniture and aspidistras. [...] The vacant spaces round the walls are filled with what-nots laden with ornaments. [...] L.C. stands a circular table, covered with a pall-like tablecloth of dull red colour with a bobble fringe."[23] The Carters' sitting-room in Clapham is not only located one step further towards the heart of the metropolis, but also

> [a]esthetically it is a step up from the parlor of the Underwoods'. The wallpaper does not assert itself and though the furniture belongs to a rather mean and indefinite period, it has been described as "this handsome suite" in its early days and has remained somewhat self-conscious ever since.[24]

Social decline in the shape of actual, though genteel poverty is demonstrated by the attic bedsit in *Tabitha*, described as

> a clean but shabby room. [...] There are heavy faded curtains at the windows, but no blinds. [...] Up R stands a cheap, narrow iron bedstead, covered with a faded travelling rug and shabby cushions. The rest of the

> furniture is just as depressing, a mixture of enamelled deal and clumsy Victorian pieces. [...] Sad looking ferns stand on the window ledges. [...] The linoleum is shabby, the rugs worn. In fact, the whole atmosphere of the room is one of faded gentility.[25]

A working-class environment may be indicated by the presence of a washing basket in the living-room, as for instance in *Power without Glory*. A set description denoting respectable poverty reads like this:

> Against one wall is a tiled brick fireplace and opposite it a rather motheaten chaise-longue. Against the other is a heavy oak dresser, covered with china plates, a teapot, various bric-a-brac and a coloured photograph of the Queen and the Duke of Edinburgh. [...] Although it is not a well-furnished room, it is not slovenly or dirty. The quality of the furniture is poor, but the place has been well looked after.[26]

While urban working-class lodgings tend towards the sordid, farm life is presented as cheerful and idyllic, for instance in the description of a Lincolnshire farm in *They Walk Alone*:

> The furniture, heavy and overmuch carved, still contrives to retain a cosy look; perhaps it is the gaiety of the curtains, the bright chintzes and flowers in the room. All these indicate the presence of "females about the house." And just as evident is the rugged air of contentment and rest: "the men" who live in it like comfort.[27]

The tendency to idealize rural life evident in this passage seems to be inherited directly from the tradition of Victorian melodrama.

Frequently the interior design of the living-room is used to introduce and characterize its inhabitants. The most obvious and least subtle instance of exposition by means of set design is introducing a portrait of the room's owner into the living-room, a practice very common in crime plays. A slightly more interesting technique of characterization by means of set design occurs in plays in which the room is supposed to have taken on character traits of its villainous owner, as for example in *This Was a Woman* and *The Third Visitor*. Thus, the audience of *This Was a Woman* is prepared for the entrance of the heroine by the following set:

> The furnishing of the room betrays either carelessness or complexity in the character of its owner. Curtains, floor coverings, chairs and couches are not particularly artistic, but the eye is startled by the presence of one or two exotic pieces, such as a Coromandel screen, a Louis commode or a Chinese lacquer chest. Against one wall dahlias riot from an enormous vase. An incongruous room, provocative rather than aesthetically satisfying.[28]

Accordingly, when Olivia Russell enters, "[h]er appearance betrays her at once as the owner of the room."[29] The villain's den in *The Third Visitor* "is a somewhat curious room, rather exotic, vaguely unpleasant, artistic, but at the same time bizarre, and a little overcolourful. There is something about it that suggests the abode of a not quite normal person."[30] "Overcolourful" and "bizarre" are fairly damning judgments, as further quotations from other plays will prove. Interior decorating may reveal yet other shortcomings of the play's protagonist. In his discussion of *The Gioconda Smile*, James Sexton notes that "the functional plate glass décor also takes on symbolic overtones. Speaking to Janet Spence, Libbard refers to Hutton's previous everyday life as equivalent to living 'behind plate-glass', and, as a result, being cut off from the ultimate reality beyond or within."[31]

If the living-room in question belongs to a family home, conclusions may be drawn as to the state of the owners' marriage. The living-room of *Black Chiffon*, for instance, "is the sort of room that has been lived in a long time – (the Christies have been married 24 years), it is mellow and a little overcrowded."[32] Apart from the duration of the owners' marriage, basic marital conflicts may also be divined from the state of the living-room. For the design of the set of *Without Witness*, Armstrong/Simpson prescribe: "The room, as far as possible, should give the impression of a pleasantly furnished home which a woman is trying to keep attractive against heavy odds in the shape of a husband who is a waster and a drunkard."[33] As the cautious "as far as possible" suggests, the authors are aware that this request may pose a problem to the set designer. In *The Crime of Margaret Foley*, the set already informs the audience that Margaret Foley has married beneath her:

> The whole room is suggestive of moneyed ease, and better taste than one would associate with Thomas Foley, the present owner of "Dunnicliffe House." On the walls are three or four family portraits in oils of the Dunnicliffe family, ranging in period from the days of James II. Over the fireplace, however, hangs a well-executed painting of a man of far coarser clay, Thomas Foley, a genial, half-smiling pig-dealer of fifty-eight.[34]

This description demonstrates the customary method of using portraits for purposes of exposition as well as the genre's predominant concern with social mobility.

In contrast to family living-rooms, rooms belonging to a bachelor are often conceived as a trap which visitors enter at their peril; they may not escape alive. In a society in which the family home is considered the norm, the bachelor apartment in itself is an anomaly: an establishment that largely escapes social control. Obviously, female and male visitors run different risks when they walk into a bachelor's room: females may be raped or shipped off to the white slave traffic, male visitors are murdered straightaway. The seductive villain of *Ten-Minute Alibi* is established as a ladies' man by the fact that "though the room is that of a bachelor, there is a modern and feminine atmosphere about it; obviously it is a room that women visit rather than men."[35] Charles Bennett goes to greater lengths to establish the atmosphere of immorality in artist Peter Hewitt's Chelsea studio:

> The apartment is decidedly striking, to say the least of it, with colouring both vivid and bizarre. [...] Eastern rugs of varied hues (these probably from the Caledonian market), a luxurious bedspread and innumerable cushions of countless shades have lent the room an almost voluptuous air. The walls have been treated with a flame-coloured distemper and all the woodwork ... the door ... the window ... the picture rail, etc ... is enamelled black. A startling frieze of nymphs and satyrs behaving quite indelicately, runs round the top of the walls. [...] Hewitt's divan-bed, piled high with cushions, is left above the fire – projecting out into the centre of the room – a rich rug below it and a crimson shaded standard lamp – which, when alight, casts a voluptuous circle of light on it, is above it.[36]

Hewitt's apartment is characterized as a villain's den both by the presence of Oriental items of furniture – Eastern rugs, a divan bed – and by being "overcolourful" and "bizarre." The heroine, however, is too inexperienced to read the obvious signs, and pays for it. The bachelor flats in *Rope* and *Trunk Crime* resemble each other to some extent. The room in Brandon and Granillo's Mayfair flat "is furnished in a luxurious and faintly bizarre manner and on no discernible model. Nevertheless, there are many good things about if you care to look for them."[37] Wolfe Guldeford's college room "is brilliantly hung and furnished. It belongs to someone eminently self-conscious, exotic, artistic and wealthy. The furniture is, for the most part, of exquisitely carved and painted Italian workmanship."[38] These rooms indicate that their owners are rich, educated, of good taste and aspiring to a high degree of individuality and originality. The fates of the hapless visitors who unwittingly walk into these rooms also resemble each other: one is killed and put into a chest, the other drugged and put into a trunk.

However, as noted above, the bachelor establishment is quite an anomaly among crime play sets; and what the preceding analysis of typical crime play sets reveals above all is the genre's predominant concern with the urban middle-class nuclear family. The family creates a visual image of itself through interior decorating, particularly of the living-room, by which it communicates the qualities and values with which it identifies most strongly to the outside world. The plot of the crime play is triggered off by some threat to the family's privacy and integrity, proceeding either from without or (more frequently) from within the family. The pronounced preference for locating crime within the family home points to one of the crime play's basic ideological assumptions: that crime is a personal, not a social ill.

Of course the set may also be occupied when the curtain rises. By merely moving about the room a character may already draw the audience's attention to certain details in set design that assume importance in the course of the action. An example of this occurs in *Distinguished Gathering*. The scene is Felix Montague's house at Hampstead.

> There is a large armchair above the fire; a smaller one below. Above hang a pair of ornamental Italian daggers. A divan is on angle before the fire; backed with a round table bearing lamp, telephone, cigarettes, etc. [...] When the curtain rises it is night; windows curtained, fire burning. The sole occupant of the room is Felix Montague. He stands on the hearthrug regarding the Italian daggers with a half-smile. Then he examines the telephone which has an auxiliary earpiece fitted. He holds out the fitting as though to another person and then replaces it. Finally he snaps off the lights, looks at the brightly burning fire and then snaps the lights on again.[39]

By means of the direction of the character's gaze and his handling of certain objects, the audience's attention is directed towards them and the following action is to some extent foreshadowed: it may be assumed that the daggers and the telephone will be used, the latter by two people at the same time, and that the lights will be switched off at a central point of the action, leaving the room in semi-darkness.

In the context of the discussion of stage props in the crime play a few preliminary remarks must be made about the presentation of objects in detective fiction. Taking his cue from Jakobson's distinction between metaphor and metonymy, Martin Priestman points out that whereas Gothic tales (particularly those of Poe) are intensely metaphorical, the operating principle in detective fiction is metonymy.

> In metonomy, by contrast, one thing stands next to another in a temporal sequence, and this in turn involves a use of synecdoche whereby an aspect or fragment of a whole "stands in" for it as an exemplary instance: realistic narratives expect us to relate these fragmentary instances back to the known wholes from which they come. This can be seen as one of the key procedures of the detective story, where our ability to deduce the whole ape or sailor from a scrap of fur or greasy ribbon is constantly put to the test. [...] The two dimensions, metaphoric and metonymic, doubtless coexist in all texts to some extent, but it is arguable that the dominance of metonymy in detective fiction makes the genre fundamentally hostile to metaphor.[40]

While in a literary text an object described in some detail is likely to assume a symbolic meaning, in detective fiction it is more likely to assume significance as a clue. While a symbol may be employed to help the reader grasp an elusive concept, a clue usually points to a very solid reality. "Mystery" in crime fiction is a mystery because the culprit tries to conceal it, not a mystery in the religious sense of being something which by its very nature cannot be understood. These observations all apply to detective fiction. In drama, however, objects already assume significance by merely being put onto the stage. At the same time, they do not invite symbolic interpretation, because they are represented by themselves. Keir Elam points out: "The theatre is perhaps the only art form able to exploit what might be termed iconic *identity*: the sign-vehicle denoting a rich silk costume may well be a rich silk costume, rather than the illusion thereof created by a pigment on canvas, an image conserved on celluloid or a description."[41] However, imagery works best the further the representation of the object is removed from the object itself – while the naming of a word conjures up a whole cluster of ideas and associations, the actual presence of an object in a space shared by the audience – if you climbed on stage you might touch it – makes it almost as uninteresting as it would be in ordinary life. Hence, the crime play as a genre may be expected to be even more hostile to metaphor than detective fiction. Unlike the clue in detective fiction, however, stage props do not only point to actions that have been performed in the past but, as in *Distinguished Gathering*, even more to actions that are to be performed in the course of the play. As has been remarked about the guns in *Hedda Gabler*, if there is a gun on stage it had better go off in the last act.

Two types of prop appear in the crime play with particular frequency: the latchkey and the letter.

That the key should be a favourite crime play prop is hardly surprising, since detective fiction has developed a specific subgenre dubbed the "locked-room puzzle." Classic examples of this formula are Poe's "The Murders at the Rue Morgue," Doyle's "The Speckled Band" and virtually anything by John Dickson Carr. The use of a key as a stage prop, however, has major consequences for set design, since it requires the presence of practical doors which convincingly look as if they could only be opened by means of a key, rather than curtains masking the entrances, which are considerably easier to install. Like the living-room set itself as the interface between private and public, the key and the letter mark points of contact with the outside world, means by which an outsider may get access to the hermetically sealed private sphere of the family. The function of the key as a clue is obvious: being in possession of a key means having access to the scene of the crime and thus being empowered to commit a crime. The importance of the key as a crime play prop is demonstrated by the fact that several crime plays use the term "key" in their title, for instance Mary Belloc Lowndes' *The Key*, Stewart Burke's *Key to Murder* and Philip Mackie's *The Key of the Door*.

In contrast to detective fiction, however, the crime play is markedly less interested in the key (representing physical access to the scene of the crime) than in the letter or, later in the century, the telephone, representing (illicit) communication between "inside" and "outside." The motif of the theft of a letter in crime fiction goes back to Poe's short story "The Purloined Letter," though it must be noted that dramatists were also quick to hit on the possibility of using the theft of a secret document as the central plot device, for instance Planché in *Secret Service* (Drury Lane, 1834) and Taylor/Lang in *Plot and Passion* (Olympic, 1853). But it was Victorien Sardou who elevated the paperchase motif to a fine art and thus revolutionized the use of so-called "hand properties." His *Les pattes de mouche,* played on English stages as *A Scrap of Paper,* showed a love letter ending up in the hands of the man least fit to read it – the recipient's husband. In the twentieth-century crime play, the theft of secret documents continues to occur as a plot motif (*Paper Chase, The Crooked Billet, The Last Hour, Black Coffee, Cottage to Let*), but it gets considerably less frequent as the century progresses. Compromising letters appear in *The Last of Mrs Cheyney, The Blind Goddess* and *The Return of A. J. Raffles*. As love letters revealing an extra-marital relationship they occur in *The Letter, Four Winds* and *Witness for the Prosecution*. Another popular type of letter is the suicide note, particularly when faked (*Dear Murderer, Someone Waiting, Something to Hide*). In a more original variation of the motif, a letter may turn out to be a murder weapon as in *Poison Pen* or, rather more elegantly, in *Mile-Away Murder*. In *Mile-Away Murder*, it is not the damaging content of the letter that is of interest but its material nature – the postage stamp is poisoned. In a variation of this idea, *Spider's Web* (1954) revolves around the theft of a rare and expensive stamp, an idea still sufficiently new to work in Stanley Donen's film *Charade* (1963), which also relies on the audience's primary interest in the content of the letter rather than its material.

In the course of the development of the crime play, the letter as a stage prop indicating illicit communication was increasingly substituted by the telephone. On stage, a telephone may serve multiple uses: as Voigts-Virchow points out, it may be used to replace teichoscopy and messenger report or as a device for re-introducing monologue[42] into the well-made play from which it had been largely banned.

> On the stage, [the telephone] became a staple of the bourgeois drawing room and a versatile plot-engineering device, prompting Marshall McLuhan to ask in *Understanding Media*: "Why does the phone ringing on the stage create instant tension?" [...] Reduced to a function in stage technique, it was thought to be most useful for establishing characters, space, and plot, and to "arrange a dramatic entrance for a star actor."[43]

In the context of the crime play, the "instant tension" created by the ringing of the telephone was clearly as important as its function as a "plot-engineering device." The above quote suggests that, with respect to plot management, the telephone is of use mainly in the exposition. However, in many crime plays, yet another aspect of telephonic conversation attained major importance. According to Ian Hutchby, the telephone achieves "intimacy at a distance," that is, "personal co-presence in the absence of physical co-presence."[44] This aspect was exploited by numerous crime plays that rely on the device of an unnoticed earwitness on the phone, for instance *Once a Crook, The Last Word* or *The Gentle Hook*. The earwitness on the phone was usually introduced to bring about a surprising last-minute reversal. On a more abstract level, telephone conversations may come to represent alienated communication. "The telephone provides a historically early example of 'technologized interaction' [...] where the physical body is absent and the splintered identity of the mediatized existence of modernity appears."[45] Hence, phone calls may serve to demonstrate "failed and manipulated communication."[46] Examples of manipulated telephone communication in crime plays include the faked alibi in *Dead Easy*, where the caller is lying about the place from which he is calling, and *Wait until Dark* and *Jumpers*, in which the caller is lying about his identity. In *Wait until Dark*, the three crooks use the telephone to make their victim believe that they are in fact policemen, while in *Jumpers*, the protagonist uses a famous *nom de plume* when complaining to the police about a noisy party in progress in his own flat: "Well, Constable, I'd like to make a complaint about a disturbance of the peace at – I'd prefer it to be an anonymous complaint. Well, do you accept pseudonymous complaints? ... Never mind, my name is Wittgenstein."[47] The telephone as a device for manipulating the victim into a vulnerable position most prominently appears in Milne's *The Fourth Wall* and Janet Green's *Matilda Shouted Fire*, a play on telephone terror best known today in its film version *Midnight Lace*.

In all of the above examples, the telephone is but a plot-engineering device. Voigts-Virchow points out, however:

> Whereas, in a realist well-made play, the telephone is likely to function metonymically, to refer simply to audience expectation of a typical drawing room inventory or to facilitate representative communication processes, it is metaphorically charged with symbolic meaning in symbolist or modernist theatre, or may even play a part in the rejection of traditional meaning-making processes in a postdramatic performance.[48]

As an example of symbolist theatre, Voigts-Virchow refers to Priestley's *An Inspector Calls*, claiming that the new technology of the telephone even serves to connect the Birlings to Divine Retribution: "Priestley refers to the idea generated by the initial fascination with bodiless voices, that of the telephone providing access to the spiritual world."[49] Voigts-Virchow's term of the "bodiless voice" points to a feasible explanation for the popularity of the device of the phone call particularly in the format of the crime play. The title and initial paragraph of the first chapter of the present study toyed with the idea of "dead" written language as being related to the dead body and "living" spoken language as related to the living body. The crime play shows a persistent fascination with the possibility of divorcing "living" language from its source, the living body; not alone in its use of telephonic communication but also in its repetitive use of sound-recording devices (for instance in *The Sound of Murder*) and audiodrama (most prominently in *Who Killed Agatha Christie?*).

A play that shows a particular deftness in the use of suggestive props is *Dial "M" for Murder*. The play starts with Tony stealing a love letter to his wife Margot from her lover Max, and then anonymously blackmailing her. This in itself is an innovative variation of the motif of the love letter used to blackmail a married woman, since normally a play would end with the husband's acquisition of a letter proving his wife's infidelity rather than begin with it. Besides, the heroine usually has to deal with the dual threat of the blackmailer and her husband, while in *Dial "M" for Murder*, these two characters are fused into one. The next prop of importance is the key. Tony hands the latchkey to his apartment to Lesgate, the killer he has hired to murder Margot. This narrows down the number of possible suspects considerably, since either Margot or Tony must have admitted the assassin into the flat. Tony then employs the telephone as a device to be both present in and absent from his flat at the same time: he wants to be an earwitness to the murder and at the same time to secure an alibi. This use of the plot motif of the phone call is also innovative, since the earwitness on the phone normally brings about the final surprise reversal and clears the innocent prime suspect. Just as the characters of the blackmailer and the husband have been fused into one, the roles of murderer and earwitness are performed by the same character.[50] After the murder plan has miscarried, Tony reacts by redistributing the props: he removes Lesgate's murder weapon, plants the love letter on Lesgate, and returns the latchkey from Lesgate's pocket to Margot's handbag. His clever manipulation of the clues leads to Margot's death sentence and completes the second phase of the play. In the last act, however, one of the props testifies against Tony; he has inadvertently mixed up the keys, since all latchkeys look alike.

Even after *Dial "M" for Murder*, Frederick Knott continued to be particularly inventive in his use of props. In *Write Me a Murder*, he employs a "bang contraption"[51] that serves to divorce the actual shot from its sound (a device probably suggested by the techniques of sound film, in which action and sound are recorded independently). In *Wait until Dark*, his "MacGuffin,"[52] the object around which the criminal plot revolves, is a musical doll filled with heroin, which means that the object desired by all three male criminals is a representation of the female human body. However, the play features yet another, more interesting representation of the female body: Roat's weapon, the ivory statue of a girl hiding a switchblade, which he fondly calls "Geraldine":

(Roat takes from his pocket a thin ivory statue of a girl. It is about five inches long and could be a small flashlight.)
Roat. Geraldine protects me. (Rising) Isn't she beautiful?
Croker. What does she do?
Roat. This! (He presses it and a thin switchblade springs out)[53]

Roat's attitude to his dagger is clearly fetishistic, but what is just as disturbing is the transgression of gender boundaries suggested by calling a "phallic" weapon like a knife by a female name – surely it should be "Gerald" rather than "Geraldine?" In terms of psychoanalytic theory, Geraldine corresponds precisely to the definition of Michael Balint's "ocnophilic object." This term refers to the single object dear to the "philobat" (the compulsive seeker of thrills), which forms part of his equipment and to which he clings in situations of high risk. This object is conceived as both phallic and as a symbolic representation of the "safe mother." As Derry has noted, *Wait until Dark* (1966) conforms to the theory laid down in Balint's *Thrills and Regressions*[54] to the letter – it might even serve as the perfect illustration of the book.

Apart from the murder weapon, a particularly significant prop in the crime play is the "MacGuffin," a term introduced by Alfred Hitchcock for the desired object that sets the plot in motion. Hitchcock himself has repeatedly issued a warning against attaching too much importance to the nature of the MacGuffin, but this has not even protected his own films against post-structuralist analyses of the significance of the objects moving to and fro between the different characters, for instance the cigarette lighter in *Strangers on a Train*.[55] The moving objects – usually stolen objects – in the crime play may also be analysed according to their symbolic significance. Most frequently, the object of theft – and hence of desire – is money, which is hardly surprising since material wealth is a highly valued goal in a capitalist society.

What is more interesting is the choice of jewels and precious stones or metals as objects of desire, which is particularly frequent during the first decades of the century,[56] and which signals the crime play's affinity to the fairy tale. Examples are the diamond in *Grumpy*, pearls in *The Last of Mrs Cheyney* and various other items of jewellery in plays from *Plunder* to *Dear Delinquent*. Allan Monkhouse's short crime play *The Grand Cham's Diamond* (Birmingham Repertory, 1918) reveals the symbolic significance of the precious jewel representing adventure and romance suddenly entering a drab ordinary life. The theft of jewellery may, however, carry the additional suggestion that women who know no better use to which to put their wealth than hanging it around their necks deserve to be robbed. In two of the crime plays mentioned above the career of a female jewel thief is cut off by an advantageous marriage – *The Last of Mrs Cheyney* and *Dear Delinquent.* Richly married, these women of course have no need to resort to stealing any longer, but the plays' endings may also be understood to suggest that stealing was only an outlet of misdirected erotic energy. In cases in which female thieves are shown to steal articles of clothing, the symbolic significance of the objects stolen is evidently more important than their material value. The most obvious example is *Black Chiffon,* in which a respectable woman is compulsively drawn to stealing a glamorous nightgown, because she feels unable to cope with her only son's imminent marriage. In *Nine Till Six,* an elegant dress that is stolen represents an entirely different life, that of a married woman, to which all the working girls in the dressmaking department aspire. And in *Breath of Spring,* the fur-coats lifted by a group of elderly ladies carry with them the promise of rejuvenation.

A discussion of set design in the crime play would hardly be complete without a brief comment on lighting. In a panel discussion on "Literary Fear" Konrad Lorenz remarked that "Be afraid in the dark!" was the most specific inborn stimulus-response mechanism he could think of.[57] LaBorde remarks on the use of darkness in crime plays: "Darkness traditionally is fearful, because it conceals the unknown; on the mystery stage darkness possesses the added benefit of allowing a murderer to work unseen."[58] His prime example is the work of Agatha Christie: "[S]he achieves her results through control of lighting rather than scenery."[59] Christie makes use of darkness not only in her most popular play *The Mousetrap* but also in *Black Coffee.* In both cases, darkness serves not merely to create an uncanny atmosphere but also, as LaBorde observes, to conceal the identity of the murderer. The only one of the crime plays discussed in this study to use darkness for nothing but eerie effect is Mason's *At the Villa Rose,* which, however, is not an original crime play but the adaptation of a previously published novel. It may therefore be concluded that Ordean A. Hagen's description of the characteristic opening of a crime play, though it reads well, is anything but representative: "Suddenly the lights go out, and there is a hushed silence. Then a shot breaks the quiet of the night, followed by a harrowing scream. When the lights come on, a body lies in full view of the audience. Now they can breathe easier again, sit back and relax. The show is on."[60] Moreover, this description does not even apply to the majority of the 117 crime plays Hagen then proceeds to list. However, Hagen's description does sound like a probable opening scenario, which makes one wonder why it was not put into practice more frequently. One reason for this may be that, since in traditional proscenium stage theatres the auditorium is dark anyway, it is not advisable to keep audiences literally "in the dark" for a too prolonged period of time.

While total darkness does not seem particularly desirable, special lighting effects are sometimes used to create a fearful atmosphere. These effects are most noticeable in *The Last Hour* with its use of magnesium "death rays" and in *The Man in Half Moon Street* which employs phosphorescent body paint. Other terrifying lighting effects occur in *Gas Light* and *The Phantom Light* – the fact that "light" actually appears in the plays' titles underlines the vital importance these effects have for the plays as a whole. Hence it may be concluded that the traditional association of darkness with fear and light with safety is reversed in a number of plays. This reversal is complete in *Wait until Dark*. Since the heroine is a blind woman attacked by three seeing men, darkness means safety to her, and an overlooked source of light signifies danger: "Roat opens the refrigerator, throwing a wide beam of light straight across at Susy at the bottom of the stairs. The refrigerator immediately starts its loud hum, which goes on and on."[61] Susy finally manages to disconnect the refrigerator, and the ensuing darkness and silence indicate Roat's death. This scene also worked very well in Stanley Donen's film version of the play.

8
Timing and Plot "Construction"

As John Russell Taylor observes in *The Rise and Fall of the Well-Made Play*, the term "plot construction" is acutely misleading, employing as it does an architectural metaphor although drama is a temporal art form. It is therefore not only appropriate to speak of "timing," "time management" or "temporal organization" of a play rather than of "construction," but also advisable to pay particular attention to the way in which the aspect of time is handled when analysing a play. Time management is also essential for the creation of suspense. Thus, Porter defines suspense as "a state of anxiety dependent on a timing device"[1] and the "French crime writers Boileau/Narcejac, in their book *Le Roman Policier* ask the question: if we call this concept 'suspense,' exactly what is it that is being suspended? Their answer is: time."[2] François Truffaut ventures the definition of the "stretching of an anticipation"[3] in his interview with suspense specialist Alfred Hitchcock.

In the same interview, Hitchcock remarks that in general, when writing a scenario, the playwright is more apt at manipulating time than the novelist.[4] It therefore seems appropriate to consider for an instant the differences in temporal organization between a narrative and a play. These differences are mainly occasioned by the fact that a play is usually watched in performance rather than read. Watching a play, the consumer becomes completely subjected to the timing employed by the production team consisting of dramatist, director and actors. A reader, by contrast, has greater control about the time management of the reading process. To begin with, he chooses his own speed. More importantly, a reader has certain techniques at his disposal that may serve to undermine the temporal organization as planned and laid out by the author. These techniques include the skipping of descriptions once they are identified as mere retardation devices – a reader for instance may easily skip Chesterton's colourful sceneries or Sayers' humorous Wimsey-Bunter dialogues without missing any essential information concerning the detective plot – or fast forwarding towards the end of the narrative in order to discover the identity of the culprit in case the suspense is getting unbearable. A reader may also interrupt the reading process whenever he chooses or go back to an earlier passage in the text, for instance in order to check clues. All these opportunities are denied to a spectator, who must wait for the interval to enjoy a break in the performance.

With respect to intervals it may be noted that they are getting fewer and fewer as the century progresses. In the early twentieth century most plays are subdivided into three acts, occasionally even four, particularly before the end of World War I. Sometimes the three-act structure is extended by adding a prologue or an epilogue. The prologue may show the discovery of the body – as in *To What Red Hell* – or contain other necessary expository information, as in *Inquest.* After World War II, plays occasionally limit themselves to two acts only, e.g. *Hostile Witness,* and from the 1970s onwards, the two-act structure becomes the norm from which only very few plays depart – the consciously Edwardian piece *The Return of A. J. Raffles* for instance has three acts. Experimental playwrights of the late twentieth century tend to discard the structuring principle of a division into acts and scenes altogether and subdivide their plays into individual scenes only. The change from three to two acts mainly indicates a change in audiences' viewing habits: spectators seem to have acquired longer attention spans than they used to have. It also means that playwrights in the late twentieth century need to keep up the audience's interest over only one major break in the action; to put it bluntly, the audience has only one opportunity to get out while the performance is in progress. Even the single interval is perceived by some dramatists to be something of a problem. In Francis Durbridge's *Deadly Nightcap,* for instance, the interval is not allowed to coincide with one of the major temporal gaps in the action, which covers about nine months altogether. Act I, Scene 5 ends with the following passage:

Cliff: He was murdered, Mr Curtis.
Geoffrey stares at Cliff, incredulously – then as the lights fade, he turns towards Sarah
CURTAIN[5]

and Act II begins:

The same. The action is continuous
As the curtain rises, Geoffrey moves away from Sarah and, still bewildered, stares at Cliff
Cliff: Mr Radford's body was discovered by Mrs Baxter, the housekeeper. It was by the pool. He'd been knifed.[6]

An accomplished writer of radio and TV serials, Durbridge finishes off a structural unit with an information creating suspense, a technique known as a "cliffhanger." A comparable instance of this technique occurs in Anthony

Shaffer's *Murderer*. Act I ends with Norman Bartholomew approaching the bathroom door in order to carry out the murder of his wife in her bath:

> He finishes undressing but retains his shorts. He tosses back the brandy and starts to ascend the stairs flexing his fingers.
> SLOW CURTAIN[7]

In Act II, the audience get what they have been waiting for:

> Act II continues where Act I stopped. Norman stands outside the bathroom door, irresolute. With sudden resolve he cautiously opens the bathroom door to its full extent. We see him approach the bath-capped head almost overtopped by bubbles, and through the swirling clouds of steam we watch the whole murder – the head pushed under, the wildly threshing legs – the water slopped over in the struggle, and then after an appreciable time, the cessation of all movement.[8]

Two pages before the actual interval, a blackout indicates the passing of several hours, yet Shaffer took care not to have the interval and the temporal gap in the succession of events coincide. For their cliffhangers both Shaffer and Durbridge make use of the climactic effect of crime – in Shaffer, it is a murder about to be committed, in Durbridge a murder recently discovered.

Because of the climactic nature of crime it seems appropriate to analyse the temporal organization of crime plays with respect to the point at which the central crime occurs: when does the crime happen in relation to overall plot development? Here it is once again useful to turn to the detective novel for purposes of comparison. Dennis Porter points out that it "is a peculiarity of detective fiction that the story of the investigation is made gradually to uncover the story of the crime which antedates it,"[9] concluding: "[a] classic detective novel may be defined, then, as a work of prose narrative founded on the effort to close a logico-temporal gap."[10] As numerous critics of detective fiction have pointed out, the plot of a detective novel only unfolds after the crime itself has been committed. The crime is thus at first absent from the text and can only be described in its entirety once the master sleuth has collected all the relevant information and reconstructed the crime. This retrospective interest of the plot structure, while appropriate to narrative, causes problems when colliding with the inevitable eternal present of the stage performance, as Erdmut August has pointed out:

> By its very nature, detective work happens *post rem*, that is, after the fact. This, however, does not pose a problem to the novelist, since the temporal structure of the novel from the very start places its narrator at a point from which he tries to elucidate an earlier occurrence from a later viewpoint in time. To the detective novelist, the period of time presented

> in the text is preceded by another period of time, the as yet opaque period of the crime. Formally this means nothing but a further deflection into the past, which anyway is the specific medium of narrative. [...] On stage, the epic character of the detective plot meets the time structure of the present that governs theatrical performance. The temporal sphere of communication of the spectators that are present in the auditorium and the actors that appear on stage logically is the present. Drama can express the past but indirectly, for instance by means of a report given by a person on something that has already happened.[11]

The above quotation reveals one of the major differences between detective fiction and the crime play: in detective fiction, someone tells us about the past, in a crime play, we watch people act in the present. It is therefore not surprising if authors had difficulties in reconciling the retrospective attitude of classical detective fiction with the temporal structure of drama which by necessity unfolds in the present. Tzvetan Todorov's seminal essay "The Typology of Detective Fiction" points to a possible solution for the crime playwright's dilemma. Todorov distinguishes between two forms of interest aroused by two different types of detective fiction. The classical whodunit causes *curiosity*: the interest is directed backwards, towards the past, asking for the reasons for the present situation – "Who done it?" The thriller, by contrast, creates *suspense*: the interest is directed towards the future – "What on earth is going to happen next?" Since retrospection, as August has pointed out, is less suited to drama than to narrative, plays more easily achieve suspense (in Todorov's narrow sense of the term) than curiosity, which means that the playwright may just as well discard the central mystery, as Wilkie Collins did intuitively when adapting his detective novel *The Moonstone* for the stage.

Another attempt to compare the principles of plot organization in detective fiction to those operating in drama was made by Dorothy Sayers in her essay "Aristotle on Detective Fiction,"[12] which applies Aristotle's observations on tragedy to the evolving genre of detective fiction. About plot development, Sayers mainly notes that a plot of detective fiction must not be episodic, giving the hardboiled thriller as a negative example, and that no loose ends must be left at the denouement. Structurally, the plot of the hardboiled thriller conforms to the format of the journey or quest. The only English crime play based on the principle of the journey is Galsworthy's *Escape*, though a number of other plays might also be called episodic, for instance *Nine Till Six*, *Boys in Brown* and *"Now, Barabbas."* These plays follow up the individual fates of several different characters brought together in the same locale. Thus the episodic nature of the plays is not caused by diversity of settings but diversity of characters, each of them being the protagonist of his/her own subplot. This model, however, was rarely realized. Another type of crime play plot that may with some justification be called episodic is the plot based on intrigue and counter-intrigue that is often found in plays dealing with espionage. Here, however, Sayers is more generous, stating that "[i]t is, however, possible, where the villain's identity is known, to make an agreeable story by showing the moves and countermoves made successively by villain and detective"[13] and giving Collins' *No Name* and Freeman's *The Singing Bone* as examples. The retrospective interest of detective fiction does not particularly trouble Sayers; she quite uncomplicatedly advocates linear temporal progression (deploring in passing the intricate time schemes of the modernist novel) and equals the three stages of the (dramatic) plot – beginning, middle, and end – with murder, detection and discovery/execution:

> The detective story commonly begins with the murder; the middle is occupied with the detection of the crime and the various peripeties or reversals of fortune arising out of this; the end is the discovery and execution of the murderer – than which nothing can very well be more final.[14]

Interestingly, a considerable number of crime plays indeed conform to the plot structure outlined by Sayers in the above quotation. The crime occurs before the beginning of the play or relatively early in the action, and the main concern of the plot is the discovery of the culprit and/or the means by which the crime was committed. The overall development of suspense may be described as follows: After the crime has been committed, tension slackens a little and remains on about the same level until the culprit is finally identified and arrested. After the discovery of the culprit, the tension is released dramatically. Local peaks of excitement occur whenever a character dear to the audience is in danger – either of becoming the next victim or of being arrested for a crime (s)he did not commit. Plays conforming to this plot type are e.g. *Inquest*, Sayers' own *Busman's Honeymoon* and *Bonaventure*. Plays employing the above time scheme do not necessarily need a "logico-temporal gap" that has to be filled, as the success of *Rope* and *Someone Waiting* shows. The crime does not have to be absent from the play's text in order to make the detection process involving, or, to put it more colloquially, a play centring on detection does not have to be a whodunit. *Someone Waiting* is of particular interest in the context of time management because in this play the retrospective attitude of the detective plot is reflected in a re-enactment of the circumstances of the crime. It might be supposed that this eminently theatrical device had been used more frequently on the stage, but it remained rare, occurring for instance in *Without Witness* and *The Crime at Blossoms* – and in the latter play the main point is precisely that the re-enactment does not tally at all with the facts of the case.

In a variation of the type of crime play described above, the crime occurs later in the course of the action, but apart from that, the temporal organization and suspense scheme conform to the model outlined above. Plays in which the commission of the crime is delayed, usually until after the first act, commonly are classical whodunits with a large cast of potential suspects all requiring introduction. Examples are *The Creaking Chair*, in which the crime occurs between Acts I and II, and the Christie adaptations *Peril at End House* and *The Hollow* (murder in II.i) and *Appointment with Death* and *Murder on the Nile* (murder between II.i and II.ii). The fact that most of the large-cast whodunits with a "late crime" are adaptations of novels suggests that this pattern of plot organization is more characteristic of detective fiction than of the crime play proper.

A group of crime plays to be distinguished from the type described above also has the crime occur early in the action, but the main concern of the plot is the effect the crime has on its perpetrator. The play therefore either shows the culprit's punishment or, more frequently, his/her attempts to avoid discovery and conviction. Accordingly, there is no major retrospective interest and no logico-temporal gap to be filled. The suspense scheme varies slightly from the first model. As in the earlier example, tension slackens after the crime has been committed, but gradually increases again as the pursuers close in on the culprit. Plays belonging to this formula are for instance *Blackmail*, *Payment Deferred* and *Ladies in Retirement*. Local peaks of excitement occur whenever the criminal is in danger of being discovered.

In a variation of this second model, the crime is again delayed, because some more extensive exposition is deemed necessary. In contrast to the large-cast whodunit discussed above, the variation of the second type is not mainly concerned with introducing numerous characters in its exposition but with establishing the criminal's motivation and his – sometimes quite elaborate – plan. This frequently causes a tri-partite structure that does not consist of crime, investigation and solution, as suggested by Sayers, but of the plan and preparation of the crime, the carrying out of the plan, and the investigation of the crime. This structure is particularly conspicuous in Anthony Armstrong's *Ten-Minute Alibi*, in which the three acts are devoted to the plan, the crime, and the investigation respectively. As far as the suspense scheme is concerned, we may imagine a rising curve with the crime as its climax, a slackening of tension and another rising curve as the danger of discovery increases, culminating either in arrest or escape. As in the case of plays on detection, the curve of suspense can be determined by the relative positions of the two major climaxes of crime and discovery/punishment.

In a third group of crime plays the crime happens comparatively late in the course of the action, because the plot is mainly concerned with the motivation of the culprit. The suspense scheme belonging to this type of play closely resembles the traditional model of dramatic construction: a slow rising action culminates in a crime that brings about a reversal and a sudden release of tension. Because the crime occurs late in the action, the problem of retrospection is avoided. This model was employed but rarely, for instance in plays like *Living Dangerously*, *The Amazing Dr. Clitterhouse*, *This Was a Woman* and *Molly*. Three of these plays depict several crimes in succession, characterized by an increasing amount of violence and culminating in a murder. *Living Dangerously* begins with drug traffic and adultery, *The Amazing Dr. Clitterhouse* with burglary, and *This Was a Woman* with the comparatively harmless attempts of the heroine to undermine her son's marriage – but in the end each of the plays' protagonists has killed another person in cold blood.

Two other types of crime play also make use of multiple crimes. The first type may be described in Sayers' terms as "an agreeable story showing the moves and countermoves made successively by villain and detective." The culprit's criminal intentions are known from the start and the play's interest resides in the battle of wits developing between the hero and his antagonist(s). The suspense scheme may be imagined as a quickly rising curve until the point at which the villain's criminal intentions are revealed, which is followed by a succession of local peaks of excitement, each bringing about a reversal of fortune for hero and villain. This series of actions is brought to a close by the final defeat of the villain after which the tension slackens. This plot model is very frequent and used for example in spy plays (*The Man Who Stayed at Home*), plays about professional criminals (*House Guest*) and crime comedies in which the crime in question is frequently theft rather than murder (*Plunder*).

For the fifth and last type it is also essential that more than one crime is committed. The culprit is frequently obvious from the start and has committed a similar crime before. The retrospective interest of solving a crime in the past thus becomes linked to the urgent necessity of preventing a crime in the present. It is helpful for the creation of suspense if the second crime is a more or less exact repetition of the first – like for instance the murder of two sisters in *The Speckled Band*. This formula turned out to be reasonably popular and brought forth plays like *The Two Mrs Carrolls, Gas Light* and *The Mousetrap*. Plays about serial killers proper also belong to this group, since they, too, create suspense by arousing fear of repetition. Since the "original" crime is usually absent from the text[15] the curve of suspense may be imagined as rising slowly but regularly in the course of the gradual discovery of clues pointing to the earlier crime and culminating in the confrontation between the criminal – who is always a murderer in these plays – and his or her intended victim. This confrontation brings about a reversal, breaking the pattern and terminating an apparently endless series of repetitions. The serial killers of these plays are to be understood as suffering from a repetition compulsion, most notably in the case of *Night Must Fall, Love from a Stranger, Murder Mistaken* and *They Walk Alone*.

These descriptions may sound very schematic, and it is perhaps more interesting to look at a number of individual plays that introduced innovative variations of the basic types of time and plot organization outlined above. *Signpost to Murder*, for example, makes use of the two last time schemes discussed: it begins like a serial killer play about a dangerous criminal on the run and then gradually changes into a battle of wits between the supposed villain and his supposed victim. Throughout the play, the audience is left in doubt as to whether Roy is the criminal and Sally the victim or vice versa.

A type of time scheme that has elicited particularly interesting variations is the variant of the second model described above. The crime is delayed because an extensive exposition is devoted to the criminal's motivation and his elaborate plan. The rest of the action consists of the actual carrying out of the crime and its investigation. Frederick Knott's *Write Me a Murder* at first glance seems to conform exactly to this formula. But then the intended victim dies of natural causes. Therefore the actual crime is postponed until III.i when a replacement victim has been found – though not by the criminal but by the dramatist: once planned, the murder must take its course. The audience's relief at the protagonist's preserved innocence turns out to have been premature. Similarly, *Not in the Book*, *Dutch Uncle* and *Murderer* revolve around inept murderers, who are forever planning a crime that never materializes. Andrew Bennett in Watkyn's *Not in the Book* is prevented from murdering his blackmailer by sudden scruples, however, he almost manages to poison a policeman by mistake. Norman Bartholomew, likewise by mistake, drowns his mistress rather than his wife, and Mr Godboy in Simon Gray's *Dutch Uncle*, despite continued efforts, does not manage to kill anybody at all.

Of these three plays, *Murderer* most thoroughly upsets the viewer's perception of the play's plot and suspense management. It seems to begin with a particularly gruesome murder carried out in full view of the audience. When the spectators are about to relax, waiting for a more leisurely investigation to follow, it is revealed that what they saw was not the actual murder being committed but only the rehearsal for one – therefore they are not in a phase of relaxation but in a phase of preparation and increasing tension. This tension is released, albeit shortly, when Norman manages to accomplish the crime and drowns his female victim in the bathtub. But this action also turns out to be incomplete since he killed the wrong victim; therefore the tension quickly rises again. What viewers are denied in this play is the alternation between mounting suspense and the release of tension. Instead, the relaxation phase is only momentary and the level of suspense rapidly rises again.

So far this chapter on temporal organization has concentrated on the management of "theatre time," the time it takes to perform the play on stage. The second half of this chapter shall now be devoted to "dramatic time," the time the play's action takes to unfold.[16] The crime plays analysed in this study cover time spans ranging from 24 hours and under to several years. Long time spans of several years occur in plays like *Justice* (three years), *Payment Deferred* (two years) and *The Winslow Boy* (two years). The maximum time span is reached in *Living Dangerously* – 12 years. What these plays have in common is that they take themselves comparatively seriously, taking their time to show character development or the effect the working of the legal machinery has on the protagonist. These plays are held together by a strong audience interest in the central character. The late twentieth century saw a number of crime plays covering even longer periods of time, the "memory plays" *The Secret of Sherlock Holmes, Killing the Cat, Normal* and *Butterfly Kiss*. These plays have abandoned linear time altogether and move about freely in time and space, often covering more than 20 years. With the plays mentioned before they share a pronounced interest in psychological development.

The average time covered by a crime play, however, amounts to between one and a few weeks. This would seem to conform to the "convenient length" prescribed by Aristotle: " 'A story or plot,' he reminds us, 'must be of some length, but of a length to be taken in by memory.' "[17] It is difficult to determine what exactly constitutes a "convenient length" – suffice it to say that the action of most crime plays unfolds within a few weeks, which seems to be the time span the average audience member can control and for which (s)he would in real life happily make plans or appointments.

Aristotelian poetics, however, traditionally defines "unity of time" as a time span not exceeding the course of one day. In the first three decades of the century a number of playwrights strove to obey this dictum and achieve unity of time in plays as diverse as *Grumpy, The Creaking Chair, The Ghost Train, Blackmail* and *The Fourth Wall*. Edgar Wallace has the action of *The Case of the Frightened Lady* unfold in less than 24 hours, whereas the action of *The Ringer, The Terror, The Squeaker* and *The Flying Squad* takes longer, sometimes considerably; dramatic time ranges from two days to more than a year. This may have to do with the fact that the latter four plays were adaptations of novels while *The Case of the Frightened Lady* was an original play that only became adapted as a novel after its success on stage. The 24-hour rule was obviously perceived as a law pertaining to drama alone. "Unity" may be useful for a crime play, since an adherence to the 24-hour rule increases tension because the various incidents must follow each other in rapid succession. Hence, time spans of under 24 hours tend to occur in crime plays consisting largely of a frantic pursuit action, like Launder/Gilliat, *Meet a Body* or Chapman/Cooney, *Simple Spymen*. The combination of unity of time and place helps to create the very tense, almost claustrophobic atmosphere of *Wise Child* and *Wait until Dark*. It should be noted, however, that many crime plays that have attained classical status disregard unity of time, for instance *Night Must Fall, The Amazing Dr. Clitterhouse* and *Dial "M" for Murder*. Yet, in general unity seemed to be an ideal crime playwrights strove to emulate, since some plays attempt to achieve unity by using a prologue or a scene of exposition which is then followed by a main action covering no more than 24 hours. During the last three decades of the century, unity seems to have become more and more unfashionable, occurring mainly in some Durbridge and in plays that parody the traditional crime play, such as Shaffer's *The Case of the Oily Levantine* or Gerald Moon's *Corpse!*

While the rule that it is advisable to limit the action of a stage play to 24 hours may appear a bit arbitrary, it is self-evident that continuity – a time scheme in which theatre time equals dramatic time – is a special case indeed. In most plays that achieve continuity, it is not an end in itself but contributes to the play's overall effect. For instance, continuity occurs in short plays like *Sus* and *The Dumb Waiter* and in crime plays in which speed is essential, in plays of intrigue and counter-intrigue that include farcical pursuit actions, such as *The Crooked Billet* or *Warn That Man!* Continuity is also realized in plays in which a large group of people needs to be kept together in the same location for some time for the action to unfold – this applies particularly to *Settled Out of Court* and *Guilty Party*, but also to *Saloon Bar* and *The Key of the Door*, which, incidentally, all imitate the format of courtroom drama. A handful of plays that are perhaps best labelled "meta-dramatic crime plays" – *A Murder Has Been Arranged, Murder in Motley, The Real Inspector Hound* – function on the basis of the principle of sharing the audience's actual space and time, being set in the theatre on the evening of performance. Finally, it should be noted that continuity may help to create a considerable amount of tension, as for instance in *Rope* and *Wise Child*. When adapting *Rope* for the screen, director Alfred Hitchcock immediately realized that unity of place and continuity of time are essential for the play's impact and decided to underline their effect by refraining from all editing of the film, a famous experiment discussed in virtually every book on Hitchcock's films. Generally, while interest in 24-hour unity is declining in the course of the century, interest in continuity is increasing – apart from the plays mentioned already it also occurs in *Jumpers, Murder in Mind, Murder by the Book, Dangerous Obsession, The Life of Stuff* and *Popcorn*. This tendency, interestingly, appears simultaneously with the opposite trend towards covering long time spans in the "memory plays" mentioned above.

Most of the crime plays analysed in the present study follow Sayers' dictum by unfolding in linear progression, that is, they show the events presented in chronological order. However, one of the crime play's favourite motifs fundamentally questions linear temporal progression, and that is the return of a character from the dead. The reappearance of a character supposed to be dead is a frequent device for the creation of suspense, occurring in about 16 per cent of the plays surveyed, figuring particularly prominently in *Ladies in Retirement, Ten Little Niggers, The Third Visitor, Conduct Unbecoming* and *Sleuth*.

Besides, the governing principle of chronological representation is sometimes disrupted by the three related mechanisms of repetition, re-enactment and flashback. As mentioned above, plays like *Without Witness* and *Someone Waiting* make use of the re-enactment of a previous action for purposes of investigation, a technique that would seem to conform well both to the retrospective attitude of the detective plot and the demands of the theatrical medium. A sinister parody of the method of re-enactment occurs in Howard Brenton's *Christie in Love,* in which serial killer John Christie is made to re-enact his first murder ever with a life-size doll operated by one of the investigating policemen. Usually, however, the earlier crime that may be about to be repeated in the course of the action is not shown but narrated, for instance in *The Two Mrs Carrolls* and *Night Must Fall.* While in these plays repetition is merely realized in the spectator's mind, a handful of other crime plays use actual onstage repetition as a structuring principle. Examples are Verner's *Meet Mr Callaghan,* in which, one by one, Callaghan blackmails and manipulates the different members of the Meraulton family, *We Must Kill Toni* and *Dutch Uncle,* which revolve around a series of unsuccessful murder attempts, and, most conspicuously, *The Unvarnished Truth,* a play that piles up female corpses with the apparent intention of reducing the audience to helpless hysterical laughter. These plays, however, do not employ exact repetition, but each instance varies slightly from the one preceding it.

An exact repetition of previously shown events occurs in *Ladies in Retirement*: the functional detectives dress up in the murdered woman's wig and dress and play her favourite tune on the piano, suggesting to the culprit that her victim has returned from the dead. The impact of this scene resides not solely in its ghostliness but also in the sense of *déjà vu* it evokes. Both the return from the dead and the *déjà-vu* effect indicate the suspension of temporality, which perhaps accounts for the strong impression created by this scene. When Tom Stoppard made use of exact repetition in *The Real Inspector Hound* he of course had the model of Beckett's *Waiting for Godot* behind him, which had set new standards for the extent to which audiences may be subjected to repetition in the theatre. Once the theatre critic in Stoppard's play has been caught up in the action, the previously shown first act begins repeating itself, and suspense is created since audiences are normally quicker than the hapless critic on stage to work out the implications of this repetition. The fear of history repeating itself had already been exploited very effectively in Walter Hackett's *The Barton Mystery* of 1916. One of the characters falls asleep and, unbeknown to the audience, the subsequent actions performed on stage are all part of his dream. On awakening, he fears that life will bring about the realization of his dream – but to the audience it is more important to find that the boundaries between illusion and reality and the linear chronological plot development have been shattered.

While *The Barton Mystery* creates a strong *déjà-vu* effect by blurring the boundaries between dream and life, the protagonist of Anthony Armstrong's *Ten-Minute Alibi* constructs a "time-loop" by merely fiddling with a clock, creating, however, a number of interesting logical problems as a consequence, all resulting from the fact that, for once, the same time happens twice over. Armstrong, however, was not particularly interested in the philosophical implications of time-loops, while J. B. Priestley was, and his concern with temporality is much in evidence in his crime play *Dangerous Corner*. Although Priestley states in *Man and Time* that a "novel or play cannot really be *about* time,"[18] he concedes that "in *Dangerous Corner* [...] I play with the idea of a loop or split in Time."[19] The most striking idea in *Dangerous Corner*, the linear development of an alternative reality parallel to what we learn in the end has "really happened," has its basis in a philosophical treatise on time which greatly influenced Priestley in the period immediately before *Dangerous Corner*, J. W. Dunne's *An Experiment with Time* of 1927. Holger Klein sums up Dunne's basic tenets as follows:

> Dunne [...] arrived [...] at the conclusion that past, present and future co-exist, and at the theory of an unending series of consciousnesses – which he called "Observers" – and of times in which they live, beyond our common consciousness which ends with the ending of our physical lives, our existence in what we normally call time.[20]

For a reading of *Dangerous Corner*, this implies that the play's main action does not represent a potential development luckily averted, but that the revelations on the characters' real relationship towards the deceased Martin co-exist, albeit in another, parallel, universe. This, rather intriguingly, would make *Dangerous Corner* a forerunner of more recent Possible Worlds Theory. *An Inspector Calls* contains a similar device: after the Inspector has departed, the imminent arrival of an Inspector is announced who may or may not be an Inspector different from the one who has just left. Thus, the ending may signal that the same plot is about to happen again (as in *The Real Inspector Hound*) or that the inquiry about to take place now will be passed with greater credit to the Birling family. It is suggested that in two parallel universes the guilt of Martin and the Birlings is simultaneously discovered and covered up – just as Radfern in *Laburnum Grove* may at once be and not be the head of a successful counterfeit ring. The theoretical possibility of acting differently in a number of alternative parallel worlds mainly serves to highlight the possibility of moral choice and the obligation to assume responsibility for the consequences of one's actions.

Possibly in imitation of Priestley, Aldous Huxley introduced a time-loop into the stage version of *The Gioconda Smile*. Dr Libbard, the functional detective of the play, tampers with a clock in order to effect the last-minute rescue of the innocent accused. By making the murderess believe that her lover has already been executed, Libbard extracts a confession from her. As in Priestley, the problem of cyclical versus linear time is related to questions of moral behaviour. This becomes evident when the innocent accused quotes Kierkegaard: "Life has to be lived forwards; but it can only be understood backwards."[21] However, the plot-management device of the time-loop and the ethical problem of how to act morally in a world in which the consequences of one's actions can only be known in retrospect are less well integrated than in Priestley or in Alan Ayckbourn's later exercise in ethics and time-travel, *Communicating Doors*.

Alan Ayckbourn's *Communicating Doors* also links a profound interest in temporal irregularities to the problem of moral choice, but Ayckbourn allows his characters to go back to the past in order to correct previous mistakes and manage earlier experiences more successfully. Learning by trial and error, Ayckbourn's heroines develop the courage to break behavioural patterns and create a better future for all of them. What attracts both Ayckbourn and Priestley to time-loops is that they may enable characters to make informed moral choices because they have had the opportunity to recognize the consequences of their actions, an opportunity not open, unfortunately, to the plays' audiences. If the moral value of an action is determined by its consequences, an individual would have to be either clairvoyant or able to move freely in time in order to make moral choices. It may be argued, however, that the theatre itself represents an alternative world in which viewers may try out diverse courses of action without any harmful consequences to their real lives, and that therefore the time-loop is mainly a metaphor for the theatrical experience itself. An interesting variation of the motif of the time-loop occurs in *Home at Seven*: here the protagonist's experience of being transported back into an earlier period in history is combined with the reverse phenomenon of a "loss" of time in the manner of Rip van Winkle. The reconstruction of his activities during the temporal gap becomes more important to the hero than the experience of reliving his past. While the heroines of Ayckbourn regain control over their lives through temporal irregularities, David Preston loses control.

The Barton Mystery is unusual among plays largely relying on a time-loop since it turns out in the end that the action shown on stage merely presented a subjective version of reality as perceived by one of the dramatic characters. Contrary to what might be expected, "[t]he device of placing the audience, without warning, inside the imagination of the protagonist has not been frequently employed in the modern comedy thriller,"[22] although it had of course been used in *Seven Keys to Baldpate* as well as in Victorian melodrama.[23] Alec Coppel's *I Killed the Count!* of 1937 also presents subjective versions of reality as dramatic actions on stage. When the various suspects are interrogated they do not merely narrate their version of the murder, but Coppel has their statements illustrated by means of a flashback. Since all of the suspects are lying, each of the flashbacks is mendacious, a practice likely to infuriate audiences as an offence against the "fair play rule" favoured so much by authors and readers of detective fiction alike. When, in 1950, Alfred Hitchcock introduced a mendacious flashback into his film *Stage Fright*, "critics scolded Hitchcock for 'cheating' his audience by inserting into the film this spurious flashback, which the viewer initially accepts as a genuine presentation of the facts, because moviegoers usually assume that the camera does not lie."[24] In drama as in film, a different degree of reliability is usually assigned to actions shown and actions merely narrated. Viewers are of course perfectly prepared to expect one of the dramatic characters to be lying, but showing a lie on stage or on screen apparently shatters audience assumptions so basic that their disruption causes spectators to feel uncomfortable in the theatre.[25] A corresponding example in detective fiction would be the unreliable first-person narrator in Agatha Christie's *The Murder of Roger Ackroyd*, a device "still debated in books and articles on crime fiction, more than half a century after the novel's first publication,"[26] as Osborne wrote in 1982. In the fictional universe of the detective story where appearances are constantly misleading and everybody is necessarily suspect, readers at least want to be able to trust in the integrity of narrative devices.

However, illustrating the evidence given by the different characters by means of flashback was not at all a new device when Coppel employed it in 1937, but had in fact been used in American courtroom drama as early as 1914, in Elmer Rice's *On Trial*. In the British crime play, however, the flashback never became a particularly popular device, though it did occur occasionally, either in plays that used a prologue-epilogue structure to frame the main action, as a local flashback illuminating the crime itself and thus closing the logico-temporal gap, or, as in *I Killed the Count!*, as a structuring principle. The practice of framing the main plot with a prologue and an epilogue and consequently presenting the central action as a flashback came into use in 1928 with Edgar Wallace's *The Squeaker*, a stage adaptation of his novel. In the 1930s this plot structure was imitated by Jeffrey Dell in his adaptation of C. S. Forester's *Payment Deferred* (1931), John van Druten in *Somebody Knows* (1932) and Percy/Denham in *Trunk Crime* (1937). And as late as 1982 Ronald Millar adopted the device for his adaptation of C. P. Snow's *A Coat of Varnish*. It is striking how many of these plays are dramatizations of novels; obviously the frame structure recommended itself as an appropriate device for rendering narrative on stage, since narrative (as August reminds us) usually looks backwards at events in the past. Of course the insertion of a flashback must be clearly signalled to the audience. Dell and Millar make use of the same technique to accomplish this: they show the same room twice, first emptied of all furniture (*Payment Deferred*) or with the furniture covered by dust sheets (*A Coat of Varnish*), that is, in a general state of desolation, and in the following scene as it looked when the house was still inhabited. Millar makes additional use of a change in season to indicate the temporal gap.

While it is hence easy to account for the use of the frame in *Payment Deferred*, *A Coat of Varnish* and *The Squeaker*, *Somebody Knows* and *Trunk Crime* require a closer look. To van Druten, the frame structure was clearly important, since he retained it even after extensive rewriting. Although the Prologue obviously left something to be desired, the structure was not abandoned altogether. What the frame does accomplish in *Somebody Knows* is an undermining of even the most basic suspense structure: by telling the audience the outcome of the action first, the playwright removes a specific "end-oriented desire" and enables his spectators to follow the whole of the action with interest without any particular end(ing) in view. This supposed intention would fit in well with van Druten's general aim of undermining the traditional structures of the detective plot. In *Trunk Crime*, we also learn about the play's outcome first, but in this case I would argue that the frame structure serves mainly to create the impression of the inevitability of the events depicted in the main plot. Looking at the chain of causes and effects from a later point in time we get the impression that the succession of events necessarily had to lead to this one particular end, whereas a series of actions presented in chronological order stresses the aspect of a possible choice between alternatives. Thus, the frame structure of *Payment Deferred* likewise lends the history of William Marble an air of fatality – Marble is doomed literally "from the start," just as Wolfe Guldeford in *Trunk Crime* is saved from the start. It is tempting to relate this concept of the preordained fate of the individual to the Calvinist doctrine of Predestination; in any case, the use of a framing device for the main action reveals a profound scepticism concerning the freedom of individual choice that is diametrically opposed to the attitude in evidence in the plays employing a time-loop.

Local flashbacks are even rarer than those employed in connection with a prologue-epilogue frame. W. Somerset Maugham uses a flashback in the third act of his dramatization of his short story "The Letter," and John Hopkins four decades later turns the entire third act of *This Story of Yours* into a flashback. In both cases, the flashback serves to close the logico-temporal gap by showing the actual crime that has so far been investigated. Actually, the flashback in *The Letter* was only introduced in rehearsal. Maugham's original third act had contained a confession speech by Leslie which he thought prudent to replace by the more theatrically appropriate device of a flashback, because he "thought it would bore the audience to listen to two long narratives in one play."[27] The stage direction introducing the flashback reads: "The lights fade to blackout. When the lights come up again, Leslie is seated by the table working at her lace. She is dressed as in Act One. It is night, and the lamps are lit."[28] The blackout is a standard device to indicate the insertion of a flashback to the audience, and of course it also facilitates necessary set or costume changes, but, as leading lady and producer Gladys Cooper feared, it may also disrupt the rhythm of the performance.

Coppel's technique of using flashback not only once but repeatedly, as an organizing principle, was imitated by a few playwrights in the late 1950s and early 1960s. Philip Mackie used the device in *The Key of the Door* (1958). Agatha Christie employed it in *Go Back for Murder*, her 1960 stage adaptation of her novel *Five Little Pigs*, and Robert Bloomfield in *Portrait of Murder* (1963). While in Mackie and Bloomfield the flashbacks serve to illustrate individual interrogation sequences, the flashbacks of *Go Back for Murder* are joint efforts by all the characters involved and it is sometimes impossible to judge whose version of events is being presented. Accordingly, the most extensive flashback is of considerable length, making up about one third of the entire play. Christie's experiment with flashback did not go down well with audiences and the play had to close after only 31 performances. Although in Christie's case the flashback technique was not favoured by audiences, it may be concluded from the repeated use of the device that by about 1960 audiences had become accustomed to dealing with multiple time shifts, possibly by the influence of cinema, where flashback is employed far more extensively. This observation suggests a general development from a standard of chronological representation with a maximum of two time shifts in the early twentieth century, via multiple flashbacks in mid-century, to the free-floating "memory plays" of the last decades.

One last aspect that needs to be discussed in a chapter on timing and plot construction is the period in which crime plays are set. The vast majority of crime plays take place in contemporary times, which corresponds to the fact that most plays are set in or around the city of performance. Both in space and in time the crime play comes as close as possible to its audience. About 10 per cent of the plays analysed, however, make use of a period setting, which can usually easily be accounted for. In more than half of the cases the plays are based on earlier narratives – such as *The Gioconda Smile, A Murder Is Announced* and *The Secret of Sherlock Holmes* – or on true crimes – like *Ladies in Retirement, The Winslow Boy, A Dead Secret* or *Molly*. Some crime playwrights, however, choose a period setting deliberately, with a notable preference for the Victorian and Edwardian periods. Examples are *Gas Light* (Victorian), *Pink String and Sealing Wax* (1880s), *An Inspector Calls* (1912), *The Late Edwina Black* (1895), *Conduct Unbecoming* (1880s) and *The Return of A. J. Raffles* (1900). Usually the period setting is chosen because of the particular social pressures and restrictions that assist the unfolding of a crime plot. Thus, both *Gas Light* and *Pink String and Sealing Wax* are based on a patriarchal family structure that puts women and children under considerable pressure, no matter whether the master of the house is presented as benign – as in *Pink String and Sealing Wax* – or as wholly evil, as in *Gas Light*. In *An Inspector Calls*, moral hypocrisy and a rigid class structure are essential for the development of the plot; in *Conduct Unbecoming*, an imperialist code of military conduct is questioned. While these two plays take themselves comparatively seriously, *The Late Edwina Black* is conceived as an entertainment, yet it also makes use of specifically Victorian laws and rules of behaviour for its plot to develop. In *The Late Edwina Black*, these laws are the rigid Victorian divorce laws, accompanied by a religious fervour that rigorously condemns suicide, and by moral hypocrisy – the latter two characteristics personified in the character of the housekeeper Ellen. These five plays, however, choose the period setting not solely for the plot. In his analysis of the image of Victorianism projected by *Gas Light*, Guy Barefoot argues:

> "Victorian" is a word descriptive of narrative, style, moral codes (and the practices underlying such codes). [...] The Victorian is identified with tastelessness, with bad art and paraphernalia, and beyond that there exists the misogyny referred to by Dickinson [Thorold Dickinson, the director of the British film version of *Gas Light*] and the "poverty, wretchedness and age" referred to by Hamilton. The association of age with negative values suggests a validation of modernity. The gloom and claustrophobia of Victorian architecture could be contrasted with the openness and light of a modernist aesthetic. The gloomy Victorian setting acts as a motivation for the action, as a motivation for both misogyny and melodrama. Male intimidation is identified with the past rather than the present.[29]

Graham Greene was clearly attracted to the period setting of *The Return of A. J. Raffles* for reasons of décor just as much as of behavioural codes, as the initial stage direction shows:

> The date is the late summer of 1900, a few months before the death of the old Queen. The scene a set of chambers in Albany, which was once described by Conan Doyle as an "aristocratic rookery." Even if we did not know it was Albany, where at that date no woman, cat or dog was allowed to reside, we would recognize a bachelor's apartment – the rather heavy furniture, the comfortable leather armchairs, the photograph on the desk of an elderly woman – a rather obvious mother – on the back wall the photograph of a cricket team and hanging on either side of it a cricket bat on which the signatures of a team have been inscribed and a red leather cricket ball dangling in a net from a nail.[30]

As usual, the initial stage direction is immensely revealing. It expresses a certain wistfulness, not only for the opulence of past times, but also for the possibility of excluding women – from certain places as well as from certain activities, such as cricket. The only woman allowed in, as will be noted, is Mother, and even she is only admitted in representation. As the rather high frequency of intertextual references suggests, however, the main attraction of the period for Greene is that it produced vintage crime fiction. For a similar reason, Gerald Moon probably chose the 1930s as the period for his 1994 crime play *Corpse!*, while Jez Butterworth was apparently attracted to the 1950s by the rock music created in that period when he chose 1958 as the year of the action of *Mojo* (1995). As the century progresses, a period setting is increasingly chosen not for reasons of plot mechanics but for atmosphere.

9
Dramatic Characters

The very first aspect of Aristotle's poetics that Sayers quotes in her essay is the primacy accorded to action before character: "[T]hey do not act in order to portray the characters; they include the characters for the sake of the action."[1] In practice, however, the distinction between the two cases – action serving characterization or characterization serving action – may not always be easy to accomplish. For in drama, more than in any other literary genre, characters are largely defined by the way they interact with each other. Accordingly, it is appropriate for a chapter on dramatic characters to begin not with the individual characters but with character configurations, which of course are but the schematic representations of recurring interactions between the characters, corresponding to habitual behavioural patterns in real life. Two basic models of character configuration derived from related literary genres offer themselves as a starting point: the typical character configuration of Victorian melodrama and the typical character configuration of the whodunit detective novel.

Critics of Victorian melodrama have isolated a recurrent character configuration, consisting of the hero, the heroine and the villain, often characterized by telling names.[2] The hero is both inordinately courageous and extraordinarily naïve; while he is largely passive, the villain is frantically active and of unusual single-mindedness. The action is frequently triggered off by the villain's desire for the heroine and his consequent attempts to separate her from the hero. The heroine is situated not solely at the centre of the amorous triangle but also at the emotional core of the play. Her main function is to suffer, as Booth somewhat exasperatedly observes:

> A cardinal rule of melodrama is that at some point, usually early in the play, the heroine begins to suffer. [...] Until the final curtain the heroine goes from one agony to the next, and some heroines do not even make the final curtain, expiring of sin and remorse in the last act.[3]

The cast of the traditional whodunit detective novel consists of a far greater number of characters. Apart from the Great Detective, who remains an outsider to some extent, there is an assortment of suspects sufficiently varied to present a microcosm of English society, as Grella points out:

> Though basically homogeneous, this society does contain variety. Its members, though roughly equal in social standing, are not of the same class, family background, or profession. Within a limited range they comprise an English microcosm. There is always at least one representative of the squirearchy, one professional man – commonly a doctor, but sometimes a lawyer, professor, or schoolmaster – a cleancut young sporting type, and a military man (never below the rank of major), usually a veteran of colonial service.[4]

The dynamics of this character configuration develops as follows. One member of this closed social group gets murdered, usually for having committed "some breach of the unwritten social or ethical code of the thriller of manners."[5] Since the victim is characterized as a somehow alien, undesirable member of the group, each of the others had a reason for wanting to murder him (or her). It is the task of the Great Detective to identify the murderer, who – by an almost perfect act of mimicry – has managed to adapt perfectly to the society around him. The story ends with the expulsion of the second undesirable member of the group, the culprit.

Considering the character configurations of melodrama and whodunit side by side, one cannot help noticing that they require vastly different cast sizes. Obviously, character configuration therefore largely depends on the size of the available cast. The smallest possible cast is required by the plot of intrigue and counter-intrigue already mentioned in the previous chapter. This type of character configuration, the "battle of wits," can be performed with no more than two leading actors. With a cast of three, a play based on the "melodrama" character configuration sketched above may be performed – although Victorian melodrama itself generally required a larger cast consisting among others of the "comic man" (the forefather of the stage detective), the "comic woman" and the "good old man." Still, the basic action of a hero rescuing the damsel in distress from the villain requires no more than three major parts. An equally, perhaps even more attractive character configuration for three actors is the "amorous triangle," which, for reasons of its very longevity has also been dubbed the "eternal" triangle. Both Carlson and Disher have commented on the configuration of the amorous triangle as one of the mainstays of drama. Disher even considers the very dominance of the motif of adultery in drama history one of the main reasons for the development of detective plays: "Very vaguely all mystery-mongers were striving to get away from sexual morality. [...] The trouble was, of course, that murder had so often been merely part of the drama of adultery."[6]

"Melodrama" and "amorous triangle" character configuration may of course be combined, as for instance in Wallace's *The Ringer* and *The Flying Squad*. In *The Ringer*, the heroine, who is deeply attached to her morally weak brother, comes under the influence of the arch villain and is finally saved by the intervention of a young heroic policeman. In *The Flying Squad*, the heroine's choice between her two suitors, the honest policeman and the dope smuggler, also entails a moral choice between Right and Wrong, turning the play into something akin to a modern morality play. The "amorous triangle" is also frequently combined with the "battle of wits" evolving between the rival lovers or between an adulterous wife or husband and his/her spouse. Casts of more than three actors can perform a play of the "whodunit" type described above, or a play in which an individual places himself (or, rarely, herself) in opposition to the whole of society. Plays of this type show the gradual decline of a criminal or the plight of an innocent suspect on the run. Larger casts are also needed to present plays in which the "battle of wits" takes place between two groups of people, a configuration particularly frequent in plays revolving around espionage. This character configuration will be called "us against them;" it is also generally found in narrative versions of the spy thriller.

The popularity of the different types of character configuration varies greatly in the course of the twentieth century. During the first three decades, they occur with about equal frequency, and it is noteworthy that the "melodrama" structure with a damsel in distress being rescued by heroic male action is still comparatively frequent. After 1930, the formula rapidly loses its interest, while the "whodunit" configuration gains in appeal: between 1930 and the end of World War II, there are about three times as many crime plays with this type of character configuration as in the previous period. This notable increase may be accounted for by the simultaneous flourishing of the classic detective novel from which Grella's observations are derived. The interest in whodunits declines a little in the following decade with about as many plays conforming to the "battle of wits" formula as to the "whodunit" configuration. After 1956, the "battle of wits" becomes and remains the most common character configuration of the crime play, a shift probably brought about by diminishing cast sizes. The most noticeable change after 1970 is a considerable increase in plays conforming to the structure of "individual versus society." The configuration of the "amorous triangle" retains its attraction throughout the century, though it mostly occurs in conjunction with other structures, usually with the "battle of wits." As this survey shows, Marvin Carlson's assumption that the typical character configuration of the whodunit is replaced by the character configuration of the amorous triangle in the period of postmodernism, though backed up by statistics to some extent, oversimplifies matters. To begin with, the decline of the whodunit configuration does not coincide with the inception of postmodernism. Rather, it is to be explained by a general trend towards the reduction of cast size – an explanation incidentally ventured by Carlson himself. Second, the character configuration of the amorous triangle has a constant appeal throughout the century and it frequently even occurs within the larger character configuration of the whodunit. Hence, it is more appropriate to conclude that the whodunit configuration is gradually being replaced by the character configuration of the battle of wits.

The first part of the present chapter has mainly dealt with what character configurations we actually find in twentieth-century crime plays. Approaching the same problem from a different, more theoretical angle, we realize that a story or play dealing with crime needs at least three main characters; the criminal, the victim and the detective, as Derry has pointed out:

> Traditionally, the popular work of art dealing with crime is composed of at least three major characters or elements: the criminal, the victim, and the detective. In most crime works, these three elements coexist in a triangular relationship [...] What is particularly significant is that in most popular works dealing with crime, one of these three elements takes precedence over the others; the works group themselves into specific varieties or (sub-)genres whose popular names often make the dominant element clear.[7]

What Derry accomplishes in the passage quoted above is not merely isolating the three basic actantial roles of any crime plot, he also suggests an approach for defining subgenres of the crime film and, as will be demonstrated in Chapter 11, the crime play. It is important to note that not all of the actantial roles must be realized in every play – the crime may not be solved, so that there is no need for a detective, or it may be sufficiently abstract to allow the author to do without an identifiable victim – income tax fraud would be a good example. Finally, it may turn out that no crime was committed after all, so that it is even possible to write a crime play technically without a criminal. This case is more common than might be supposed; examples include *Libel!*, *Murder without Crime*, *Without Witness*, *To Kill a Cat* and *A Lady Mislaid*. The triangular structure may also be upset by having two different actantial roles fulfilled by one actant, for instance if the victim dies by her own hand (*Appointment with Death*, *The Late Edwina Black*) or if the intended victim himself or herself undertakes investigations for his/her own protection, as in *Love from a Stranger* or, perhaps most blatantly, in *Sleuth*, where the victim of Andrew's cruel game dresses up as a Police Inspector to indicate his transformation from the actantial role of the victim to the actantial role of the detective. A blurring of the roles of detective and criminal may be said to occur if the crime is solved not by the detective but by the culprit revealing himself and thus fulfilling the narrative function of closing the logico-temporal gap. LaBorde disapproves strongly of this device:

> That type of solution has numerous shortcomings. First, it is arbitrary. At some point in the script, the author decides to have the murderer reveal himself, although the villain has gone to great lengths until then to conceal his identity. Two thousand years ago Aristotle labeled such a method of disclosing new information "inartistic." [...] Second, that approach to the solution violates the trust the audience places in an author who it assumes is playing fair.[8]

What is remarkable about LaBorde's statement is the amount of confusion and irritation obviously occasioned by the blurring of actantial roles within a recognizable formula. Carlson comments on the aspect of confusion engendered by the re-allocation of actantial roles:

> We are trained, not only in detective fiction but in all traditional narratives, to seek a pattern of relationships in which the individual characters, or actants, continue to relate to one another, and to the narrative as a whole, in fairly predictable ways. In the comedy thriller, we might say that actantial roles are cut loose from the actants, so that one can never be sure, from scene to scene, who will be playing any particular actantial role.[9]

While the confession in a whodunit brings about a blurring of the actantial roles of criminal and detective, the character configuration of the amorous triangle permits a blurring of the actantial roles of culprit and victim. The formula of the "battle of wits" even consists entirely of a succession of role reversals between criminal and victim. Hence, the gradual decline of the "melodrama" character configuration with its fixed and stable actantial roles, and its replacement by the "battle of wits" formula means that a blurring of actantial roles is becoming increasingly likely as the century progresses.

Keeping in mind that not all actantial roles must be realized in every play, and that actantial roles may become blurred, it is yet appropriate to investigate how the basic actantial roles are most typically realized. Of the three roles, Sayers and Grella have most to say about the victim, which is surprising since surely not a lot of activity is required from a mere victim. Sayers states that, according to Aristotelian poetics, a violent reversal – such as a fortunate, wealthy person being reduced to the state of a corpse – must be brought about by a fatal flaw in that person's character. "The defect may be of various kinds. The victim may suffer on account of his unamiable character, or through the error of marrying a wicked person, or through foolishly engaging in dubious finance, or through the mistake of possessing too much money."[10] Grella is more definitive in his statements, declaring right at the start that the victim must be "an exceptionally murderable man."[11] To define what constitutes murderability, Grella also resorts to character configuration, defining the best victim as a "blocking character" of the type frequently employed in comedy, often the negative father or mother figure "who opposes a marriage, makes an unfair will or refuses to act his or her age, all actions which cause distress to the young."[12] He explicitly singles out the squire, who "being elderly, stupid, and irascible, makes an excellent murderee."[13] Alternatively, a good blocking character is "the ineligible mate, whose impending marriage to the decent young girl would interfere with her natural preference for the eligible young man."[14] This character's ineligibility is defined by his acting "like a cad," which consists of cheating at gentlemanly sports and games, advocating free love and other sexual misdemeanours, such as taking advantage of a working-class girl (John Dickson Carr, *The Problem of the Wire Cage*) or being a gigolo (Dorothy Sayers, *Have His Carcase*). Grella does not comment on the obvious prohibition of cross-caste love relationships implied by the last two examples. A similar prohibition applies to upward social mobility in general: a victim may effectively damn himself by "posing as a gentleman" or otherwise aspiring to a rank that is not his by birth: "Because he may have risen from humble beginnings to achieve a doctor's high status, the professional man – generally the only character who actually works for a living – is also a potential victim."[15] Grella's pervasive use of the personal pronoun "he" is clearly not accidental, since he states explicitly that women, particularly young women, rarely get murdered in detective fiction, unless they, too, hide an unacceptable secret, most commonly related to sexual morality, for instance being an adulteress or an actress. Adultery in particular is a favourite offence in detective fiction bringing about instant retribution.

> One other damning trait is a fatal un-Englishness. Foreigners lead dangerous lives in detective novels, and men otherwise exemplary die for merely ethnic reasons. The Mediterraneans of the detective novel never seem true gentlemen, perhaps a result of the Italian villainy of Elizabethan revenge tragedies and Gothic fiction.[16]

W. H. Auden agrees with Grella that "the best victim is the negative Father or Mother Image."[17] He also explains why the victim should not be a criminal: "because he could then be dealt with by the law and murder would be unnecessary," adding, however, that "[b]lackmail is an exception."[18]

To a considerable extent, the observations of Sayers, Grella and Auden also apply to the crime play; however, some modifications will be necessary. As indicated by Grella, the largest group among victims consists of characters that have committed some breach of the unwritten code of conduct of their society; approximately one in two crime plays uses a victim guilty of a misdemeanour (or, as Sayers so politely puts it, of "an unamiable character"). About 50 per cent of these misdemeanours are sexual in nature. The pervasive influence of the adultery motif in melodrama and detective fiction alike has already been pointed out; and indeed a great number of crime play plots are triggered off by sexual infidelity. Female victims may also be prostitutes, particularly in crime plays of the first half of the century (*Somebody Knows*) or sexually promiscuous lower-class girls (*Someone Waiting*). *The Shot in Question* reveals an ambiguous response of simultaneous attraction and repulsion when introducing its intended victim with the words "She has that touch of the gutter about her that some men find distasteful, and others irresistible"[19] and the first victim of *Someone Waiting* complements her loose morals with being a foreigner. Male sexual misdemeanours range from attempted rape (*Dangerous Corner*) and seduction (*Ten-Minute Alibi*) to homosexual inclinations (*Wise Child*). The seducer and pimp of *Ten-Minute Alibi* combines dubious morals and foreign origin:

> He is half English, half Latin-American, slim, dark, tall, and handsome, and about thirty-three years old. He is obviously very attractive to women, possessing the ability to combine masterfulness with an air of passionate admiration in his dealings with them. Under his suave and fascinating exterior there is, however, a calculating ruthlessness. Women and pleasure are the dominating notes in his life. He is dressed in a dark grey suit – too well cut.[20]

This description fits in almost perfectly with Grella's observations on the ineligible mate, although the character in question is not Italian but Latin American. The above quotation is from the stage direction that introduces the character, and its third sentence already indicates a built-in reversal of audience perceptions. This type of comment, suggesting that a character is not what (s)he seems, is very common in stage directions introducing new characters and would seem to conform to detective fiction's interest in revealing the true character behind the social mask. Incidentally, it should be noted that male characters die for sexual misdemeanours quite as frequently as female ones, even though conservative sexual morality has usually tended to be stricter towards the female.

Dominant father and mother figures, though also frequent, make up less than 10 per cent of crime plays' victims. Again, the number is equally divided between the sexes. *Appointment with Death* contains a perfect description of the "bad mother": Mrs Boynton "is a vast obese woman, rather like an idol, with an expressionless face. She moves her head and eyes, but not her body. A stick is beside her chair. Her family are grouped round her like courtiers round a queen."[21] This stage direction creates an image of a woman both powerful and immobile – the perfect "blocking character" – and contains advice for acting ("She moves her head and eyes but not her body") as well as for blocking ("Her family are grouped round her"). The victim's immobility, which is also indicated by the stick lying by her side, is to be related to the type of the invalid victim which is to be discussed below, but also to the crime play's recurrent interest in "death in life" and "life in death" – the logical opposite of Mrs Boynton is the "busy body," the dead body that is surprisingly mobile. The most interesting instance of a "bad father" is Graham Greene's choice of the Marquess of Queensberry as a victim – though only of theft, not of murder – in *The Return of A. J. Raffles*, since Queensberry is mainly known to posterity as a blocking character preventing his son's love affair with Oscar Wilde. Queensberry is described as "a small man with curly red hair, heavy eyebrows and long whiskers, and a heavy ugly lower lip"[22] and clearly also fits the description of the murderable squire.

Other crime play victims that are in some way socially unacceptable include social climbers (*The Crime of Margaret Foley*), alcoholics or drug addicts (*Grand National Night*), foreigners (*I Killed the Count!*) and tramps (*Double, Double*). One might be tempted to read the defensive attitude towards tramps as a remnant of ancient laws against vagrants, but the choice of a tramp as a victim is clearly informed by necessities of plot: the culprit wants to play dead and collect the insurance money and therefore needs a replacement corpse. A tramp is ideal for the purpose because he will not be missed. The female alcoholic victims of *Grand National Night* and *A Clean Kill* are described in extremely similar terms and also serve as blocking characters, since their presence prevents their husbands' happiness with more suitable younger girls.

Of the crime play victims that become targets of aggression because they possess too much money, only five are actual country squires, and two of their respective plays are stage adaptations of novels by Agatha Christie. In the second half of the twentieth century, the squire is no longer a prominent figure in the crime play. Instead, murder between professional rivals is becoming more and more popular, occurring for instance in *Settled Out of Court*, *The Real Inspector Hound* and *Murder by Misadventure*. In a variation of the motif of the murdered squire some crime play victims get themselves killed because of their political influence rather than their wealth, being eminent politicians as in *Meet a Body*, or scientists engaged in politically sensitive research as in *Cottage to Let*. In about 10 per cent of the crime plays analysed, however, the victim is rich and nothing but rich, often an heir or an heiress, which seems to imply that they have done nothing to deserve their wealth. Generally, the source from which the money came originally remains remarkably obscure; there is no particular interest in revealing the dubious financial practices that Sayers mentions and that play such an important part in American hardboiled detective fiction. Contrary to Auden's assumption, a large number of crime play victims are criminals themselves, making up a group almost as large as that of the idle rich. Most frequently an accomplice or collaborator gets killed by the prime criminal

(*Wait Until Dark*) or a blackmailer is removed, usually by one of his victims (*Interference*). In a handful of plays the victim himself (or herself) is guilty of (attempted) murder, this applies for example to *Ten Little Niggers, Dial "M" for Murder* and *Nightcap*. The frequency of the blackmail motif, which was already noted by Auden, reveals society's dominant concern with the keeping up of appearances, of a respectable façade. *Distinguished Gathering*[23] provides an impressive list of disreputable secrets thought too terrible or shameful to be revealed. Two of the women among the *dramatis personae* are hiding an illicit love affair while the third one was compromised by her brother's cowardly behaviour in war. Of the three men, one is a cocaine addict, one forced a rival firm into bankruptcy and one attempted suicide. What is most remarkable is that each of these secrets seems a sufficient reason for murder.

A fourth group of victims is characterized not by antisocial behaviour or inordinate wealth but by sheer vulnerability; a possibility discussed neither by Grella nor by Sayers or Auden. This group can be subdivided again into helpless females or "damsels in distress" (*Gas Light*), invalids (*The Gioconda Smile*) and children (*Wild Justice*). The helpless female in *Portrait of Murder* is introduced by the following words: "She has a slightly defensive air, like a sensitive child in a strange and possibly hostile world."[24] The stage direction obviously strives to create the greatest possible contrast between the "defensive," "sensitive" heroine and her "strange," "possibly hostile" environment. The decisive point, however, is the comparison between an adult woman and a child; being compared to a child automatically marks a character for an appropriate victim, as in *The Bargain* where we learn about the victim: "When he is enthusiastic about some new adventure – as he is now – he can no more restrain it than a schoolboy."[25] While childish grown-ups make good victims, the genre for some time hesitated to use actual children as murder victims. The earliest play in which it is at least suggested that a child may have been murdered is *The Day's Mischief* of 1951. Even earlier, in 1936, Alfred Hitchcock had risked showing the killing of a child in *Sabotage*, his film version of Joseph Conrad's *The Secret Agent*, and regretted it afterwards as an error in judgment.[26] It is only in the 1990s that crime plays revolving around the murder of children appear with increasing frequency. If Sayers and Grella are right and the identity of the victim reveals a latent hostility within a society, hostility towards children must have reached a peak in 1990s' Britain.

According to the same line of reasoning, there must be a pervasive strain of hostility towards disabled people in British society, since in one crime play in 15 a disabled person is killed or attacked. What is particularly noticeable is that in these plays the culprit is often allowed to escape. In *Night Must Fall, Bonaventure* and *Molly* some punishment for the crime occurs, whereas in *The Unexpected Guest, Loot* and *Dead Guilty* the murderer goes free. In plays revolving around a mercy killing the culprit generally escapes punishment

(*The Sacred Flame, Beyond Reasonable Doubt, Rough Justice*), although the murderous husband of *Beyond Reasonable Doubt* commits suicide in the end. The two latter plays are courtroom dramas, which suggests that by the late 1980s and early 1990s it had become possible not only to publicly discuss euthanasia but even to claim it as a moral right. In *The Sacred Flame* matters are slightly different, since the invalid man is an obvious blocking figure in Grella's sense, preventing his wife's happiness with another man. As in *Rough Justice*, the victim's mother carries out the actual murder, and as in *Rough Justice*, her moral right to do so is asserted: by her act, she secures the happiness of her family and, accordingly, she eventually escapes punishment. In four other plays the blocking character is an invalid wife, who must be removed so that her husband can find happiness with another woman. For a while it seems that one member of the happy couple will be arrested for the murder, but it then turns out that the wife died of an accident (*Murder Happens*) or committed suicide (*The Late Edwina Black*). A particularly elegant solution for granting a happy ending while at the same time denying the murderer (or, in this case, murderess) any benefit from the crime is found in *The Gioconda Smile* and *Verdict*. The amorous triangle is here extended into a foursome, with one of the husband's lovers kindly removing the superfluous wife and afterwards herself, so that he can live happily ever after with his other mistress, whom he truly loves. Incidentally, *The Gioconda Smile* is the only play with an invalid victim that is the dramatization of a narrative, which would seem to suggest that the murder of a handicapped person is a special concern of the stage play. Drama expresses itself through the bodies of actors, and the genre accordingly attaches great value to physical perfection. Sayers and Grella do not mention the disabled victim in particular, because the detective novel takes a different attitude to physical handicap – even the Great Detective may be disabled; blind like Max Carrados, the hero of the novels of Ernest Bramah Smith, or so fat that he is virtually immobile, like Rex Stout's Nero Wolfe. The film thriller, by contrast, also makes extensive use of the handicapped victim, as Georg Seeßlen has noted about postwar *film noir*:

> The first thing one notices is that in several films the murderer is said to direct his aggression against women who are invalid and hence useless. Besides *The Spiral Staircase, Sorry, Wrong Number* might be mentioned. [...] Most films that tell the story of a murderer threatening a woman close to him are subconsciously governed by the obsessive idea of destroying the "defective."[27]

Film as a genre is clearly as obsessed with physical perfection as drama, perhaps even more so. It is therefore hardly surprising that physical imperfection would seem to amount to an offence against unwritten codes of conduct. One notable exception in drama and film ought to be mentioned: Susy, the blind heroine of *Wait until Dark* manages to defend herself against three criminals, but it is only after she has defeated her oppressors single-handedly that her rescuers arrive. Susy is allowed to survive once she has proved that though visually impaired she is not *dis*-abled but perfectly capable of taking care of herself, and hence not a burden to the healthy people around her.

Like most of her previous argument, Sayers' conclusions concerning the characterization of the culprit are drawn from her reading of Aristotle:

> Aristotle takes the modern, realistic view, as when he says, for instance, that the plot ought not to turn on the detection and punishment of a hopelessly bad man who is villainous in all directions at once – forger, murderer, adulterer, thief – like the bad baron in an Adelphi melodrama; but rather on that of an intermediate kind of person – a decent man with a bad kink in him – which is the kind of villain most approved by the best modern writers in this kind. For the more the villain resembles an ordinary man, the more shall we feel pity and horror at his crime and the greater will be our surprise at his detection.[28]

The "ordinary" man as a culprit is moderately popular among crime playwrights, featuring for instance in *Payment Deferred, Laburnum Grove* and *Not in the Book*. "Ordinary" here means first of all "middle-class," but also physically insignificant. The "ordinary" criminal of *Not in the Book* for instance is described thus: "He is in his early 50s, of medium build, largely unnoticeable. His patient, dog-like face, not helped by glasses, attracts however through its essential modesty and, in the unlikely event of people looking at him twice, they would like him."[29] And the protagonist of *Someone Waiting* is introduced as "an insignificant-looking boyish man in his early 40s,"[30] the term "boyish" suggesting, as argued above, that he would also make a good victim – which of course is what he does in the end. The stage direction characterizing the criminal protagonist of *Laburnum Grove* slightly varies the formula, stating that George Radfern

> is a man about fifty with nothing remarkable about his appearance, though even at first there should be a certain quiet assurance and authority visible beneath his easy manner. At this hour, he is very much the suburban householder at ease, wearing slippers and an old coat, and smoking a pipe.[31]

This description suggests that the character of the suburban householder may be but an assumed role, and it already prepares the major reversal within the play by ascribing to Radfern "a certain quiet assurance and authority."

Initial stage directions characterizing the culprit frequently contain similar phrases that already indicate the intended reversal in character perception, for instance in *A Clean Kill*, where the culprit is "what everyone describes, without second thoughts, as 'a nice girl.' But she possesses, as will be seen, more brains and determination than her good-looking face reveals at first sight."[32] It is tempting to view this type of built-in character reversal as a function of the whodunit plot structure, ensuring the final surprise at the revelation of the identity of the culprit. However, similar introductory stage directions also occur in crime plays that are not whodunits, plays in which the identity of the criminal is never in doubt, as the following examples will show:

> His personality varies considerably as the play proceeds: the impression he gives at the moment is one of totally disarming good humour and childlike unself-consciousness. It would need a very close observer to suspect that there is something wrong somewhere – that this personality is completely assumed.[33]
>
> Not a conventionally good looking man, he gives the impression of being mild, rather ordinary, pedantic, self-deprecating, but his face is capable of assuming other expressions which can be inexplicably dangerous.[34]

The conventional reversal from the apparently ordinary human being to a dangerous criminal is itself reversed in the initial stage direction of *Christie in Love*:

> Christie's first appearance is in the Dracula tradition. Happy horror, creeps and treats. He rises from the grave luridly, in a frightening mask. It looks as if a juicy evening's underway, all laughs, nice shivers, easy oohs and aahs. But that's smashed up. The lights are slammed on, and the mask is seen as only a tatty bit of papier maché. Off it comes, and what's left is a feeble, ordinary man blinking through his pebble glasses.[35]

The fact that deceptive appearances are such a constantly recurring motif of the crime play suggests that the concern with human personality and the social mask is not a function of the whodunit structure, but that the gap between appearance and reality is a primary concern of crime play and detective novel alike, and that it only produces the whodunit formula as one means of expressing itself. Grella's description of the ideal criminal also stresses the difference between the culprit's real character and his assumed social role:

> Usually the culprit is a more acceptable person than his victim, because he comprehends the elaborate social ritual well enough to pose as an innocent. This sustained pretense of innocence culminates, however, in the unveiling of the murderer's true character; the detective exposes him as an impostor, the *alazon*, a familiar comic figure who, Northrop Frye says, "pretends or tries to be something more than he is."[36]

It is doubtful, however, if the culprit really is a more acceptable character than the victim. In the crime play, "good" criminals frequently belong to the same social groups as "good" victims, and occasionally, the culprit may have a more shameful secret to conceal than the victim, as the paragraph on "abnormal criminals" below will prove.

Frye's definition particularly applies to social climbers, who, as we have seen, make good victims, but even better culprits, appearing for instance in *Murder on the Nile, Dial "M" for Murder* and *Deadly Nightcap*. Most frequently the criminal social climber is a murderous husband who has married a rich wife. All these plays reinforce the prohibition of cross-caste marriage noticed, though not expressly mentioned, by Grella. Criminals who have risen from humble surroundings tend to inadvertently betray themselves. Crozier in Durbridge's *House Guest*, for instance, "wears a blazer and carefully pressed cavalry twill trousers. There are times when his public school accent does not ring quite true."[37] Like Julian in *Communicating Doors*, he is marked as a social climber by his accent and by being immaculately dressed, which automatically seems to suggest that the man in question is not quite a gentleman.

In the context of social class it should be noted that a considerable proportion of crime play culprits are lower-class, that is, either poor or dependent on their victims. Examples include *At the Villa Rose, Power without Glory, Murder Story* and *Drummers*. In crime plays, underprivileged criminals are more frequent than conspicuously middle-class culprits – the "ordinary" criminal mentioned above – or upper-class criminals, as personified in the "gentleman-thief." This is surprising since neither melodrama nor traditional British detective fiction considers lower-class characters eligible as culprits. The favourite culprit of melodrama is the "bad baron" mentioned by Sayers, and in traditional British detective fiction servants are usually automatically exempt from suspicion. S. S. Van Dine, an author frequently criticized for his snobbishness, for instance remarks in his "Twenty Rules for Writing Detective Stories:" "A servant must not be chosen by the author as the culprit. [...] The culprit must be a decidedly worth-while person – one that wouldn't ordinarily come under suspicion."[38]

Considering the crime play's culprits under the aspect of profession we immediately notice the abundance of doctors. The reason Grella gives for the medical man making a good victim of course also makes him a good criminal. In fact, in crime plays he is even more popular as a culprit than as a victim, appearing for instance in *The Ghost Train, The Amazing Dr. Clitterhouse* and *The House by the Lake*. What makes a doctor such a suitable culprit is that he has access to means of murder that are out of reach for ordinary people, but also that he tends to be associated in people's minds with pain, fear and – as TV serials show – power over life and death.

Another favourite culprit is the immoral artist, an umbrella term for homicidal writers (*Sleuth*), painters (*The Two Mrs Carrolls*) and actors (*The Case of the Oily Levantine*). The *topos* of the criminal as artist is a recurrent motif in detective fiction, which G. K. Chesterton has pointedly summed up in the formula: "The criminal is the creative artist; the detective only the critic."[39] The quality that initially associates the criminal with the artist is their firmly anti-bourgeois stance, as Sweeney has pointed out:

> Literature often depicts the writer as a criminal – a brooding, Byronic figure, exiled from society and exempt from its laws. According to this Romantic view, art itself is a kind of crime, a secretive act in which man presumes to create as if he himself were God. Fictional criminals, then, unlike their real-life counterparts, are usually cunning, cultured artists manqué, who execute their crimes for aesthetic pleasure as much as for more mundane motives.[40]

A self-referential interest of the crime play in the performing arts, however, only evolves with the postmodern comedy thriller. Marvin Carlson notes that many comedy thrillers contain characters that are actors while others, though not actors, "consciously 'theatricalize' their situations by appearing in disguise, committing false murders with blank guns, staging elaborate hoaxes, and so on."[41] Examples of conscious theatricalization follow (from *Murderer* and *Stage Struck*), but Carlson himself admits to bafflement as to why the theatre as a setting for a crime play has remained comparatively unpopular, particularly considering that the major writers of British Golden Age detective fiction were all "interested in the theatre and utilized it to some extent in their work."[42] A concern with metatheatricality seems to have remained an interest of individual playwrights in the genre, most notably Anthony Shaffer and Tom Stoppard, rather than a general trend. About the immoral artist as a painter it should be noted that he also makes a good murderee, for example in *Blackmail* and *Go Back for Murder*. Foreigners likewise make good victims but even better culprits. But while the ideal victim hails from Spain or Latin America, the most popular country of origin for a culprit is Germany (*The Man Who Stayed at Home*, *Cottage to Let*, *Warn That Man!*, *The Third Visitor*), particularly around the middle of the century, and for obvious historical reasons.

While most characteristics that mark a dramatic figure for a suitable victim also qualify that character for the position of the criminal, one large group of culprits is clearly meant to be criminal and nothing but criminal. This is the type of culprit that will be called the "abnormal criminal," which covers psychological aberrations from sexual ambiguity to downright insanity. The earliest example of a crime play with a culprit of an ambiguous sexual orientation is Benn Levy's *A Man with Red Hair*, followed by *Rope*, *Loot*, *Entertaining Mr Sloane* and *The Return of A. J. Raffles*. *Wise Child*, *Corpse!* and the rather obscure *The Murder Game* (which never reached the West End) make use of the initial confusion engendered by a male actor appearing in drag, and *Suspect* and *Rebecca* both include the character of a devoted chambermaid apparently homosexually attracted to her mistress. Since homosexuality was clearly a major taboo, it may be doubted if Grella is right in assuming that the criminal is generally a more "acceptable" character than the victim.

While the homosexual criminal, however, is a minor phenomenon, the emergence of the homicidal maniac marks a major new phase in the development of the English murder mystery play, as Rahill has noted. According to Rahill, the appearance of the homicidal maniac marks the transition between the psychology of melodrama, which characterizes the villain as "evil" and the modern psychology of the crime play, which diagnoses him as "ill." What is even more remarkable is that by its mere presence, the character of the homicidal maniac questions one of the basic tenets of classic detective fiction, namely that a crime may be traced back to its perpetrator and his motives by a chain of logical reasoning. If the killer is mad, however, no amount of reasoning will help the detective to uncover his motives. In order to contain the insecurity engendered by the device of the homicidal maniac, most crime plays offer some kind of pseudo-scientific explanation for the culprit's insanity. Some plays present medical explanations for mental illness, ranging from a simple childhood head injury (*What Happened Then?*) to the suggestion of degeneration in an old aristocratic family (*The Case of the Frightened Lady*). Other plays blame the environment of the criminal. The murderers in *The Wrecker, The Amazing Dr. Clitterhouse* and *Ten Little Niggers*, for instance, have become obsessed with their profession and resolved the tension between private self and social role by subordinating all other considerations to their professional duty, finally regarding even murder a means to be justified by its end. The culprits of *The Mousetrap, A Man with Red Hair* and *Love from a Stranger* were physically abused as children and consequently began to derive pleasure from inflicting pain on others. While Crispin's elaborate theory of the sensations in *A Man with Red Hair* is largely derived from the source text, the novel by Hugh Walpole, Bruce's emotional aberrations were only added to the source (Christie's short story "Philomel Cottage") by the adaptor Frank Vosper. Bruce recalls being punished at school:

> I can remember so well, standing in my headmaster's study at school, waiting for a beating, and seeing the hands [of the clock] move on – nothing I could do could stop them. [...] I can remember the funny mixture of sensations it used to give me ... terror, and yet with it a strange sort of delight.[43]

Institutionalized education teaches Bruce that intense sensations can only be experienced in situations of domination and submission, and in later life he carries this discovery to its logical conclusion. Towards the end of the century, crime plays increasingly featured murderers that had been seduced by the heroization of the mass murderer by the mass media, for instance *Dutch Uncle, Murderer* and *Popcorn. Murderer* and *Popcorn* both criticize the mass media's readiness to satisfy the audience's craving for violence and at the same time proceed to satisfy that same craving, a technique that Anthony Shaffer, referring to *Sleuth*, jokingly called "to have my cake and eat it." *Murderer* and *Popcorn* are the most extreme plays offering an extrinsic explanation of homicidal mania, because they locate the source of the murderous instinct not vaguely in society in general but more specifically in the audience, as if to say: "This is what you have come to see, and because you want to see it, it is going to happen."

A number of crime plays featuring a psychopathic killer, however, deliberately refrain from offering a reassuring explanation for the genesis of homicidal mania, for instance *Rope, Night Must Fall, Christie in Love* and *Normal*. Of all the choices open to the playwright, this is probably the most disquieting one. Howard Brenton in his production note for *Christie in Love* prescribes a style of production that leads to a "kind of dislocation, tearing one style up for another, so the proceedings lurch and all interpretations are blocked, and the spectator hunting for an easy meaning wearies, and is left only with Christie and his act of love."[44] And Michael Clayton Hutton in *Power without Glory* explicitly reveals the motif of the homicidal maniac itself as a comforting device when he has the young murderer's father declare: "No, Cliff's not mad, Mum. People who do those sort of things aren't mad. We say they are 'cos it makes us feel better. If we say they are then it means we couldn't do the same sort of thing."[45]

A handful of plays that refuse to give a rational explanation of homicidal mania instead revert to a pre-scientific, magical understanding of mental illness as demonic possession. The murderess of *They Walk Alone* is obviously both a witch and a werewolf – her fits of homicidal frenzy are announced by the howling of a dog. In *Conduct Unbecoming*, the folklore motif of possession by an evil spirit is hidden under a thin veneer of psychoanalytic jargon:

> [I]t is reasonable to suppose that in Roach's unconscious mind, the idea was formed that Scarlett had died in his place, that he owed Scarlett a life. Unconsciously he has allowed Scarlett to come back to life by occupying his body and mind. But Scarlett is the stronger personality, Roach the weaker. And in the battle for occupation of the one available mind and body, Scarlett is winning. He becomes stronger every day, and Roach less and less in control.[46]

The concept of the insane as an individual gifted with an inordinate access to the spiritual world is varied in *Trunk Crime*, where the psychiatrist remarks that "the abnormal is like a crack in the wall. It lets in the light and you can get a glimpse through of the truth on the other side."[47] The play which follows, however, does not fulfil its initial promise of showing the mentally deranged as gifted with a superior existential awareness: the action that unfolds depicts an ordinary mental breakdown.

It is not accidental that the discussion of the figure of the detective has been left for last. For in the crime play, the detective is clearly of lesser importance than in the detective novel. The character of the Great Detective familiar from the traditional English detective novel appears in no more than eight plays, six of which are stage adaptations of novels. Christie and Sayers each wrote one crime play featuring their most successful serial detectives, Hercule Poirot (in *Black Coffee*) and Lord Peter Wimsey (in *Busman's Honeymoon*), and afterwards abandoned the experiment. Christie, as has variously been noted, even replaced Poirot by other detective characters in later dramatizations of her novels. But it is not alone the Great Detective of the English tradition who is absent from the crime play – the hardboiled dick of the American thriller is equally rare. Mr Callaghan in Gerald Verner's adaptation of Cheyney's *The Urgent Hangman* is the only character to conform exactly to type; there are a few female private detectives (*The Wrecker*) and some detectives employed by insurance companies (*Policy for Murder*), but by and large the investigation is left either to the police (in about 70 plays) or to an amateur personally involved in the case (over 100). The functional detective may be personally involved because, in contrast to the detective novel, the crime play does not feature a serial detective; with the sole exception of Lily Piper, who appears in two successive plays by Jack Popplewell. The fact that in most plays the detective is personally involved means that he can never be the impartial, objective super-reader familiar from detective fiction.

In a number of crime plays the detective is drawn into the investigation because (s)he has married or fallen in love with the criminal. This applies for instance to *The Two Mrs Carrolls, Double Image* and *Dear Delinquent*. In most of these plays the functional detective is prepared to avert the course of justice for the sake of his (or her) beloved, at least for some time. In some cases they even use the crime as a means of ensuring the culprit's devotion, most conspicuously so in *Entertaining Mr Sloane*, but also in more conventional plays like *The Last of Mrs Cheyney, The Sound of Murder*, and arguably even *Night Must Fall*. The absence of an impartial detective abolishes the clear-cut distinction between the investigating subject and the object to be investigated that governs traditional detective fiction. According to William W. Stowe, the shift from the concept of the detective as a "super-reader" outside the "text" of the crime he investigates to the concept of the investigation as an interactive process may be related to a paradigm shift in science "from semiotics to hermeneutics."[48]

Another popular type of functional detective likely to be compromised by his association with the culprit is the lawyer, advising either the culprit (*Witness for the Prosecution*) or the victim (*Towards Zero*). The lawyer is a suitable choice because he is a semi-official investigator, yet even lawyers are sometimes tempted to shield their clients, as in *Living Dangerously* or *The Amazing Dr. Clitterhouse*. The description of the lawyer-detective in *A Clean Kill* sums up the characteristics of the type of the gifted amateur investigator:

> Mr Schofield is a solicitor of the post-war type. That is to say, he is about 40, spent the whole of the war in an active branch of the armed forces, and is now just about beginning to catch up with the Law. This mixed upbringing has broadened his mind if it has foreshortened his professional knowledge. Meeting him casually you would suppose him to be almost anything from a gentleman farmer to something in the City. Not "clever" but, as will appear, shrewd enough.[49]

This description shows an appreciation of physical courage (going back as far as to Bulldog Drummond) accompanied by a profound scepticism towards formal education. The ideal detective is no longer an obsessive specialist of the Sherlock Holmes mould but broadminded, and, what is perhaps most important, not immediately recognizable as the detective.

In a number of crime plays the authors go to considerable lengths to conceal the true character of their functional detectives, in the same way they go about concealing the identity of the culprit. The detective of *Night Must Fall*, for instance, "is a subdued young woman of 28, her hair tied severely in a knot, wearing horn-rimmed spectacles; there is nothing in any way remarkable about her at the moment."[50] The final clause prepares a reversal of audience perceptions similar to the one prepared for the culprit. A particularly good example of a functional detective undergoing a major change before the audience's eyes is Rupert Cadell in *Rope*. He is introduced as a typical crime play misfit:

> He is a little foppish in dress and appearance, and this impression is increased by the very exquisite walking-stick which he carries indoors as well as out. He is lame in the right leg. He is enormously affected in speech and carriage. He brings his words out not only as though he is infinitely weary of all things, but also as though articulation is causing him some definite physical pain which he is trying to circumvent by keeping his head and body perfectly still. His sentences are often involved, but nearly always syntactically complete. His affectation almost verges on effeminacy, and can be very irritating, but he has a very disarming habit, every now and again, of retrieving the whole thing with an extraordinarily frank, open and genial smile.[51]

Cadell has all the characteristics of a good victim or culprit: he is too well-dressed, physically disabled, affected and effeminate. It seems that Hamilton, after making the identity of the two murderers plain from the start, wanted to introduce some amount of suspense concerning the true character of his functional detective. LaBorde describes the way in which the spectator's perception of Cadell changes in the course of the play, noting that Hamilton

> conceals his other major character, the functional detective Rupert, among the typed minor figures, thereby creating a false initial impression that Rupert adheres to type and is as vapid, bored, and insensitive as the characters with whom he is grouped. Only as Rupert reveals more of his real character does it become apparent that he belongs instead in a separate class as a more than worthy adversary for the murderers.[52]

In spite of the intended surprise, Cadell does belong to a recognizable type of fictional detective, the detective posing as a "silly ass," represented for instance by Sayers' Lord Peter Wimsey in fiction, and, in the genre of the crime play, appearing for instance in *The Man Who Stayed at Home* and *The Ghost Train*.

In *The Ghost Train*, the detective posing as a silly ass is an actual policeman, and generally during the period, the police are very fond of disguising themselves. The tradition of the detective disguising himself goes back to Sherlock Holmes in detective fiction and Hawkshaw, the detective in *The Ticket-of-Leave Man*, in the crime play. Policemen adopt the role of criminals (*At the Villa Rose*), of an alcoholic (*The Terror*) or, more respectably, a retired major (*The Mousetrap*). The technique of concealing the detective is applied most pervasively in the plays of Edgar Wallace, particularly in *The Squeaker*, where the final surprise relates to the identity of the detective, not of the culprit. Disguise of course also works the other way round, with criminals pretending to be policemen. This motif was most popular early in the twentieth century, appearing in *Bulldog Drummond*, *Paper Chase* and most notably in *The Ringer*. By mid-century, the character of the bogus policeman had lost a little in appeal, though of course Christie made use of it in *The Mousetrap*, and Priestley in *An Inspector Calls*. Both Christie and Priestley presumably chose the motif deliberately because it was already such a conventional, hackneyed device.

The phenomenon of policemen dressing up as criminals and vice versa of course fits in with what was said above about the cutting loose of actantial roles from their actants. Besides, it serves to undermine the audience's faith in the institution of the police. The police are portrayed as ineffectual (*A Lady Mislaid*) or, if they manage to discover the true culprit, they must let them off nevertheless (*Ten-Minute Alibi*). The convention of the "stupid police" is of course familiar from detective fiction, where it serves as a foil to celebrate the gifted amateur.[53] In crime plays, too, the policeman frequently receives substantial help from amateurs, for instance in *Dial "M" for Murder* and *Dead Easy*. Apart from clever amateurs, the police are often aided by surprise witnesses and other agents or acts of poetic justice, for instance in

Once a Crook or *Corpse!*. Policemen who finally solve the case may belong to the type called the "plodder" by Julian Symons, for instance in *Four Winds*, but more frequently they resort to methods that are not wholly legal and engage in games of cat-and-mouse with the criminals, for example in *Mile-Away Murder, Dear Murderer, Dead on Nine* and most conspicuously in *Who Saw Him Die?*, where the retired policeman kidnaps the master criminal in order to enforce a confession. In other, particularly late twentieth-century plays the police are portrayed as corrupt, racist or violent (*Dutch Uncle, This Story of Yours, Loot, Sus*), and even basically decent policemen are shown to be corruptible through sentiment (*Deadly Nightcap*).[54]

As these observations show it would be inappropriate to speak of the "police procedural" as a subgenre of the crime play, as LaBorde does. The term "police procedural" is borrowed from the field of detective fiction and refers to the type of novel, written for instance by Ed McBain, which "concentrates upon the detailed investigation of a crime from the point of view of the police, and in the best examples of the kind does so with considerable realism."[55] A list of plays filed under "police procedurals" by LaBorde will immediately show that they do not conform to the above definition: *The Creaking Chair, The Case of the Frightened Lady, The Hollow, I Killed the Count!, Murder on the Second Floor, Spider's Web, Towards Zero* and *The Unexpected Guest*. In the Christie plays, the focus is firmly on the family of the murder victim, in *Murder on the Second Floor* on the playwright inventing the detective plot. The cast list of *The Creaking Chair* lists the policemen at the very bottom, two of them as "utility." *The Creaking Chair* was not the only crime play to treat policemen as "utility." The production note for in *Christie in Love* for instance expressly prescribes: "The Constable and the Inspector are not 'characters,' apart from the facts of their age and rank. [...] They are stage coppers."[56] Shaffer describes what the stage copper should look like: "His hair is balding, and he wears cheap round spectacles on his fleshy nose, above a greying moustache. His clothes, dark rumpled suit, under a half-open light coloured mackintosh occasion no surprise, nor does his pork pie hat."[57] His clothes occasion no surprise, because Milo has donned them with the express intention of embodying the prototypical stage copper, but also because they were previously immortalized by Peter Sellers' Inspector Clouseau.

10
Violence, Crime and Punishment

One of the most remarkable features of British Golden Age detective fiction is its lack of violence. Though the crime is most frequently murder, neither the pain of the victims nor the pain of their relatives and friends is particularly indulged in; for a genre dealing predominantly with violent death, the traditional British detective novel is surprisingly pain- and bloodless – the latter, as some critics would say, in both senses of the word. As Dorothy Sayers points out in her introduction to *The Omnibus of Crime*: "The victim is shown rather as a subject for the dissecting-table than as a husband and father. A too violent emotion flung into the glittering mechanism of the detective story jars the movement by disturbing its delicate balance."[1] American hardboiled detective fiction, by contrast, has – as the name suggests – always been eager to depict violent actions, for instance in the recurring scene of the tough detective being beaten up by thugs, which serves to demonstrate the hero's power of endurance and perseverance. A particularly good example of excessive violence in classic American detective fiction is Dashiell Hammett's *Red Harvest*, in which the detective hero "cleans up" the corrupt city of "Poisonville," killing a major proportion of its population in the process. However, excessive violence did not become the trademark of Hammett's writing, as in the case of Mickey Spillane, who has become known as the main representative of the detective novel of extreme cruelty. The spirit of Spillane is invoked in David Hare's crime play *Knuckle*, in which the protagonist, arms dealer Curly Delafield tells the audience:

> I was once stranded in Alaska for ten days with a single copy of G. E. Moore's *Principia Ethica*. And one copy of *My Gun Is Quick*. The work of Mickey Spillane. I was able in this period to make comparisons under scientific test conditions. The longest word in *Principia Ethica* is "contrahydrapallotistic." The longest word in Spillane is "balloon." Moore wins outright on length of sentence, number of words and ability to contradict yourself in the shortest space. Spillane won on one count only. It burns quicker.[2]

The competing moral universes of Moore and Spillane are humorously compared and discarded, but Delafield proves susceptible to the specific language of hardboiled fiction, when he laconically remarks in parting: "It burns quicker." *Knuckle* is the rare example of a crime play consciously modelled on American hardboiled detective fiction, and to an even greater extent on its cinematic equivalent, *film noir*. Hare, however, abstains from showing acts of violence onstage in *Knuckle*, although stage censorship had by then been abolished.

Graphic violence shown on stage is a matter altogether different from violence explicitly described in a literary text, and attitudes to violence on stage have greatly varied in the course of history. In English theatre history, excessive violence on stage is mainly associated with Elizabethan and Jacobean revenge tragedy, for instance *The White Devil* with its numerous and varied murder methods. For most of the twentieth century, however, stage censorship prevented the display of excessive violence[3] on the assumption that, rather than cathartically purging the viewer from aggression, the representation of violent actions on stage leads to their imitation in real life; an argument still conducted today, though mostly in the context of videos and computer games.

The trend towards the control and containment of aggression was countered by the opposite movement towards the excessive display of violence. In 1920, for instance, theatre manager Jose Levy opened a Grand Guignol theatre at the London Little Theatre, producing a number of plays characterized by extreme cruelty, starring Sybil Thorndike in her pre-Saint-Joan days.

> Sybil Thorndike suffered every torment that the British Grand Guignol could invent, being crushed by a collapsible ceiling; murdered and stuffed into a trunk; blinded and strangled; terrified as her husband was tossed to a ravenous band of wolfdogs; and encased in a block of plaster.[4]

Some 40 years later, Peter Brook imported another type of French theatre of violence to the British Isles when he produced a season of "Theatre of Cruelty," based on the theoretical writings of Antonin Artaud. Artaud's *Le théâtre et son double*, written between 1931 and 1937, had been translated into English in 1958. On the English stage, Artaud's theories were put into practice most extensively by the in-yer-face theatre of the 1990s. A comparison between Aleks Sierz' definition of in-yer-face theatre[5] and Antonin Artaud's *Le théâtre de la cruauté* shows that imagery and vocabulary are remarkably similar. Like Sierz, Artaud confirms the need for a theatre "that wakes us up heart and nerves,"[6] that respects no taboos and accepts no limits of the representable, but shows "everything in love, crime, war and madness."[7] The theatre envisaged by Artaud employs shock effects to create "a show aimed at the whole anatomy"[8] – like in-yer-face theatre, theatre of cruelty is experienced bodily by the spectator, affecting the whole organism, not (or not only) the intellectual faculties. On the English stage, theatre of cruelty probably found its most perfect realization in the works of Sarah Kane, which initially met with violent and unqualified critical disapproval, but were soon hailed with equally violent and unqualified critical praise. While the responses of professional drama critics may perhaps have been informed by their fear of failing to recognize the new Osborne or the new Beckett, audience reactions are more puzzling. For the curious thing about in-yer-face theatre is that it is popular, that audiences seem to actually enjoy being taken "by the scruff of the neck" and shaken into awareness. This response can probably most easily be accounted for by assuming the existence of a pervasive state of emotional numbness and alienation among viewers, who prefer feeling horror and disgust to feeling nothing at all. It may also be explained by a general hostility to intellectual analysis, by being tired of competing and increasingly complex theories about life. This hostility towards intellectual activity seems the most disturbing aspect of theatre of cruelty and in-yer-face theatre alike. In his selection of in-yer-face plays, Sierz includes the crime plays *Butterfly Kiss*, *Normal*, and *Mojo* as well as Irvine Welsh's *You'll Have Had Your Hole*. *You'll Have Had Your Hole* in particular displays a positive relish in cruelty and torture and received unanimously scathing reviews from West End critics.

Frequently, however, crime playwrights created shock effects not through excessive sensations but through an entire lack of sensation. Geoffrey Kerr's World War II spy play *Cottage to Let* contains a scene in which the protagonist invites the other characters to stick needles into his anaesthesized body for purposes of scientific demonstration. A disturbing lack of feeling is also in evidence in the particularly cold-blooded murders in *The Fourth Wall* and *Once a Crook*. The stage direction for the murder of Estelle in the latter play reads as follows: "A moment's pause – and the Duke – calmly and deliberately – shoots Estelle. She remains where she is – sitting on the settee – she slowly turns her head to Charlie and smiles. Then, with no other movement, her head sinks forward."[9] Neither the act of killing nor of being killed is performed with any expression of emotion. Like an excess of feeling, a lack of feeling turns out to be a recurrent concern of the crime play.

It is not quite as simple to establish a play's degree of violence as might be supposed. Audience responses, for instance, are notoriously difficult to measure, and even in the case of *Love from a Stranger* or *Afore Night Comes*, which reportedly sent spectators into fainting fits, it cannot be proved beyond any doubt that it was the perceived violence on stage that caused them – viewers may lose consciousness in the theatre for a variety of reasons (including faulty air condition). It seems therefore advisable to move away from the precarious field of audience reception studies to the safer ground of a study of the plays themselves. In order to gauge the degree of violence, three qualities of a crime play should be considered: the type of crime committed (e.g. murder, theft, espionage), the number of crimes committed, and the treatment of violent actions – are they graphically displayed on stage, or merely narrated?

As in the case of detective fiction, the crime play's favourite crime is murder. W. H. Auden and S. S. Van Dine explain why murder is the ideal crime for crime fiction. Under the subheading of "Why Murder?" Auden explains: "Murder is unique in that it abolishes the party it injures, so that society has to take the place of the victim and on his behalf demand atonement or grant forgiveness; it is the one crime in which society has a direct interest."[10] And Van Dine decrees: "No lesser crime than murder will suffice. Three hundred pages is far too much pother for a crime other than murder."[11] While Van Dine argues in the Aristotelian tradition, requesting that the action be of a certain magnitude, Auden argues that after the removal of the victim, justice can only be done if society as a whole takes action. In a crime play, "society" may be substituted by "audience," and accordingly there resides a strong direct appeal to the audience in the very crime of murder. It is therefore hardly surprising that only one crime play in five manages to do entirely without murder. The majority of crime plays without murder revolve around a case of theft, and it will be found that Bentley's dictum: "Comedy is very often about theft, exactly as tragedy is very often about murder"[12] applies in a large number of cases.

Returning to the violent crime play of murder, we find that usually one murder is deemed sufficient to keep up audience interest. However, the crime play plot formulas of the "battle of wits" and serial crime generally require more than one criminal action. Crime playwrights escape the accusation of gratuitous violence by logically relating the individual crimes to each other – theft leads to murder in *The Creaking Chair*, rape to murder to blackmail in *Blackmail*, and in *Murder on the Nile*, the second murder is committed to silence a blackmailer who witnessed the first. In order to create an increasing tension, playwrights tend to order the crimes according to a rising degree of violence, a technique for instance in evidence in *Dangerous Corner*, *Recipe for Murder* and *Play with Fire*. From the mid-1950s onwards, there is a noticeable increase in multiple-crime plays, mostly occasioned by a greater complexity of plot. Frequently, the various crimes are committed by different characters – in accordance with the increasing popularity of the "battle of wits" formula – and escalating in violence. Blackmail prompts its victims to murder in *Something to Hide*, *Not in the Book* and *Kill Two Birds*, and rape provokes manslaughter in *This Story of Yours*. The use of multiple crimes does not always create the impression of extreme cruelty but may also be experienced as hysterically funny, as in the farcical *The Unvarnished Truth*, *A Spot of Bother* and *Simple Spymen*, and indeed it has already been frequently noted that violence is a necessary ingredient of farce. This observation may serve as a timely warning against a too mechanistic approach to the problem of onstage violence. As the example of farce shows, violence on stage may serve not only to shock the viewer into a state of heightened awareness but also to allow him to vicariously enjoy the release of previously pent-up aggression.

The trend towards multiple crimes, greater plot complexity and a greater relish for onstage violence persists after 1970, accompanied by a growing interest in professional crime. Crime plays about organized crime with a liberal lacing of violence from the last three decades of the century include *Who Saw Him Die?*, *Mojo* and *The Life of Stuff*.

A few words need to be said about organized crime in fiction. In hard-boiled American detective fiction, the excessive amount of violence is usually caused by its background in organized crime, as sketched by Raymond Chandler:

> The realist in murder writes of a world in which gangsters can rule nations and almost rule cities, in which hotels and apartment houses and celebrated restaurants are owned by men who made their money out of brothels, in which a screen star can be the finger-man for a mob, and the nice man down the hall is a boss of the numbers racket.[13]

The traditional English detective novel, by contrast, clearly favours domestic crime; Van Dine for instance declares that a "professional criminal must never be shouldered with the guilt of a crime in a detective story. [...] The motives for all crimes in detective stories should be personal."[14] The English crime play generally follows the example of English detective fiction in preferring domestic to professional crime. Plays around organized crime make up no more than about 15 per cent of all crime plays, with peak times of popularity towards the beginning and the end of the century. In the stage directions of Attiwell/Price's *Once a Crook*, it becomes evident that organized crime and the mythical figure of the gangster were immediately and inevitably associated with the American nation. Describing the American master criminal the Duke, Attiwell/Price can think of nobody more appropriate than Edward G. Robinson, who had created the character of the American gangster on screen in *Little Caesar*. This national stereotype is parodied in Alec Coppel's American crime play *The Gazebo*, which features an English master criminal introduced as an "Old Etonian from England known as the Dook."[15]

So far only the quantity of violence contained in any one crime play plot has been considered. But it is even more important to look at the manner in which violent actions are presented, distinguishing first and foremost between violence displayed onstage and violence that is supposedly happening offstage and is merely narrated or suggested. As Saroyan/Cecil have their master criminal Mr I state in *Settled Out of Court*: "The trick is to keep violence both concealed and unexploited, but always a very real possibility."[16] As is to be expected, violence graphically performed onstage increases markedly after the abolition of censorship. Before 1970, between 20 and

30 per cent of crime plays show violent actions onstage, with the percentage slightly but steadily increasing. After 1970, one in two crime plays shows graphic violence, sometimes excessively indulging in it, even before the advent of in-yer-face theatre in the "nasty 90s."

The most extreme act of violence in any of the crime plays surveyed occurs at the very beginning of Anthony Shaffer's *Murderer*. The play starts with a stage direction specifying that "[t]hirty minutes should elapse between the start of the play and the first spoken words."[17] In these 30 minutes of dumb show, the audience watches Norman drug and strangle his girlfriend and then witnesses the undressing of the corpse and the removal of teeth and limbs by a variety of implements: pliers, a knife, a saw and finally an electric drill. Norman then

> picks up a chopper from the wood basket and returns to the bathroom where he sets to work to cut off the head. The blows are deliberate and sickening. After three or four he straightens up, holding the head by its hair. He examines it closely, then kisses it playfully on the nose.[18]

After he has burned the torso and head in a stove and cleaned his workspace and himself, finally the police arrive. The stage direction describing the killing and dismemberment of the victim is incredibly meticulous; it runs for four pages and is almost unbearably detailed and realistic to read – let alone watch. Without being given any background information the audience is plunged into a variety of conflicting emotions: initially feeling horror at the murder, then suddenly finding themselves siding with the murderer when it seems that he will be surprised at the process of disposing of the corpse, and sensing a little morbid amusement at the sight of Norman kissing the severed head on the nose and amusement mixed with disgust when he contemplates eating a sandwich that has become soaked in blood in the course of the cutting up of the dead body. This initial sequence was obviously thought to be the pivotal scene of the whole play by publisher Marion Boyars, since the cover features a photograph of Norman strangling his girlfriend. The scene evidently corresponds to Sierz' description of in-yer-face theatre to the letter. It appeals to the nerves and sensations rather than the mind, employs shock tactics and challenges the moral norm of a dignified treatment of the dead body. It is remarkable that the major shock effect is not created by showing a prolonged killing but the graphic dismemberment of a body already dead, a technique also employed in Joe Orton's *Loot*. Dismemberment of a dead body is an act particularly abhorrent to the public, as the reaction of British juries in murder trials has proved. A survey of famous English murder trials in the recent detective novel *Original Sin* by P. D. James ends with the recommendation that "men wishing to be acquitted of murder should avoid dismembering their victims, a practice for which British juries have long demonstrated their distaste."[19] A deep-rooted "distaste" for the practice of dismemberment is in evidence for instance in the history of legislation concerning the practice of anatomical study. From the reign of Henry VIII until 1832, the only bodies available for dissection were those of executed murderers. Dismemberment was obviously considered a desecration of the dead human body appropriate for the most infamous of criminals only. Recently, the debate on the dignified treatment of the dead human body has been fuelled again by Professor von Hagens' exhibition *Body Worlds*, which exhibited dead bodies to the public for purposes of entertainment and, even more, by the public autopsy performed by von Hagens on Channel 4, 20 November 2000. The way in which a dead body may be presented to the audience in an acceptable manner also constitutes a major problem for the crime play.

A specific way of presenting the dead body on stage is the device termed the "busy body" by LaBorde:

> One other mystery device complicates comic mysteries almost exclusively: a vanishing corpse or "busy body," as it was labeled by Donald E. Westlake in the title of one of his comic mystery novels. What would ordinarily be a loathsome, repugnant object becomes a comic prop instead. The humorous use of a corpse in stage mysteries begins with *Seven Keys to Baldpate*, in which a body disappears and then reappears toward the end of the drama (II.871), and continues in most of the comic mysteries.[20]

One of the risks of employing the "busy body" motif is that it can easily cease to be funny, as LaBorde points out two pages later:

> For example, George Batson warns about the body in *Ramshackle Inn*. He includes the prohibition that the corpse "should not be too realistic" (II.54) He fully realizes that even a remotely human body carted about the stage can easily become grisly; therefore, he cautions against exceeding the perilously close limits of unseriousness.[21]

Ramshackle Inn, however, premiered in 1944, and since then, the standards of what can or cannot be shown onstage have changed rapidly. LaBorde's previous remark about the "comic prop" is problematic in two respects: first, because the motif occurs not only in comedy, but also in other types of crime play, and secondly, because, technically, the "busy body" is often represented not by a prop but by an actor, particularly if the victim was first seen alive (as in *Spider's Web*) or if the corpse is to revive later on,[22] as in *Meet a Body* and *Who Saw Him Die?*, or both – as in *The Unvarnished Truth* and *Corpse!*. The most frequent device for presenting a dead body on stage is a living body.

The picturesque arrangement of a dead body in a bedroom in the style of Victorian *tableaux vivants* is described in the initial stage direction of Robinson's *To What Red Hell*:

> The Scene is the rather poorly furnished but cheaply dainty bedroom of a first floor flat in Kennington. [...] The outer passage is well lit, but the room itself is in darkness but for the light that comes from doorway, and the little gleam from between the curtains. [...] Down R.C. is an ottoman, chintz spread along it – on it the flashy garments of a woman – a pair of gaudy shoes stand in front of ottoman, while on the floor behind it is a nightdress-case, with a nightdress half out of it. An incandescent gas bracket is above window. The bed itself is C. and on it, covered to the chin by the disordered bedclothes, lies the body of a woman. Her head is hanging a little over the R. top end of bed, with the back of head towards floats, her hair streaming.[23]

This stage direction is very particular in prescribing not only the use of indirect lighting and the specific distribution of furniture and garments but also the exact position of the corpse, mocking sleep and at the same time presumably intended to evoke associations with Fuseli's famous painting of *The Nightmare*, with its deliberate reference to the head hanging a little over the top end of the bed and the hair streaming. Thus, the play opens on a "tableau vivant" that on closer inspection turns out to be a still life – a "nature morte." The living body of the actress represents a dead body which is treated as an aesthetic object.

If a dummy is used to represent the corpse it will hardly bear closer inspection, which is why the dummy is sometimes revealed to be a dummy at second glance, and not a corpse as previously supposed. This device is employed for instance in *Murderer, Stage Struck* and *Who Killed "Agatha" Christie?*. This indicates a shift in the attitude towards representational realism on stage. In *Ramshackle Inn*, the audience is expected to accept a non-naturalistic corpse as a part of the contract between performers and spectators, but this willing suspension of disbelief can no longer be taken for granted in the English crime play of the 1970s. In *Loot*, the problem is solved by a double logical somersault: the audience is expected to accept a dummy as the representation of a dead body, which within the world of the play is then passed off as a dummy to the investigating police officer – successfully of course, since it looks a lot like a dummy anyway. Since a dummy always looks remarkably like a dummy, authors of crime plays have sometimes resorted to having the corpse represented by its container, thereby avoiding the risks of both the grisly and the ridiculous. LaBorde mentions the device of placing the body in a bag or sack "for easier handling"[24] in *Ramshackle Inn* and *The Gorilla*, to which *Brush with a Body* and *Jumpers* may be added. In *Rope* and *Trunk Crime*, the container is a trunk; *Night Must Fall* most inventively uses a hatbox supposedly containing the head of Dan's previous victim. As in *Night Must Fall*, the whole body is represented by one of its members in *Who Killed "Agatha" Christie?* (a dummy head), *Loot* (false teeth and a glass eye) and *Brush with a Body* (a shoe suggesting a foot). Apart from being easier to handle, these props also transgress the taboo against dissection mentioned above.

Among murder methods, shooting and poisoning are the most frequent, followed by stabbing and strangling. The advantage of shooting and poisoning is of course that the murderer does not have to come into close physical contact with his victim – does not have to kill, so to speak, with his own claws and teeth – but apart from that, the use of a gun or of poison usually means that the crime has to be premeditated, which is clearly an aspect attractive to the crime playwright, since the elaborate plotting of a crime is a fascinating activity to watch.[25]

The use of varied and inventive murder methods is a characteristic feature of *The Last Hour* by Charles Bennett, which develops a positive relish for describing various murder methods and the way they may convincingly be presented on stage. Killings are performed by means of strangling, shooting and burning alive – the latter being accomplished theoretically by a "death ray machine," and in practice by blinding the audience momentarily with a flash of magnesium and meanwhile replacing the actor playing the victim by his double, already made up as a scorched corpse. It was not until 65 years later that another crime play author showed a comparable interest in the different methods of transporting people from life to death. In Simon Donald's *The Life of Stuff*, however, the various killings are mostly narrated, Donald does not share Bennett's enthusiasm about trying to make scenes of violence work on stage.

During the more squeamish period between *The Last Hour* and *The Life of Stuff*, inventive murder methods featured in *Dear Murderer*, where the victim is made to cooperate in his own gassing and *Trunk Crime*, in which the victim is intended to die of asphyxiation, which is a murder method with the additional benefit of making the murder reversible. The severing of heads occurs in *Who Killed "Agatha" Christie?* and *The Case of the Oily Levantine* – once the ban of censorship had been lifted – and Stoppard employed a sensational example of "death from a great height" in *Jumpers*.

Discussing the relationship between the crime play and true criminal trials, Roy Vickers has noted a significant difference between the treatment of crime in drama and the principles of criminal procedure in court as regards the relative importance of opportunity and motive. The crime play stresses motive:

> Motive becomes the first essential, facts of time and place being little more than an explanation that the suspect could have "dunnit" if so minded. The author's main task – as opposed to that of prosecuting counsel – is to show that the suspect *could* have been so minded.[26]

Since the crime writer is specifically concerned with motive, it is not surprising that F. Tennyson Jesse has devoted a whole monograph to the subject. In her book *Murder and Its Motives*,[27] she distinguishes between six major categories: "for gain," "from revenge," "for elimination," "from jealousy," "from lust of killing" and "from conviction." Although this system of classification indubitably makes sense, it is not particularly helpful for an analysis of the crime play. Wargrave in *Ten Little Niggers*, for instance, kills just as much "from lust of killing" as "from conviction," and Tony in *Dial "M" for Murder* attempts to kill his wife both "from jealousy" and "for elimination" – arguably, all murders are committed "for elimination," with the result that the vast majority of murder plays may be grouped in this category. Therefore, a slightly adapted method of categorization will be adopted in the present study, keeping the categories of "gain" and "revenge," and subsuming "lust of killing" and "conviction" under the heading of "insanity." There will be an extra category for plays in which murder is the inevitable outcome of the configuration termed the "amorous triangle," which will contain murders "from jealousy" as well as some committed, in Jesse's terms, "for elimination." Finally, there will be the new category of "self-defence," which includes not only actual self-defence against violence, but also murder committed to protect one's reputation, since the murder of a blackmailer is a crime by no means infrequent in drama.

As might be expected, gain is the most frequent motive for murder, occurring in about one quarter of all classifiable murder plays. It is particularly popular in the postwar period between the end of World War II and the mid-1960s. What is most remarkable is that a large number of murder plays motivated by greed are adaptations of novels, for instance *At the Villa Rose* and *Payment Deferred*. Among original crime plays, the motive of financial gain figures particularly prominently in *Busman's Honeymoon* and *A Dead Secret*. The amorous-triangle ending in murder is almost as frequent and also reaches its peak in popularity between 1945 and the mid-1960s. Revenge as a motive for murder is less frequent, occurring in approximately one sixth of the plays surveyed, distributed evenly through the century. About as often as revenge, insanity features as an explanation of the culprit's motivation. As stated in the previous chapter, the homicidal maniac is a character typical of the crime play proper; in this category, adaptations of novels are rare. The motif of insanity is especially frequent in crime plays from the late 1920s to the end of World War II and reaches a second peak around the time of the abolition of stage censorship, when the limits of permissiveness were increasingly being tested. While actual self-defence is a motive rarely employed in crime plays, gradually getting even rarer as the century progresses, murder for the protection of one's reputation is slightly more frequent, occurring in about 10 per cent of the plays surveyed. It is most common between the late 1920s and 1945 and notably occurs in original crime plays only.

The different frequency of the various murder motives in adapted and original crime plays suggests that the crime play prefers what may be termed an "interactive" murder in which the victim contributes to his/her own demise. Particularly in the amorous-triangle plays of the 1950s with their interlocking triangles it is often perfectly unpredictable who will turn out to be the murderer and who will end up being the victim. Besides, murders in original crime plays are more likely to be committed spontaneously; crimes that require prolonged planning and plotting often occur in crime plays that are adapted from novels.

As has already been stated above, the most frequent non-violent crime in crime plays is theft. Plays about theft may be serious case studies of crime, justice and punishment, as in the works of Galsworthy, in *The Winslow Boy*, *Murmuring Judges* or *Drummers*, or they may be lighthearted comedies, like *The Last of Mrs Cheyney*, *Plunder* and *Dear Delinquent*. Both types are more likely to represent a portrayal of society as a whole, its norms and its stratification, than the average murder play. While the crime of murder illuminates a specific relationship, the crime of theft illuminates what society thinks desirable (in terms of objects) and acceptable (in terms of behaviour). Thomson has persuasively argued that theft is a particularly suitable crime to investigate the values of a society, stating that "Crook plays, dwelling on theft, explicitly insist on the social and political meaning of the crime and investigate American materialism from a Progressive viewpoint."[28] It is debatable, however, if the playwright has to be politically progressive, since even a conventional entertainment like *Dear Delinquent* ends up delivering the positively Brechtian message that there is hardly any difference between the professional crook and the respectable citizen, when the author has Sir George welcome a notorious family of burglars as prospective in-laws with a generous "Well, dammit, I'm a business man too!"[29] The crime play applies a similarly relaxed attitude to other crimes directed against property, for instance *Laburnum Grove*, *Dusty Ermine* and *The Man with Expensive Tastes*, which deal with forgery, and *A Spot of Bother* and *Mrs Inspector Jones*, which are concerned with smuggling. Both forgery and smuggling are apparently sufficiently abstract crimes to allow the perpetrator to retain audience sympathies, since there is no perceivable individual victim. Forgery is considered a talent rather than an offence, while in the case of smuggling, much depends on what it is that is being smuggled – transporting arms or drugs across the border, as in *The Ghost Train* or *The Flying Squad*, often results in violence and hence is perceived as definitely unacceptable and distinctly unfunny.

Early criticism of detective fiction[30] has tended to assume that the appeal of the detective story resides largely in its promise of justice being done; the culprit will be identified and punished in the end. In her article "Getting Away with Murder: An Analysis,"[31] Nancy Wingate contests this widely held view, claiming that the "satisfaction of the traditional mystery comes not from the reader's certainty of the immanence of justice but from his certainty of the immanence of truth."[32] While identification of the culprit is essential, readers do not necessarily insist on seeing the criminal punished. Wingate discovers four basic types of novels that let the criminal escape: texts that employ "poetic justice," those that apply "detective's justice," novels revolving around the character of "the charming rogue" or around the theme of "the perfect crime."

Examples of poetic justice are *Payment Deferred* by C. S. Forester, the source of the crime play by Jeffrey Dell, and *Malice Aforethought* by Francis Iles. Both novels are indubitably instances of poetic justice, since the murderer, while remaining unpunished for a crime he did commit, is later arrested and sentenced for another crime of which he is innocent. "Detective's justice" is said to occur in texts in which the detective successfully discovers the identity of the culprit but decides to let him (or her) escape, because (s)he considers the crime justifiable. Cases in which the detective knows who committed the crime but cannot prove it are also counted as examples of "detective's justice" – apparently they no longer qualify as "perfect crimes" once the criminal has been found out. The largest group in Wingate's essay are "perfect crimes," occurring for instance in Christie's *Curtain* and "The Witness for the Prosecution" – the short story, not the play version. In a footnote, Wingate shows that she is aware of the shift from "perfect crime" to "poetic justice" accomplished by the ending of Billy Wilder's film version, but she fails to note that the change was actually introduced by Christie herself in the 1954 stage version. The difference between the two endings must therefore be accounted for by the difference between fiction and drama (supposing film to be closer to drama), rather than by a change in authorship from Christie to Wilder. Taking issue with Auden's statement that, ideally, the detective novel ends with the execution of the criminal,[33] Wingate concludes "from a wide reading of twentieth-century detective fiction, the inference is clearly that traditional books are just not concerned with the punishment for the crime."[34] As will be shown in the following, this statement also applies to the crime play. Wingate's essay boasts a useful appendix listing the endings of Agatha Christie's novels, excluding spy thrillers, grouped under the following headings: "Murderer gets away with it (4)," "Murderer kills self or is killed by another who also kills self (15)," "Murderer goes 'over the edge,' may not come to trial (6)," "Murderer killed while trying to kill final victim (4)," "Murderer arrested (19)," "Murderer revealed but no further information given (14)."[35] Obviously, Wingate had to modify her previously established categories as soon as she had to deal with a specific canon.

For the following discussion of crime play endings, yet another method of categorization seems advisable, distinguishing between:

A. Plays that indicate that the culprit is being punished according to the Law;
B. Plays that end with the (impending) arrest of the culprit;

C. Plays that achieve justice outside the legal apparatus, either by acts of poetic justice or by allowing the culprit to commit suicide;
D. Plays that do not indicate that the culprit will be punished in any way;
E. Plays in which it is finally revealed that no crime was committed after all, or in which an intended crime is averted.

As will be noted, A and B correspond to Wingate's fifth category, "murderer arrested"; C contains the second, third and fourth categories of Wingate's essay with the murderer getting killed by self or other, or going insane, and D conforms to the first and last categories of Wingate's list – getting away plus "no further information given." Category E (not in Wingate) is a delicate matter, since this type of ending defies one of Van Dine's cardinal rules: "A crime in a detective story must never turn out to be an accident or a suicide. To end an odyssey of sleuthing with such an anti-climax is to hoodwink the trusting and kind-hearted reader."[36] According to the theories of John G. Cawelti, however, this type of ending makes perfect sense, since Cawelti argues that the main concern of the reader of traditional whodunit detective fiction is to see their favourite suspect cleared of guilt, a function obviously fulfilled by type E solutions.[37]

In accordance with Van Dine's criticism, type E, the "revelation of innocence," is the least frequent ending occurring in less than 10 per cent of the crime plays analysed, most notably in *Blackmail, Tabitha* and *Knuckle,* in which it is finally revealed that no murder was committed after all, and in *Trunk Crime* and *Communicating Doors,* where courageous women prevent the crime from being brought to its horrid conclusion. Endings of type A, indicating what kind of punishment awaits the culprit, are also rare – slightly over 10 per cent – and mainly occur in plays based on true criminal cases (*People Like Us, Christie in Love, Cause Célèbre*) or plays that specifically intend to criticize the judicial system (*Justice, Murder Story, Murmuring Judges*). Apparently, crime playwrights do not expect their audiences to feel any particular satisfaction at the idea – even less the sight – of the culprit being punished and hesitate to devise a severe punishment for a fictional character in an entirely invented plot. If a play is based on a celebrated criminal trial, the matter is different since the public will be familiar with its outcome anyway. The most popular conclusions of crime plays accordingly are arrest, poetic justice and escape, each making up about one quarter of crime play endings altogether. Arrest – performed or imminent – for instance ends the action of such classic crime plays as *The Fourth Wall, Rope, Dial "M" for Murder* and *Sleuth,* though it is sometimes, most economically, merely suggested by the sound of police sirens approaching.

Endings of the "poetic justice" type are particularly satisfactory if the criminal is killed in the same manner as his victim, which happens for instance in *The Speckled Band, Love from a Stranger, Dutch Uncle* and *Murder by Misadventure.* The formula of "plots fall'n on the inventor's head" is obviously still very attractive. Cases of suicide have been counted among acts of poetic justice, because they, too, indicate the final recognition of a moral law quite independent from institutionalized justice. Suicide of course might also be grouped under ways of escape, since it is traditionally thought more gentlemanly to end one's life oneself than to be executed. Suicide as the more gentlemanly way out is allowed to the murderers in *To What Red Hell, Alibi, The Two Mrs Carrolls* and *Beyond Reasonable Doubt.* With the abolition of the death penalty and the general decline of a gentlemanly code of conduct, suicide as a means of escape from justice necessarily becomes less and less popular in the course of the century. Two crime plays establish justice by means of distinctly metaphysical intervention: *A Murder Has Been Arranged,* in which the ghost of the murdered man returns to bring his murderer to justice, and *An Inspector Calls,* which ends with a prophecy of World War I as another deluge meant to purge a society characterized by thoughtless cruelty and moral hypocrisy.

Though crime plays that end without any punishment of the guilty frequently revolve around less serious crimes (as in *Paper Chase, For the Love of Mike* or *Breath of Spring*), some plays literally show the culprit "getting away with murder," for instance in *The Ringer, Suspect* or *Dead Guilty.* Of these plays, *Suspect* is the most interesting, since it does not deal with a crime that would seem justifiable by any moral standards; as will be remembered, the protagonist killed her parents with an axe. Some crime plays do not merely deny their spectators the certainty of the imminence of justice but even the certainty of the imminence of truth.[38] Either the play ends before the climactic act of murder is finally committed (*The Sound of Murder, The Dumb Waiter*) or the identity of the culprit remains forever unrevealed (*Somebody Knows, Jumpers*). These two possibilities are cleverly combined in *Murder by the Book*: if Selwyn intended to poison Imogen in earnest, he will drop dead any minute (which would then constitute an ending of poetic justice), if he used placebos, he will survive – which would be an instance of type E, the "revelation of innocence." However, since the curtain falls before the outcome is revealed the audience is somewhat paradoxically left without the revelation of truth, but with the certainty of the imminence of justice. Since crime fiction is a genre dominated by end-oriented desire, it is extremely rare for a detective novel to deny its readers closure to the extent to which this happens in the crime plays listed above. There are, however, comparable cases in the subgenre of antidetection, associated with writers like Jorge Luis Borges and Alain Robbe-Grillet. Anticlosure accordingly is a phenomenon of highbrow crime fiction, and this also applies to the genre of the crime play – as will have been noted, two of the five plays listed above were written by critically acclaimed dramatists: Pinter and Stoppard. "Middle-brow" dramatists like van Druten, Fairchild and Greenwood/ King, who catered for a different type of audience that mainly went to the theatre to be entertained, accordingly were running a considerable risk in departing from the established formula; a fact that should be noted and appreciated.

In the course of the century, the relative frequency of the various types of ending underwent considerable shifts. During the first three decades, arrest and escape were equally popular, followed by poetic justice, punishment and innocence revealed. The period from 1930 to the end of World War II showed a marked increase in plays being brought to a close by acts of poetic justice, while arrest and particularly punishment proportionally lost in appeal. Faith in institutionalized justice had apparently reached its lowest point: of 55 crime plays, no more than 15 ended with arrest or punishment. After the end of World War II, arrest once again became more popular than trusting in Divine Retribution, and until 1970 was the most frequent type of crime play ending. Criminals were now as likely to be arrested and punished as to be left alone or to poetic justice. After 1970, the belief in the efficiency of the organs of established justice declined again, and together poetic justice and escape became the most frequent types of conclusion. Moving from the question of the faith in the efficacy of the legal apparatus to the problem of the "certainty of the imminence of justice," we find that while before 1945 and after 1970, one crime play culprit in three could count on escaping justice (institutionalized or poetic) altogether, the odds were more strongly against criminals in postwar times, when only one in five escaped both arrest and the workings of Providence. This would seem to reveal a strong reactionary trend and a profound longing for law and order to be reestablished among postwar middle-class theatre audiences.

Bookstores catering for crime fiction addicts, such as London's "Murder One," usually devote an extra shelf to "true crime." This indicates that, although a reader of detective fiction may occasionally enjoy a report on a true criminal case, the two genres are by no means interchangeable and satisfy different readers' appetites. While classic detective fiction is mainly celebrated for its deftly constructed puzzles, a true crime documentary is hardly ever a whodunit – if it has received any amount of media coverage, "who done it" will already be widely known – and it certainly never offers the pleasures of following up artfully planted clues or demolishing elaborate alibis. What true crime does satisfy, on the other hand, is the reader's craving for the sensational, and, above all, for realism. Since the crime play generally strives for verisimilitude, it is hardly surprising that stage plays frequently took their cue from true criminal cases. Stokes has noted that British thrillers of the inter-war years "continually drew upon real-life cases."[39] The phenomenon, however, is not limited to that period – as pointed out in Chapter 2, Victorian melodrama had already exploited recent criminal cases, and true crime again became a very popular source for the stage play in the late twentieth century, after the abolition of stage censorship. While censorship was in operation, relatives of a convicted criminal could demand that a theatrical production be banned if it offended their sensibilities. This happened for instance to Frank Vosper's *People Like Us* which prematurely terminated its run following a complaint from a relative of Edith Thompson's. Accordingly, crime playwrights tended to avoid easy identification of the real-life sources. Rodney Ackland for instance declares in his preface to *A Dead Secret*: "The play was originally evoked by some of the more vivid circumstances of a celebrated case. The whole of the play and all the characters are in every other way fictitious and have no resemblance to any person concerned in the real-life story."[40] Yet J. C. Trewin manages to identify the Seddon case as the model of Ackland's work, pointing out the following similarities: "a querulous old woman's hoard, arsenical fly-papers, a pauper's funeral, a commission on the coffin."[41] Interestingly, Vickers considers the Seddon case particularly unpromising material for stage adaptation, in spite of the trial's indisputable appeal to the public:

> The case was colourful because it revealed in Seddon an incredible callousness and cupidity. [...] On the stage, the callousness and cupidity would require a formidable build-up, and the total result would merely be to present a particularly detestable murderer. [...][A] completely detestable murderer is unhandleable.[42]

The tracing back of crime plays to their real-life models, though an entertaining occupation in itself, hardly yields any significant insights. However, a survey of true criminal cases dramatized suggests that specific types of crime particularly recommend themselves for stage adaptation. A substantial number of crime plays are based on the exploits of serial killers: of John Christie (*Christie in Love* and *Dutch Uncle*), Peter Kürten (*Normal*), Myra Hindley (*Foreign Lands*) and George Joseph Smith (*The Mysterious Mr Love*). Another major group of attractive true crimes conforms to the type dubbed "the thriller of murderous passions" by Charles Derry: "[T]he thriller of murderous passions is organized around the triangular grouping of husband/wife/lover. The central scene is generally the murder of one member of the triangle by one or both of the other members."[43] An American *cause célèbre* that prototypically shaped the formula was the crime of Ruth Snyder and Judd Gray, which inspired James M. Cain's novel *The Postman Always Rings Twice*, variously adapted for the screen by Luchino Visconti, Tay Garnett and Bob Rafelson. Among famous British trials, the Bywaters-Thompson case, the Greenwood case (which inspired Huxley's *The Gioconda Smile*) and the Rattenbury trial, which formed the basis of Rattigan's *Cause Célèbre* and Gray's *Molly*, conform to the formula. What makes these cases so attractive for stage adaptation is not merely the combination of murder and illicit sexuality. There is also the lack of closure: the execution of Edith Thompson has been widely regarded as a judicial error, and the fictionalized treatments of the Greenwood case by Huxley and Agatha Christie (in *Verdict*) suggest that not Greenwood himself but one of his mistresses committed the murder. In the Rattenbury case, it turned out extremely difficult to establish the individual guilt of wife and lover. The combination of murder, sexual transgression, and residual doubts about the justice of the outcome of the trial has also contributed to the notoriety of the Leopold/Loeb case, which inspired *Rope*, and the Hulme/Parker case that formed the basis for Denham/Orr's *Minor Murder*. Due to the particularly abhorrent nature of the crimes – the apparently motiveless murder of a child by Leopold and Loeb, the murder of Parker's mother by Hulme and Parker – and the inordinate youth of the guilty couples, these cases have retained their appeal for the mass media and were recently adapted for the screen as *Murder by Numbers* and *Heavenly Creatures* respectively.

The relationship between the crime play and true crime has itself become a popular theme of the crime play. *Murderer, Dutch Uncle* and *Mindgame* in their various ways all criticize the public's craving for accounts of true murder cases. In *Murder in Motley*, what seems to be a real murder disrupting a theatrical performance turns out to be a crime play after all, while in *The Real Inspector Hound*, the apparently fictional crime play turns into a real murder once the critics have lost their aesthetic distance and been drawn into the action. A comparable instance is *Who Killed "Agatha" Christie?*, where the elderly drama critic is also made to lose his critical detachment and an uncommonly convincing crime play suddenly turns into an effective murder weapon. However, the two plays that object most violently to the exploitation of true crime by the theatre and the media are *The Crime at Blossoms* and *Popcorn*. But while in the earlier play the dramatized version merely fails to represent reality accurately, the later one claims that the audience's hunger for sensational crime is itself partly to blame for an increasing crime rate. The play, however, suffers from a certain lack of integrity, because it itself exploits the sensationalist strategies it claims to criticize, and, more importantly, because it locates the action at a comfortable distance in America.

11
Genre, Generic Development and Subgenres

As the preceding four chapters have demonstrated, crime plays throughout the century share a number of recurrent characteristics. It is debatable, however, if these characteristics are sufficient to claim that the crime play is a distinct dramatic genre. This largely depends on the concept of "genre" we wish to apply. Particularly from the 1980s onwards, genre studies based on the concept of the "loose family group"[1] have gained in influence:

> Contemporary theorists of genre tend to follow the lead of the philosopher Ludwig Wittgenstein, who thought of genre in terms of "family resemblance," a set of similarities some (but by no means all) of which are shared by those works classified together. Viewed this way, genre is a convenient, though arguably loose and arbitrary, categorizing and descriptive device that provides a basic vantage point for examining most historical and many modern and contemporary works.[2]

Among the crime plays discussed in this study, family resemblances are indeed very conspicuous. Stanley Richards for instance declares that a certain family resemblance exists between crime plays as various as *Witness for the Prosecution*, *Dial "M" for Murder*, *Sleuth*, *The Letter*, *Gas Light*, *Dangerous Corner*, *Ten Little Niggers*, *Hostile Witness*, *Night Must Fall*, *An Inspector Calls* and *Ladies in Retirement* by grouping them all together in his two collections of "Mystery and Suspense Plays." Moreover, the impression of family resemblance (and hence of the crime play as a distinct genre) is produced by a web of intertextual references connecting individual plays to each other. A particularly conspicuous example of an intertextual reference assisting in the constitution of genre is the identification of Arnold Ridley as the "author of 'The Ghost Train' " on a poster announcing *Peril at End House*.[3] The poster identifies the play as a "crime play" in a twofold manner: as a play based on a novel by Agatha Christie with Hercule Poirot – and hence "like" Christie's Poirot novels – and as a play adapted by Ridley, and hence "like" *The Ghost Train*. Apart from authors specializing in crime plays (most prominently

Agatha Christie and Edgar Wallace), recurring actors and producers also contributed to the reinforcement of intertextual links between groups of plays. Among actors, for instance Gerald du Maurier, Emlyn Williams and Gordon Harker might be mentioned; among producers, Agatha Christie's producer Peter Saunders comes to mind.

Besides, crime plays reinforce their link to the genre by metafictional reference to the genre of detective fiction – as in *Murder on the Second Floor, Not in the Book* or *Sleuth* – and to the genre of the crime play – for example in *Amber for Anna, Whodunnit* and *Who Killed "Agatha" Christie?*. The most interesting instance is the intertextual reference in the title of *Who Killed "Agatha" Christie?*, since it promises to be both an example of the crime play ("Who Killed?") and the ultimate demolition of the genre since the play anticipates the killing of Agatha Christie.

Another form of intertextual reference that strongly contributes to the creation of genre is the imitation of particularly conspicuous plot structures in successful plays. Innes has noticed the similarities between Orton's *Loot* and Travers' *Plunder*. Other conspicuous borrowings include the suicide made to seem murder in *The Late Edwina Black* (from *Appointment with Death*), the enforced re-enactment of a court trial in a private setting in *Guilty Party* and *Settled Out of Court*, the basic character configuration of *Verdict* (from *The Gioconda Smile*) or the ordinary citizen hired by the Secret Service in *Out of Bounds* (from *Simple Spymen*). The most remarkable example of this practice is David Ellis' *Make Me a Widow*, which borrows its entire basic plot mechanism from Williams' *Someone Waiting*, merely changing the gender of its protagonist.

According to contemporary genre criticism these multiple family resemblances would probably be sufficient to speak of the crime play as a genre. However, the category of the "crime play" is clearly of greater heuristic and interpretative value if we can – as in a more traditional, strict definition of "genre" – identify distinct characteristics that apply to all crime plays. A number of distinct characteristics were already part of the initial definition of the subject-matter of the present study. All the plays surveyed were written in the same century for the commercial London theatre, which imposes the doctrine of stage naturalism. And they all deal with crime, which carries the implicit promise of suspense and surprise. The combination of these features produces a type of play characterized by a strong tension between stabilizing (meaning also "comforting, reassuring") and de-stabilizing (meaning also "disquieting") tendencies. Stabilizing features are the adherence to the doctrine of stage naturalism, which mainly manifests itself in the preference for the single living-room set, and the use of a linear, chronological plot. These are characteristics the crime play shares with the Well-Made Play. Another stabilizing feature is the society depicted in the crime play, usually a "well-ordered" society characterized by fear of social mobility, of foreigners, and of disease and the diseased. While the linearity of the plot is comfortingly traditional, the crime play departs from the conventional plot structure of classical tragedy by accommodating more than one major climax and reversal. There tend to be at least two: the commission (or discovery) of the crime and the demasking of the criminal. Most crime plays, however, contain multiple climaxes and surprising reversals. As Thomson remarked about the "crook play:" "Reversal and surprise are the constants."[4] Another de-stabilizing feature with respect to plot design is the comparative irrelevance of punishment for the guilty. On the level of characterization, the crime play is de-stabilizing because it tends to question the notion of "character" itself, of a stable, "rounded" and invariable "identity," by making use of built-in character reversal and the blurring of actantial roles. Finally, a crime plays is de-stabilizing because it confronts the audience with the spectacle of the dead body. The presentation of a corpse may work as a powerful *memento mori*, particularly when the dead body is represented by a living body "playing dead." While a prop representing a dead body highlights the materiality of the corpse, the living body representing a dead body stresses continuity: it takes so little to transform a living body into a dead body. In *Rosencrantz and Guildenstern Are Dead*, Tom Stoppard playfully suggests that being subjected to the repeated spectacle of actors "playing dead" may trigger off similar reflections in the spectator: "Do you ever think of yourself as actually *dead*, lying in a box with a lid on it?"[5]

Within this general framework of opposing stabilizing and de-stabilizing forces, there exists a wide variety of crime plays. In this respect, the crime play resembles the genre of detective fiction. As Tzvetan Todorov has shown in his "Typologie du roman policier," although it makes sense to speak of the Detective Novel as a distinct class of texts – "*le* roman policier" – the main generic features only emerge once the genre has been divided up into suitable subgenres. A typology of the crime play is equally to be desired, and attempts in this direction were already undertaken in Chapters 8 and 9 of the present study, by pointing out various types of plot structure and various types of character configuration. However, neither the plot structures nor the character configurations are particularly helpful for creating a sensible system of categorization of crime plays into subgenres or "formulas."[6]

In his dissertation, Charles LaBorde attempts to create a typology of the crime play. His basic distinction is that of "Standard Formulas," which include "Confined Mysteries" and "Less Restrictive Formulas," and "Variations" and the "Comic Mystery," which are understood to be derived from the standard formulas. "Confined Mysteries" are again subdivided into "Murder-House Mysteries," "Police Procedurals," and "Psychological Thrillers." "Less Restrictive Formulas" include "Inverted Mysteries" and "Had-I-But-Knowns." As variations of the standard formulas, LaBorde lists the "Supernatural Mystery," the "Collective-Detective Mystery," the "Environmental Mystery" and the "Propagandistic Mystery." "Comic Mysteries" are understood to derive from the standard formulas because they are parodies of "straight" crime plays.

This system of classification is problematic in several respects. By borrowing some of his key terms from detective fiction studies, LaBorde overlooks the crucial difference between detective fiction and drama. The category of the "Had-I-But-Known" is a case in point. This is a derogatory term first coined by Ogden Nash to deride the novels of Mary Roberts Rinehart and applicable to detective fiction only, since it obviously refers to first-person narratives told in retrospect. Both retrospection and a first-person narrative voice are notoriously difficult to realize in drama – there is no "had" in drama, nor is there an "I."

However, LaBorde's typology suffers from yet more substantial problems than a mere unfortunate choice of terms. The main trouble is that his categories are not mutually exclusive, because from category to category, he applies different criteria for classification. Categories are defined according to setting ("Murder-House Mystery"), according to central character ("Had-I-But-Known"), according to dominant themes ("Supernatural Mystery"), to plot construction ("Inverted Mystery") or mode ("Comic Mystery").[7] Hence, it is hardly surprising that plays are capable of belonging to two or more categories at the same time – or to none at all. One of the main requirements of a typology of the crime play should therefore be that the individual categories must be complementary, that is, mutually exclusive and at the same time covering as many different texts as possible. Such a model of classification will be introduced in the course of this chapter. This typology will be partly inspired by Symons' *Bloody Murder.* Symons' basic distinction is the one between the Detective Novel, which deals with the discovery of a crime, and the Crime Novel concerned with the commission of the crime. The genre of the Detective Novel is subdivided into the subgenres of the "Classic" or "Formal" British Detective Novel, the American Hardboiled Thriller and the Police Procedural. Types of the Crime Novel include the Inverted Detective Novel (founded by Iles) and the Gangster Novel (typically represented by W. R. Burnett's *Little Caesar*). These categories and subcategories are to be distinguished most quickly and easily by their choice of central character. In the Detective Novel, the protagonist is the investigator, in the Crime Novel, it is the criminal. However, the identity of the protagonist is not the only distinctive feature – as the term "hard-boiled" and the derogatory label "cozy" for the formal British detective novel suggest, the difference is just as much one of mode as of central character. Hence, in the following typology, the two categories of mode and central character will be applied simultaneously.

Subtitles of crime plays occasionally reveal an interest in mode: Comic plays call themselves "farces" (*Plunder*, *Simple Spymen*) or "comedies," sometimes indicating the criminal context as in "criminal comedy" (*Well-Caught*), "an immoral comedy" (*Laburnum Grove*), "a detective comedy" (*Busman's Honeymoon*) or "a felonious comedy" (*The White Sheep of the Family*). Plays that take themselves more seriously are announced as "a drama" or "a tragedy," and plays that promise their audiences mainly suspense and excitement are advertised as "thriller," "adventure" or "melodrama." Finally, a number of crime plays employ compounds with "play" as their subtitles, among which "mystery play" is the most common, promising, above all, the pleasures of the puzzle (*Black Coffee, Towards Zero*). This leaves us with four different types of crime play: a comic one, a sad one, a thrilling one, and an intellectual puzzle with the respective appropriate audience responses of laughing, crying, trembling, and thinking. A similar principle of classification is suggested by Patrick Hamilton in his "Preface on Thrillers," introducing *Rope*:

> Laughter and tears and horror, after all, are the three essential factors which keep the theatre alive; and just as a man might have gone all out to write an emotional drama, to make you cry; or a farce, to make you laugh; so in "Rope" I have gone all out to write a horror play to make your flesh creep.[8]

Categories based on the intended effect of a literary work are particularly appropriate for drama, as Paul Hernadi has convincingly explained:

> Perhaps because our public reaction to performed plays is more observable and less idiosyncratic than our private reactions to printed written texts, drama criticism has always been very concerned with the effect of works on an audience. [...] [A] congenial performance can impose additional "unity" on the represented action, and live audiences further homogenize the response of individual spectators: who would not be somewhat embarrassed to laugh when most others are on the verge of crying and *vice versa*?[9]

There remains the question of what the four different modes of the crime play are to be called. In accordance with the decision of most of their authors, comic crime plays will henceforth be called Comedies (although the term "farce" may be more appropriate for some) – but serious crime plays will not be termed "tragedies."[10] Tragedy traditionally is said to arouse both terror and pity, whereas within the genre of the crime play, it is usually possible to distinguish clearly between plays that mainly aim at creating terror ("thrillers") and those that mostly provoke pity. In popular film as well as in modern everyday usage, an entertainment mainly intended to offer the audience the pleasure of "a good cry" is usually called "melodrama." The use of the term "melodrama" in the context of this study, however, may be slightly confusing, because it refers to a genre distinct from (though indebted to) Victorian Melodrama, which has already been mentioned and discussed several times as an important generic predecessor of the twentieth-century crime play. To distinguish the "tear jerker" variety of the crime play from Victorian Melodrama, it will therefore be dubbed (somewhat awkwardly) "Twentieth-Century Melodrama." Finally, plays that mainly appeal to their audiences because of their intricately plotted central mystery will be called crime play "Puzzles."

As Todorov has declared, new genres are made out of old ones: "A new genre is always the transformation of one or several old genres: by inversion, by displacement, by combination."[11] In the case of the crime play, the influence of several traditional genres may be traced. The impact of detective fiction and Victorian Melodrama on the genesis of the crime play has already frequently been remarked on. A further genre that needs to be mentioned in this context is the Well-Made Play, a dramatic subgenre that has also suffered critical neglect and contempt. The striking parallels between the well-made play and the crime play (particularly of the whodunit variety) become even more obvious in the following definition of the well-made play by Beckson/Ganz:

> The plot of a well-made play regularly revolves about a secret known only to some of the characters; revealed at the climax, it leads to the downfall of the villain and the triumph of the hero. The action, which centres on a conflict – especially a duel of wits – between the hero and his opponent, builds with increasing intensity through a series of reversals which culminates in the climactic revelation scene. Misunderstandings, compromising letters, precisely timed entrances and exits, and other such devices contribute to the suspense. The denouement is always carefully prepared, and, within the framework of the manipulated action, believable.[12]

As G. C. Ramsey has noted, the entire definition might also be taken to apply to a Detective Novel of the traditional British type.[13]

According to John Russell Taylor's *The Rise and Fall of the Well-Made Play*, the English well-made play was derived from the *pièce bien faite* of Eugene Scribe and his follower Victorien Sardou and influenced, albeit to a lesser extent, by the more mechanical farces of Feydeau and Labiche. These French models were imitated by English dramatists like Tom Robertson, who represented the realistic strain in Victorian melodrama. Robertson's major contribution to greater stage realism was that he abolished stock casting and expected every part to be written and played as an individual character. Other representatives of the late nineteenth- and early twentieth-century drama conforming to the standards of well-madeness were the plays of Henry Arthur Jones, whose *The Silver King* of 1882 was mentioned in Chapter 2 as an important nineteenth-century crime play, and Arthur Wing Pinero. Pinero wrote problem plays and farces, both characterized by an easy flow of dialogue and an inconspicuous plot construction. After Pinero, a serious and a comic variety of the well-made play continued to exist side by side, but after Shaw's attack on the well-made play as an inappropriate vehicle for the discussion of "ideas," Galsworthy remained the only representative of the serious problem play. The comic variety is represented by the "comedy of manners" of W. Somerset Maugham and the drawing-room comedies of Noël Coward and Frederick Lonsdale.

The four varieties of the crime play may be said to relate to these different subgenres as follows: The thriller is most closely related to Victorian melodrama, particularly in its fear-inducing quality. Twentieth-century melodrama may also be supposed to be influenced by its Victorian predecessor, particularly its interest in depicting suffering and arousing pity, but also by the problem plays of Galsworthy. Although fear and pity are traditionally linked to the concept of tragedy, Eric Bentley claims that they should more correctly be associated with melodrama.[14]

> Pity for the "hero" is the less impressive half of melodrama; the other and more impressive half is fear of the villain. Pity and fear: it was Aristotle in his *Poetics* who coupled them, and tried to give an account of the total effect of tragedy in these terms. It seems an oversimplification. In tragedy, most of us now feel, more is involved. Is more involved in melodrama? Is not working on the audience's capacity for pity and fear the alpha and omega of the melodramatist's job?[15]

The division of fear and pity into two separate subgenres helps to create greater unity of effect, and traditionally, melodrama has been said to strive for "monopathy." "Monopathy" is the term coined by Robert B. Heilman for the "singleness of feeling that gives one the sense of wholeness."[16] One of the pleasures of melodrama, according to Heilman, is therefore that it does away with ambiguity of character, mixed emotions or dubious value judgments. Dividing fear from pity arguably would serve to create an even greater "singleness of feeling," though the twentieth-century-melodrama variety of the crime play is saved from becoming an unadulterated "tear jerker" by the injection of seriousness and realism from the problem play.

The comic variety of the crime play has its source in drawing-room comedy, though some of the more farcical treatments of the theme are clearly indebted more directly to the French tradition. Taylor for instance remarks that the virtues of well-madeness are largely missing from the Aldwych farces of Ben Travers (among them *Plunder*), which "ramble all over the place, picking up and dropping characters without rhyme or reason, the perfect dramatic embodiment of an era which prided itself on believing that everything went."[17] The crime play puzzle also profited from drawing-room comedy with respect to stagecraft, a fact much in evidence for instance in Milne's *The Fourth Wall*. For the plotting of the puzzle, this subgenre was of course strongly indebted to the contemporary detective novel.

Apart from a categorization according to mode, it is of course also possible to classify crime plays according to their central actantial role as suggested by Charles Derry.[18] The greatest technical difficulty for the purposes of the present study is how to establish which of the dramatic characters is the "central" one. The following procedures have been adopted: It has generally been assumed that a play's title character is also its central character, as for instance in *The Amazing Dr. Clitterhouse, Carrington, V.C., Miss Mabel* or *Meet Mr Callaghan.* Problems arise in the case of more general titles (e.g. "Night Must Fall") or in cases in which the English and the American title of the same play feature different central characters – *The Case of the Frightened Lady*, for instance, became *Criminal at Large* in America. In these cases, a part has been supposed to be a leading part if it managed to attract a star actor. Apart from the aspects of title character and star part, the amount of time a character is actually on stage has been considered. Hence, in *Ten-Minute Alibi* and *Dial "M" for Murder*, the central character is clearly the culprit, since all his actions are witnessed by the audience from beginning to end, while other characters spend some of their time offstage. Uninterrupted active stage presence of a dramatic character tends to manipulate audience sympathies in favour of that character, and so does the casting of a star actor or actress. Choosing the criminal as the central character subjects spectators to a state of considerable moral ambiguity, or, one might argue, allows them the vicarious pleasure of imaginatively indulging in forbidden activities. In Victorian times, Henry Irving had already discovered that criminal activities do not necessarily alienate audiences from a dramatic character, and hence he played a long series of repentant sinners. Criminal protagonists that retain audience sympathies are Dr Clitterhouse, Carrington, Mrs Cheyney and Andrew Bennett (*Not in the Book*), while Dan (*Night Must Fall*), Ellen Creed (*Ladies in Retirement*), Julian Fanshaw (*Double Image*), Norman Bartholomew (*Murderer*), Leslie Crosbie (*The Letter*), Mr Sloane or Tony Wendice (*Dial "M" for Murder*) evoke a mixed response of fascination and repulsion.

The criminal is by far the most popular protagonist of the crime play, taking centre stage in about one half of the plays surveyed. In about half as many cases, approximately 25 per cent of plays altogether, the hero of the crime play is the detective. As has been explained in Chapter 8, the choice of the detective as the central character has far-reaching consequences for the construction of the plot, since the crime is usually only illuminated in retrospect. While this approach is favoured by detective fiction, the crime play has considerable difficulties with this procedure. Accordingly, the proportion of adaptations of previously written novels and short stories is particularly high among detective-centred crime plays. Apart from seven Christie adaptations and dramatizations of novels by Edgar Wallace, this applies to *Dead Ringer, Rebecca, A Coat of Varnish, The Phantom Light* and *Alibi for a Judge*. Famous serial master sleuths of detective fiction were also sent onto the stage, including Bulldog Drummond, Slim Callaghan and, of course, Sherlock Holmes. As early as 1899, William Gillette proved that a detective may be turned into the hero of a successful crime play and spent the rest of his working life impersonating the great detective. Gillette, however, had subjected his source, the Holmes short story "A Scandal in Bohemia," to a most thorough rewriting in order to make it suitable for the stage. Of the original crime plays, only three resemble classic British detective fiction, two of which are by Christie and Sayers: *Black Coffee* and *Busman's Honeymoon*. (The third one is *Bonaventure* by Charlotte Hastings.)

This leaves the crime play revolving around the victim. This subgenre is very rare, making up no more than approximately 10 per cent of the crime plays surveyed. Usually, the crime play with a victimized central character also requires a strong culprit (see for instance *Gas Light, The House by the Lake* or *Dead Guilty*), since the special attraction of this type of play resides in the interaction between the criminal and his victim, and in some cases it may even be debatable who the real protagonist is, the cat or the mouse. *Love from a Stranger, The Two Mrs Carrolls* and *Gas Light* recognizably conform to the formula called the "Had-I-But-Known" by LaBorde, the type of play revolving around the "innocent, trusting, foolish, or even stupid" heroine. Within the genre of the crime play, this is the most conspicuous case of gendering: crime plays revolving around the character of the victim frequently have a female lead. Although the subgenre of the "Had-I-But-Known" would at first glance seem to be inherently sexist, it definitely does provide actresses with splendid material, more than any other subgenre of the crime play. What can be done with the part of a victimized woman was proved by Ingrid Bergman when her performance in *Gas Light* earned her an Academy Award.

As has been pointed out above, the proportion between crime plays centring on the criminal, the detective, and the victim respectively is approximately 4 : 2 : 1. How do we account for these remarkable differences in frequency? That is, what quality is it that makes a criminal a better protagonist than a detective, and a detective a more popular one than the victim? A reasonable suggestion would seem to be that it is the degree of activity and initiative displayed by a character that makes him (or her) a good central character. As observed by Chesterton in his famous dictum on the criminal as the creative artist and the detective as the mere critic, the initiative is always in the hands of the criminal, the detective merely reacts and follows the traces left by the criminal. The victim frequently does not even have time to react, but may be taken by surprise or already be dead when the curtain rises. To approach the question from another angle: What other quality, then, would be able to account for the different proportional relations in crime fiction? In crime fiction, the most popular protagonist is the detective, followed by the criminal (in inverted and gangster novels) and the victim (in Had-I-But-Knowns). Here, popularity would seem to correspond to the degree of intelligence: the detective must be more intelligent than the criminal, or he would not be able to catch him, and the criminal must be smarter than the victim or he would not be able to pull off his crime. While detective fiction is obsessed with intelligence and knowledge, the crime play is predominantly concerned with vitality.

Mathematically inclined readers will have noted that so far percentages do not add up to a full 100 per cent, and indeed there is still one group of plays missing. These are plays without a clearly identifiable central character, or with a central character who is neither criminal, nor detective, nor victim. Plays without a clearly identifiable central character invariably are whodunits that lack a strong detective figure. Since the victim is usually killed early in the action and the culprit must not be revealed until the very end, the audience's interest is transferred to the fourth group of people involved in a crime, "those threatened by the crime but incapable of solving it."[19] This applies to the thrillers *The Terror, The Creaking Chair, The Mousetrap* and *Whodunnit* and the puzzles *Poison Pen, Murder on the Nile, The Hollow, Four Winds, The Key of the Door, The Shot in Question* and *Amber for Anna*. This list includes three adaptations of Agatha Christie's fiction, and her decision to dispense with the services of her master detective led to the peculiarity of her stage whodunits having no star parts. This was also reflected in casting policies – when *Witness for the Prosecution* opened at the Winter Garden in 1953, *Plays and Players* specifically recommended Christie's ability to fill the theatre "without stars."

An instance of a central character who is neither criminal, nor detective, nor victim is the protagonist of the formula dubbed "Innocent on the Run" by Derry. In a number of crime plays, the central character is an innocent man suspected of a crime he did not commit, in most cases he has been deliberately framed by another. Since his ostensible crime places him outside the law, the innocent suspect resembles the criminal, and he often participates in criminal activities (*They Came by Night, Guilty Party, The Bargain*) or for some time even suspects himself of having committed the crime he is accused of (*Libel!, Home at Seven, Signpost to Murder*). However, if he has been framed deliberately, the innocent suspect may also be conceived as the crime play's victim, most notably in *The Gioconda Smile, The Day's Mischief* or *Hostile Witness*. The protagonists of *The Gioconda Smile* and *The Day's Mischief*, however, are the only ones to accept the role of victim quietly, while most innocent suspects take considerable action in order to clear themselves and thus assume the functional role of detective. *Guilty Party, Settled Out of Court, Hostile Witness* and *Suddenly at Home* revolve around a suspect's – frequently very original and ingenious – attempts to discover the truth. Arguably, plays revolving around the character of an innocent suspect may therefore be grouped under each of the actantial roles, because it is capable of fusing all three major parts – criminal, detective, and victim – into one.

After the categorization of crime plays according to central character, some attention shall now be devoted to the second criterion, that of mode.[20]

As in the case of central character, it is sometimes difficult to establish if a given play is a "thriller" or a "melodrama," a "puzzle" or a "comedy." To some extent, nearly all crime plays exploit the contrast between different modes, most notably in their use of "comic relief." Moreover, generic hybridity may be considered a specific characteristic of popular drama in general, as Thomson has pointed out:

> Generally, the popular play cannot be separated entirely from progenitors or unerringly demonstrate all the traits of one single genre. The assertion might be true of all plays: Aristotle is hard put to get far beyond *Oedipus*. Popular theatre, however, is especially unconcerned with the niceties of a crystalline distinction between genres. The potpourri characteristic can be itself a characteristic of the popular, and often such plays carry ambiguous subtitles indicating conglomerate genres such as comedy-drama or comedy-melodrama.[21]

While it is therefore necessary to keep in mind that some amount of blurring of generic distinctions occurs in the crime play, it is yet possible to isolate the typical dynamics of each crime play mode. It must also be remembered, that while thriller and puzzle require a crime to trigger off the action, melodrama and comedy can do without it. It will therefore have to be discussed what particular uses melodramatic and comic crime plays make of the motif of crime.

We will, however, start the discussion of the modes of the crime play with the thriller, the subgenre bent on the creation of maximum fear. Discussing the emotion of fear in the context of Aristotle's theory of tragedy, Bentley remarks:

> In his *Rhetoric,* Aristotle explains that pity and fear have an organic relation to each other. An enemy or object of terror is presupposed in both cases. If it is we who are threatened, we feel fear for ourselves, if it is others who are threatened, we feel pity for them.[22]

However, audiences in the theatre will hardly ever feel personally threatened by the actions performed on stage. What the above quotation would therefore seem to suggest is that the reaction of fear requires an almost total audience identification with a dramatic character in danger, whereas the response of pity indicates that, though the spectator may sympathize with a dramatic character, he still perceives him as an individual separate from himself. To put it differently, in a thriller, we react as if we were the dramatic characters threatened by the criminal, but in melodrama, we sympathize with the victims while keeping our emotional distance. This view is supported by a reading of Aristotle's *Poetics*: "[P]ity is aroused by someone who undeservedly falls into misfortune, and fear is evoked by our recognizing that it is *someone like ourselves* who encounters this misfortune."[23]

Bentley also engages with the problem of why people voluntarily subject themselves to the feeling of extreme fear, that is, with the question of whether catharsis is a pleasurable experience.

> Do people enjoy being shaken and upset? The question is part of the larger one: is there pleasure in pain? To which the answer is: on certain terms, yes. Nor are the terms necessarily those of masochism. Being shaken by a tragi-comedy of Ibsen or Chekhov is a pleasure, because it is a shaking into life.[24]

Michael Balint's *Thrills and Regressions* addresses the same problem, starting off with the question of why people enjoy funfair rides. In his concise summary of Balint's theory, Derry isolates three major characteristic attitudes of personalities who are attracted to these kinds of controlled thrill: "first, some amount of conscious fear, or at least an awareness that a real external danger does exist (even if only in a marginal, controlled form), second, a voluntary exposure to this danger and to the fear aroused by it, and third, a confident hope that the fear will be tolerated or mastered, the danger will end, and a return to safety will follow."[25] Balint points out that the same dynamics provide the basis of many children's games "like tag, hide-and-go-seek, or musical chairs." As Balint points out, all these games basically consist of (1) a safe zone and an external danger represented by the catcher or seeker, (2) the other players' being forced to leave the zone of security, which is often called home, and to accept exposure to the danger, in (3) the hope that they will reach security again.[26]

In the following, Balint identifies three different basic types of thrill – that of exposure (as in the children's games), that of high speed and that of unfamiliar forms of satisfaction, "such as new foods, new customs, and new sexual experience."[27] To the children's games listed above, one form of playful, voluntary exposure to fear must be added: the dare as an initiation rite. A stage or a film thriller presents itself as a kind of dare, asking: "Can you bear watching this?" This applies in particular to the subgenre of the "splatter," and it would seem to be the equivalent of the intellectual "Challenge to the Reader"[28] presented by detective fiction. Passing through a state of intense fear accordingly promises the following rewards: greater respect from one's peers, higher self-esteem, and – if Bentley, Sierz and Artaud are to be believed – a heightened sense of being alive. Balint calls the thrill-seeking personality addicted to this experience the "philobat," and the thrill-avoiding, "clinging" personality the "ocnophil." As Derry and Seeßlen point out, the "three-act philobatic drama," in analogy to the children's games cited above, shows the ocnophilic protagonist (with whom the spectator is asked to identify) in a state of safety, but under constant pressure to avoid danger and fear, then passing through a number of extremely dangerous and fearful situations, and finally emerging on the other side as a more confident and more fully realized individual.

In crime play thrillers, all of the above elements are sometimes to be found, though occasionally in a modified form. The element of the dare for instance is already announced in some of the plays' titles, for instance "The Touch of Fear" or "Play with Fire." These titles only partly refer to the action presented on stage but more directly to the intended audience response. The drama of a liberation from a regressive way of living through the experience of extreme fear and pain is most fully developed in *A Man with Red Hair*. In a number of plays, the central character indeed emerges from the ordeal a changed person, most notably in *Wait until Dark*, *Libel!*, *The House by the Lake* and *Gas Light*. The last sentences spoken by Mrs Manningham in the latter play are a particularly good example of regeneration through fear:

Rough: I came in from nowhere and gave you the most horrible evening of your life, didn't I? The most horrible evening of anybody's life, I should imagine.

Mrs. Manningham: The most horrible ... Oh, no ... (With a sort of proud defiance.) The most wonderful ... Far and away the most wonderful.[29]

The playful aspect of exposure to fear is stressed in such late twentieth-century thrillers as *Sleuth* and *Who Killed "Agatha" Christie?*, the latter showing the negative example of an ocnophilic spectator who was unable to survive the challenge of "Can you bear watching this?"

The thrill-seeking personality of the philobat of course also makes a particularly good crime play protagonist. The first and foremost example is Sapper's Bulldog Drummond, who is said to have started fighting crime not because of any inherent righteousness, but because he "found peace dull." Other crime play protagonists who engage in high risk activities are the murderers in *Rope* and *Murderer*, who mainly kill for the excitement of it. The amazing Dr Clitterhouse is a borderline case: he still clings to the respectability of his medical profession but is drawn increasingly into crime. The same process is dramatized in thrillers revolving around the motif of "the worm that turns," like *Stage Struck*, *The Business of Murder* and *Dangerous Obsession*. It is not merely crime itself that is attractive, it is also the philobatic criminal. The seductive powers of the criminal are announced by crime play titles like "Dear Murderer" or "Dear Delinquent," and they emerge most clearly in plays in which an ocnophil comes in the thrall of a glamorous philobatic criminal, as in *Night Must Fall* and *Normal*. The types of the philobat and the "clinging" ocnophil, the adventurer and the miser, also confront each other in Wallace's *The Green Pack*, and audiences are expected to approve of the murder of the latter by the former.

While the basic mechanism of the philobatic drama is in evidence in most of the crime play thrillers surveyed, the three basic types of thrill listed by Derry and Balint are rarely to be found. The experience of high speed of course is difficult to convey on stage, but it should perhaps be noted that a number of stage thrillers have car accidents and hit-and-run killings as their starting point (*Something to Hide, A Shred of Evidence, Dead Guilty*) or revolve around a series of accidents involving other vehicles (*The Wrecker, The Phantom Light*). Besides, the philobatic protagonist is sometimes characterized by his fondness for high speed and high risk sports, as in Dorothy and Campbell Christie's *Grand National Night* and in Mackie's *The Big Killing*. The thrill of exposed situations is also rarely presented on stage, also for practical reasons. In Balint's theory, exposure is associated with great heights and empty spaces, which may of course be depicted in film much better than on stage. Instead, individuals in stage thrillers are often trapped in enclosed spaces as in *The Mousetrap* and of course in all of LaBorde's "Murder-House Mysteries." "Home," which is the safe place of the children's games in Balint's list, is not a safe place any longer in the stage thriller, an observation that recalls Freud's explanation of the term "unheimlich" (usually translated as "uncanny") as the familiar yet frightening. Unfamiliar forms of satisfaction, which might be expected to be represented on stage, are also largely absent from crime play thrillers. Instead, protagonists are sometimes shown to invent their own *alter ego*, and in acting out their alternative personality, they find (sexual) gratification that has been denied to them before (*Double Image, Hapgood*).

The "melodramatic" mode of the crime play has been defined above as the mode granting the viewer the pleasures of "a good cry." A good cry is a "good" cry, however, only as long as we are not seriously distressed. Therefore, melodramatic crime plays often employ distancing devices that prevent the action from being taken too seriously. Stereotypical action and language that clearly mark the play as a work of fiction and a typical example of its genre may help to maintain the necessary distance. In performance, a certain "wooden" acting style may also serve as a distancing device. Unlike the thriller, melodrama usually does not challenge the viewers to test the limits of what they can bear, but, just as in the case of the thriller, audiences seem to enjoy subjecting themselves to the experience of watching plays that arouse pity. First, weeping in the theatre (or the cinema) may be enjoyable because it relieves stress: instead of having to cope with the innumerable ordinary frustrations of daily life which never allow an emotional release through weeping, audiences vicariously experience large scale disasters that permit emotional outbursts. The experience of pity offers yet another secondary gratification: "Because we feel that our pity is virtuous we begin to enjoy it," writes Bentley,[30] and Johann N. Schmidt speaks of "the pleasurable feeling of cultivating one's capacity for pity."[31] Bentley continues his argument by pointing out that pity is a somewhat dubious emotion, since it is an attitude that belittles others: "Then we start to look for potential objects for our pity. These objects then become our *victims*."[32]

The use melodrama makes of the motif of crime is explained by the thematic formula of melodrama identified by Denis Salter, who recognizes the following "conventional melodramatic pattern – the expiation of guilt followed by a kind of secularized apotheosis or revolution of the spirit."[33] This is the precise formula of the repentant-sinner plays of Henry Irving, though most twentieth-century crime plays do not stress repentance but concentrate on the gradual decline and mental anguish of the culprit. Two particularly good examples of plays depicting the decline of the criminal are *Payment Deferred* and *A Dead Secret*. Vickers has claimed that this type of play, which he calls the "inverted crime drama" in analogy to inverted crime fiction, is related to classical tragedy: "The inverted crime drama is in the tradition of classical tragedy, in which sympathy is engaged for the hero, who is driven by fate to become the instrument by which evil triumphs over good, thus involving his own destruction."[34] The comparison is debatable. What prevents crime plays on the life of downgoing men from being tragedies in the classical sense becomes most obvious when comparing them to a tragedy like, say, *Macbeth*. It is not merely that these plays do not show "the fall of princes"; the protagonists of *Payment Deferred* and *A Dead Secret* are not even shown as capable of making a moral choice; nor do they intimate their qualms of conscience to the audience. Rather, they are presented as animals with conditioned reflexes caught up in an infernal machine.

In a number of crime plays, this infernal machine is no other than the machinery of institutionalized justice itself, and indeed it is remarkable how many melodramatic crime plays concentrate on the defects of, or historical changes within, social institutions. Cawelti recognizes "Social Melodrama" as the type to which nineteenth-century melodrama converted in the hands of twentieth-century best-selling novelists. Social melodrama

> synthesizes the archetype of melodrama with a carefully and elaborately developed social setting in such a way as to combine the emotional satisfaction of melodrama with the interest inherent in a detailed, intimate, and realistic analysis of major social or historical phenomena. [...] [T]he social setting is often treated rather critically with a good deal of anatomizing of the hidden motives, secret corruption, and human folly underlying certain events or institutions; yet the main plot works out in proper melodramatic fashion to affirm, after appropriate tribulations and sufferings, that God is in his heaven and all's right with the world.[35]

Cawelti's discussion of typical features of social melodrama is followed by a historical overview of the development of melodramatic prose writing from Dickens and Beecher-Stowe to Harold Robbins and Irving Wallace. What this historical outline shows above all is that melodrama focuses on historical shifts in moral values and social arrangements. This makes melodrama "the most time-bound of all the major formulas."[36] Within the genre of the crime play, melodrama mainly discusses the institutions of the law and law enforcement. Plays like Galsworthy's *The Silver Box, Justice* and *Loyalties* deal with the law, while mid-century crime melodrama tends to concentrate on the institution of the prison (*Boys in Brown, Now, Barabbas, Murder Story*). Late twentieth-century crime plays show the police as a contested institution (*This Story of Yours, Sus, The Colour of Justice*) or deal with the problem of re-integrating a former convict into society (*The Awakening, Foreign Lands*).

All definitions of melodrama agree on the importance of the concept of poetic justice to the genre: in the end, the good are rewarded and the bad are punished. As Cawelti puts it:

> intense sympathy must be aroused by placing admirable (or hateful) characters in situations where they are threatened with the wrong fate. Then, the emotion must be released either by removing the threat of an inappropriate fate, or by explaining and justifying it from some higher perspective.[37]

Some of the crime plays surveyed in this study that would else fit the subgenre of melodrama perfectly seem to violate the rule of the punishment of the wicked. A closer look, however, reveals that a criminal usually goes unpunished in a melodramatic piece of writing either because his crime is justified, since it was an act of self-defence (*Dangerous Corner*) or a mercy killing (*The Sacred Flame*), or because the culprit is punished instead by perpetual mental anguish, as for instance in *The Letter*, which ends with the following exchange between Joyce and Leslie:

Joyce: It's not easy to live with a man you don't love. But you've had the courage and the strength to do evil; perhaps you will have the courage and the strength to do good. That will be your retribution.

Leslie: No, that won't be my retribution. I can do that and do it gladly. He's so kind and good. My retribution is greater! With all my heart I still love the man I killed.[38]

This quotation seems a particularly apt illustration of Cawelti's dictum on "explaining and justifying an apparently inappropriate fate from a higher perspective." Finally, in a handful of melodramatic crime plays, the universe of the play is conceived as hell, and the most appropriate punishment is to let the wicked remain where they are – this applies for instance to *Wise Child* and *Breaking Point*. Only three crime plays in the melodramatic mode altogether affront the audience's demand for poetic justice: *Somebody Knows, Dusty Ermine* and *The Colour of Justice*.

Clearly this is not the place to develop a comprehensive theory of comedy and the comic; rather, the following remarks will wisely limit themselves to the discussion of the various opportunities for creating comic effect the crime play has at its disposal. Some amount of humour is indispensable for any crime play, since mounting tension must always alternate with comic relief – you cannot keep the audience on the edge of their seats for an entire three hours. In fact, finding the appropriate mixture of humour and suspense is probably one of the secrets of a crime play's success. Bentley has explained the conspicuous link between the theme of theft and the genre of comedy as follows:

> The desire to live is not merely love of living. It is also greed. Comedy deals with the itch to own the material world. Hence its interest in gluttons who imbibe part of this world, and misers who hoard another part. And from *devouring* and *clutching*, human nature makes a swift leap towards *grabbing*. In how many comic plots there is theft or the intention of theft![39]

As even a cursory look at crime plays written in the comic mode reveals, a lot of them indeed centre on theft; most overtly *Grumpy*, *The Last of Mrs Cheyney*, *Plunder* and *Breath of Spring*. Moreover, many of them are comedies of greed in the strict sense of the word, since they do not present a single greedy character in the tradition of Molière's character comedy but a whole society that is greedy; the most striking examples being *Plunder*, *Paper Chase*, *For the Love of Mike*, *Miss Mabel* and *Loot*.

Besides comedies of greed, the crime play produces what may be termed "comedies of public embarrassment." Public embarrassment of course is an effective means of creating comic effect, as any viewing of slapstick comedy or comic circus or vaudeville acts will reveal: in this type of entertainment, humour is created almost exclusively by making individuals perform degrading or embarrassing acts in public – often they are literally "caught with their pants down." A lot of crime play comedies revolve around the danger of being caught while performing an illegal action and the consequent danger of losing one's social status. This applies to the above-mentioned *The Last of Mrs Cheyney*, *Plunder*, *For the Love of Mike* and *Breath of Spring* as well as to several other plays, for instance *Mrs Inspector Jones*, *Not in the Book*, *Brush with a Body* and *Spider's Web*.

Comic effect also ensues if a character is inappropriate for the actantial role he has to perform, which usually takes the form of the inept criminal or the inept detective, though there may also be the case of the uncooperative victim, as in *Meet a Body* and *We Must Kill Toni*. Of a number of inept criminals, Andrew Bennett (in *Not in the Book*) and Mr Godboy (in *Dutch Uncle*) are perhaps the most striking ones. Inept detectives are slightly less frequent, appearing for instance in *Simple Spymen* and *Out of Bounds*, but the characters played by Gordon Harker in *Warn That Man!* and *The Phantom Light* and Irene Handl in *Busybody* and *Dead Easy* also belong to that formula, as does the character of Miss Marple as performed by Margaret Rutherford. Apart from an inappropriate "casting" of the wrong individual as the criminal or the detective, an inappropriate response to the fact of crime is another well-established technique of creating comic effect. In this type of play, a reversal of values takes place and crime is perceived as normal. Examples include *Laburnum Grove, Before the Party, The White Sheep of the Family* and *Loot.* Finally, the subformula of parodic crime plays must be mentioned. This is the type of play described by LaBorde as the "Comic Mystery." Among the crime plays achieving comic effect by parodying the genre, *Murder on the Second Floor, My Giddy Aunt, The Real Inspector Hound* and *Corpse!* are to be mentioned, but this type of play is far less common among comic crime plays than supposed by LaBorde.

This leaves the subgenre of the crime play puzzle to be discussed. As pointed out above, "puzzle" is the term which will be employed for a crime play that mainly appeals to the audience's intellectual faculties, as opposed for instance to the modes of thriller and melodrama which mainly appeal to the emotions. It is not quite clear what constitutes the appeal of the puzzle for the spectator. The view that consumers really try to "solve the puzzle" has been variously contested even in the field of detective fiction, where this would be a more practicable proposition, since a reader may choose his own speed of reading or return to earlier passages in the text, options barred to the spectator in the theatre. Cawelti has argued that readers (and, by inference, audiences) become emotionally involved in detective puzzles because they wish to see their favourite suspect cleared of suspicion. Obviously, this explanation applies most perfectly to whodunits; yet there are crime plays that appeal first and foremost to the viewer's intellectual faculties and yet are no whodunits. The best description of the audience's pleasure in watching a puzzle is to be found in Carlson's *Deathraps,* even though it is originally intended to refer to the postmodern comedy thriller only:

> The audience must know the rules of this complex game [...] and must utilize these rules to construct continually a series of hypothetical explanations for the strange events on stage, paradoxically deriving their great pleasure from discovering that their hypothetical constructions are again and again subverted by the evolving action.[40]

The intellectual challenge to the spectator in puzzles generally is to make sense of the actions performed on stage and try to recreate the hidden pattern – even if they do not manage to "solve the puzzle" in its entirety. In order to transport the audience into a state of heightened attention it is sufficient to suggest that there is a hidden design behind the apparently random actions displayed on stage, and that every detail may be meaningful.

An intricately plotted crime may be just as arresting as an intricately plotted solution, which also accounts for the popularity of "big caper movies." This means that even a crime play of the puzzle type may be enjoyed twice if we sufficiently admire the author's clever plot and the criminal's clever plotting. The importance of the mystery element in crime fiction has clearly been overestimated, particularly by critics who insist on attaching "spoiler warnings" to their writings if they reveal the identity of the culprit. The vast majority of crime play puzzles revolve around a particularly ingenious central idea, frequently an elaborate murder plan (*Ten-Minute Alibi, I Killed the Count!, The Fourth Wall, Busman's Honeymoon*), or another original idea, like a suggestion for hiding a purloined letter (*Black Coffee*) or a violation of a convention, such as the Least-Likely-Suspect rule (*The Hollow*). Though it has been stressed above that not all puzzles are whodunits, the formula is of course frequent, occurring in about 10 per cent of crime plays altogether, mostly in plays by Christie or stage adaptations of her novels. Puzzle plays with smaller casts of characters make use of the convention of the innocent suspect in order to ensure audience involvement. Most noticeably this happens in *Bonaventure, Dial "M" for Murder* and *Hostile Witness*. Both in the whodunit and in the innocent suspect formula the emotional dynamics described by Cawelti are in operation: audience sympathies are engaged by a character who might be the culprit but – hopefully – is not.

Of the four different modes of the crime play, the thriller is the most common (about one third of plays altogether), followed by the melodramatic mode (approximately one quarter), comedy and puzzle (about one fifth each), with comedy being slightly more popular than the puzzle. The frequency of a specific type thus seems to be related to the amount of audience identification generated – the greater the audience involvement, the more popular the formula. The importance of audience identification for a crime play's success was pointed out persuasively by crime playwright Reginald Denham:

> The people in the theatre must say, "This could happen to me, or to my family or to my friends." In that way the impact on them is intensified. They do not feel they are merely looking from afar at some academic case history. Instead the particular dramatic happening is something that could occur to themselves the moment they leave the theatre, or perhaps it might actually be going on in their own homes while they are looking at the play. [...] There must be some spot where the public exclaims vocally. In *Angel Street*, when the detective left his hat behind in the room just as the villain was about to enter, the entire house shouted nightly, "Your hat! Your hat!"[41]

Why it makes sense to argue that the thriller formula invites stronger audience identification than melodrama has already been explained above.

The emotional distance between spectator and character in comedy is obviously greater than in melodrama, since melodrama aims at arousing pity whereas comedy exploits public embarrassment and the accompanying response of *Schadenfreude*. Finally, a puzzle play would even be impaired in its impact if audiences sympathized too strongly with one particular character – as Sayers has convincingly argued – and in the case of the whodunit, viewers may even hesitate to attach their sympathies to a dramatic character who may be guilty after all.

Since, as suggested above, any crime play is to be defined by the two aspects of central character and mode, each subgenre is constituted by a specific combination of these two factors. Theoretically, therefore, 12 subgenres of the crime play are possible. Some combinations, however, were so rarely realized that they cannot be said to constitute a proper subgenre or formula.

The most popular form is the melodrama centring on the criminal, making up about one fifth of crime plays altogether. These plays mainly deal with the suffering consequent to the commission of a crime. Bentley's shrewd observations on the patronizing attitude inherent in pity suggests that the twentieth-century melodrama of the criminal has indeed managed to "extinguish the dubious glory of the criminal" (as Foucault phrased it), turning him from an object of terror and admiration into an object of pity and condescension. Twentieth-century melodramatic crime plays no longer present stage villains that are conceived on a scale larger than life. Lionel Trilling has explained the demise of the Victorian Villain "by the modern tendency to locate evil in social systems rather than in persons."[42] Though this may be so in theory (and hence in literature), though we may profess to blame the system, not the man, penal law subsists, punishing not the system but the man. Accordingly, a great number of melodramatic crime plays focusing on the criminal uneasily question the relationship between individual guilt and a penal system that is also perceived as faulty. The legal apparatus is particularly visible in this formula. Crime plays conforming to this type are for instance *Justice, The Letter, People Like Us, Murder Story* and *A Dead Secret.*

The second formula is a little less common: about one crime play in six is a thriller with the criminal as its protagonist. These plays often make extensive use of the mysterious and uncanny, sometimes approaching the realm of the supernatural tale of horror (*A Murder Has Been Arranged, Ladies in Retirement, Signpost to Murder*). The criminal is often mentally deranged and in that respect more closely resembles the monster of Gothic fiction than a fully developed human character (*A Man with Red Hair, Rope, The Case of the Frightened Lady, Night Must Fall*). Alternatively, the protagonist may be the type of the "heroic" criminal, of whose crime the audience is meant to approve (*The Ringer, The Green Pack, The Amazing Dr. Clitterhouse*). In any case, the criminal in this subgenre is gifted with superhuman powers.

In the formula of the comedy revolving around the character of the criminal, which makes up about 10 per cent of the plays surveyed, the criminal is an enviable figure because of his (or her) freedom from social and moral constraints. As suggested by Grella, the criminal in these plays understands the codes of conduct of his/her society better than anyone else which allows him (or her) to emerge victorious – and to retain audience sympathies. Functional detectives and other would-be representatives of justice are often shown to be taken in and condone the crime. Nancy Wingate's term of the "charming rogue" hence is most applicable to this formula. Examples include *The Last of Mrs Cheyney*, *Laburnum Grove* and *The Return of A. J. Raffles*.

Detective comedy and detective puzzle are slightly less frequent, each making up about one twelfth of the plays surveyed. The detective in comedy is a practical, down-to-earth character, without the esoteric knowledge of a Sherlock Holmes or the uncanny reasoning powers of the Great Detective of fiction, and is perhaps best represented by Lily Piper, the Cockney charwoman of Jack Popplewell's *Busybody* and *Dead Easy*. There is a strong strain of self-parody (as suggested by LaBorde in his chapter on "Comic Mysteries") and a surprisingly high proportion of politically motivated crimes, which apparently are conceived as less serious than personal ones. The detective puzzle is the subgenre of the crime play that most closely resembles classic detective fiction, which is hardly surprising since it is often adapted from the latter. By mid-century, however, the Great Detective had largely been written out of the formula, which suggests that the detective is the subgenre's problem, not its solution.

The subgenre of the detective thriller also resembles a traditional subgenre of detective fiction, that of the hardboiled novel. Accordingly, many plays of this type are adaptations of novels: *Bulldog Drummond*, *The Flying Squad*, *Meet Mr Callaghan*. Audience sympathies are unambiguously with the heroic detective protagonist, and there are comparatively many action sequences consisting of stage fights and chase and pursuit. Among original crime plays, *The Man Who Stayed at Home*, *The Last Hour* and *The Wrecker* illustrate the characteristics of the formula particularly well. Contrary to what one might expect, the detective thriller is not gendered: the detective may also be a woman, as in *The Wrecker*, *Murder on Arrival* or *Hapgood*. The formula is not particularly frequent: about one crime play in 15 is a detective thriller.

The formula of the victim thriller occurs with approximately the same frequency. This subgenre closely resembles LaBorde's "Had-I-But-Known," and the plays usually end in a stunning reversal with the supposed victim turning the tables on his – or more frequently her – aggressor. The high degree of audience identification with the weakest character creates considerable suspense, which accounts for the reported fainting fits among the original audience of *Love from a Stranger*. While this subgenre was strongly gendered during the first two thirds of the century, the victim protagonist was no longer necessarily female in the last three decades. Female victims appear in *Gas Light*, *The House by the Lake* and *Wait until Dark*, male victims for instance in *Stage Struck* and *Popcorn*. All the other subgenres centring on the character of the victim are extremely rare.

This leaves two minor formulas to be discussed: the criminal puzzle and detective melodrama. The criminal puzzle resembles "inverted" detective fiction and is usually based on the intricate plotting of a crime. The playful attitude of the criminal (and the police!) towards the crime is indicated by the grace with which they admit defeat, a feature that also recalls the concept of the "charming rogue." Chief examples are *Ten-Minute Alibi, I Killed the Count!, Dial "M" for Murder* and *Beyond Reasonable Doubt*. About 7 per cent of the plays surveyed belong to this formula. Detective melodrama is a rare subgenre (less than 4 per cent) that puts its detective protagonist at the centre of a decidedly bleak universe. There are no pursuit actions, and the heroes themselves are tainted, or at least sceptical about the possibility of establishing justice. In mood, this formula is closely related to *film noir* or the "entertainments" of Graham Greene. Examples are *An Inspector Calls, The Winslow Boy, Conduct Unbecoming, A Coat of Varnish* and *Sus*.

Two minor formulas that evade neat classification in the scheme outlined above are the whodunit without a central character (*Amber for Anna*) and innocent-suspect plays (*The Whole Truth*). Together, these two formulas make up a little less than 10 per cent of the plays surveyed.

To sum up: the classification of crime plays according to subgenres produces three major formulas – criminal melodrama, criminal thriller and criminal comedy – and five minor formulas: detective puzzle, detective comedy, victim thriller, detective thriller and criminal puzzle. All major formulas put the criminal at the centre, but each of the three formulas advances a different theory of crime. In melodrama, crime is conceived in moral terms, as "evil," even though the source of evil may often be located in the system rather than the individual. In the thriller with its abundance of insane murderers, crime is identified as a "disease," and in comedy, crime frequently is an accomplishment that frees its perpetrator from social pressures.

The dominance of the character of the criminal on stage is remarkable because histories of detective fiction tend to stress the fact that the criminal as hero of broadsheet and ballad is replaced by the detective as hero in late nineteenth- and early twentieth-century detective fiction. This shift has usually been attributed to changes in penal practice.[43] The results of the present study, however, strongly suggest that even in the twentieth century, the criminal could still be conceived as a hero, albeit on the stage rather than the page. It might be argued that drama is a more subversive art form than fiction, and indeed evidently was thought so, since it was more rigorously monitored by censorship. On the other hand, stage censorship effectively prevented any genuinely subversive ideas from being openly proclaimed on stage. Alternatively, one may suppose that the replacement of the criminal by the detective in fiction is not (or not alone) indicative of a major cultural shift but also prompted by the sheer literary efficacy of the device. The detective as a reader surrogate is a particularly attractive character in fiction and obviously considerably less appealing on stage. The dominance of the detective in fiction is related to the structure of narrative rather than to shifts in cultural practices – in order to prove this statement, in the following, detective novels will be compared to their respective stage versions. Thus, the different poetics of detective fiction and the crime play will emerge more clearly.

Part III
The Crime Play and Detective Fiction

12
Stage Adaptations of Agatha Christie's Detective Stories

As the previous chapters have shown, the crime play has developed a set of conventions quite distinct from those governing the genre of detective fiction. In the two final chapters of this study, the question of how the crime play differs from detective fiction and to what extent it is similar shall be approached from a new angle; by examining stage adaptations of detective fiction in comparison with their respective source texts – mostly novels, in some cases detective short stories. Among British detective novelists, Agatha Christie is particularly prominent as an author whose works have proved consistently popular for transformation into stage plays. The present chapter will therefore concern itself with adaptations of Christie's detective stories, written either by herself or by other playwrights.

Before embarking on a comparison and analysis of the texts themselves, however, it may be useful to consider the specific difficulties facing an adaptor of a novel in general and a detective novel in particular. The major generic difference between prose fiction and drama is that in drama there tends to be no instance of narrative mediation, while the novel usually has a more or less personalized narrator. Although there have been several attempts in twentieth-century drama to introduce a narrator figure into a stage play, this practice is obviously at odds with the conventions of stage naturalism imposed by the well-made play. Discussing "Point of View in Modern Drama," Edward Groff points out that plays that are set in the central character's mind – such as Strindberg's *Dream Play*, O'Neill's *The Emperor Jones* or Miller's *Death of a Salesman*,

> are not really at home in the structure that has been provided for them. Therefore, the audience is made to feel that they are somehow "unreal"; for though, as spectators in the theater, they have been conditioned to believe that they have the superhuman sight to penetrate the wall of a room, they are now asked to observe an action which could not have taken place in any room they know anything about. [...][T]he playform

> has been made *subjective*, whereas the stage on which it is presented remains chiefly concerned with the representation of *objective* reality.[1]

Since the practice of adapting detective novels for the stage was particularly popular during the heyday of the well-made play, adaptors were confronted by serious difficulties. At the time the bare stage and the memory play that moves about freely in time and space had become fashionable, the adapted crime play was practically extinct.

Although writers and critics of detective fiction have tended to favour the ideal of a largely "invisible" narrative technique,[2] narratologists have not failed to note the importance of narrative perspective for the genre. Their observations are concisely summed up by Peter Hühn:

> The plot of the classical detective novel comprises two basically separate stories – the story of the *crime* (which consists of action) and the story of the *investigation* (which is concerned with knowledge). In their narrative presentation, however, the two stories are intertwined. The *first story* (the crime) happened in the past and is – insofar as it is hidden – absent from the present; the *second story* (the investigation) happens in the present and consists of uncovering the first story. [...] Conventionally, the coupling of these two stories is presented to the reader in a specifically involved manner (invented by Poe and standardized by Doyle) [...]: the story of the crime is mediated in the discourse of the detective's investigation; and the story of the detective's investigation, in its turn, is mediated in the narrator's discourse (for instance in Dr. Watson's uninformed written account of Holmes's detection). In both cases the story is hidden for the most part so that the reader is doubly puzzled – trying to make out the mysterious crime story by way of the almost equally mystifying detection story."[3]

The reference to Holmes and Watson suggests that these observations apply mainly to classic British detective fiction that makes use of the narrative device of the "idiot friend" (this uncharitable term is by Christie herself, who used the device in a number of her detective novels).

A striking lack of awareness concerning the importance of narrative technique is in evidence in the earliest stage adaptation of a Christie novel, Michael Morton's *Alibi*, a dramatization of *The Murder of Roger Ackroyd*. With this novel, Christie had not merely scored her first major success but also triggered off a lively critical controversy:

> Arguments raged, friends fell out, and the newspaper columns vied with each other in giving the book gratifying publicity. The air was rent with cries of "foul," "unfair," and "cheat," enunciated with a degree of passion which only a nation of sportsmen can muster.[4]

The "unfair" device Christie had employed of course had been to make the Watson character the culprit. This causes interesting complications in terms of narrative technique, as Hühn points out:

> Agatha Christie's famous novel *The Murder of Roger Ackroyd* provides an interesting example of both an extreme deviation from, and a close adherence to, the narrative scheme of the classical detective novel. Christie identifies the central agents of both stories with each other: the murderer (the author of the first story) is also the narrator of the second story (Poirot's investigation of the murder). The murderer's *intention* in narrating the second story is to chronicle, for once, a failure on the part of the great detective. In other words, the criminal tells the second story to prove the unreadability of the first. The fundamental self-contradiction inherent in his intention, namely, to make public that the meaning of the first text can never be made public, seems to be the root of his eventual self-defeat.[5]

This complex narrative strategy obviously could not be transferred to the stage. A reader excludes Dr Sheppard from the group of suspects because he is the narrator and the criminal "must not be anyone whose thoughts the reader has been allowed to follow," as Father Knox categorically declared in his "Detective Story Decalogue."[6] On stage, Sheppard is not automatically exempt from suspicion on grounds of genre conventions – he is solely exempt on the basis of plot logic, since he has an apparently watertight alibi: the voice of the dead man was heard by several witnesses long after Sheppard had left the house. This is obviously the reason of why Michael Morton retitled the play *Alibi*.

Apart from its narrative perspective, *The Murder of Roger Ackroyd* is a fairly conventional detective story with a particularly long list of minor mysteries. Of the original six secondary mysteries, Morton removed three (concerned with footprints, a blackmailer and an illegitimate child) and retained three. The most important of the minor mysteries concerns Roger Ackroyd's niece Flora. She steals money from her uncle's dressing table and later pretends that she saw him alive in his study as late as 9.45 in order to cover up the fact that she had been in his bedroom. Thus, her evidence initially seems to support the all-important alibi of the story. In the novel, Flora's evidence is quoted in Chapter 5. In Chapter 15, Poirot asks Flora and the butler Parker to re-enact their actions of the evening and notes that Parker did not actually see Flora leave Ackroyd's study, but he only reveals his conclusions in Chapter 19. In the play, Flora's actions are openly presented to the spectator in I.ii. It thus becomes immediately obvious to the spectator that Flora's testimony is worthless for fixing the time of death. Poirot realizes this fact some 16 pages later, when he has Flora and Parker re-enact the scene. Hence, in the novel, Poirot knows more than the reader for about four chapters, but in the play, the spectator knows more than the detective for the equivalent of 16 pages. This divergence points to the different conventions of detective fiction and drama. In detective fiction, the detective knows more than the reader, which produces surprise, while in drama, the audience knows more than the characters, which creates suspense. These two conflicting conventions pose an interesting dilemma for the crime playwright.

Like the minor mysteries, the amorous subplots are less developed in the play than in the novel, but in return Poirot himself is provided with a love interest. Osborne remarks about the adaptation of the character of Poirot for the stage:

> Mrs Christie much disliked Morton's first suggestion which was to take about 20 years off Poirot's age, call him Beau Poirot, and have lots of girls in love with him. With the support of Gerald Du Maurier who produced the play, she persuaded the adaptor not to change the character and personality of Poirot, but agreed to allow Caroline Sheppard to be turned into a young and attractive girl, in order to supply Poirot with romantic interest. Mrs Christie's agreement was reluctant. She resented the removal of the spinster Caroline, for she liked the role played by this character in the life of the village, and she liked the idea of that village life being reflected through Dr Sheppard and his sister.[7]

The concept of "Beau Poirot" is still much in evidence. The stage-adapted version of Poirot is French, not Belgian, and represents the prototype of the stage Frenchman almost to the point of caricature. He kisses the ladies' hands a lot, uses inordinately ornate language and speaks English with a pronounced accent, saying "Mees Flora" and "leetle reunion." While the novel's Poirot allows Sheppard to escape by means of suicide out of consideration for Caroline, the Poirot of the play version is motivated by love for Caryl and even goes so far as to pretend that he has for once failed to solve the case. This behaviour would be completely out of character for Christie's Poirot, with whom the passion for knowledge always supersedes the erotic impulse. Osborne's regret at the substitution of Caroline by the more conventional "bright young thing" Caryl is understandable, since Caroline is one of Christie's most successful character creations. Osborne is also right to complain about the removal of the presentation of village life – as will be shown in the course of this chapter, village life seems to be particularly difficult to portray on stage.

Alibi opened at the Prince of Wales's in 1928 and saw 250 performances in London, followed by a tour. In 1932, it premiered in New York under the title of *The Fatal Alibi*, but closed after only 24 performances, a good example of the difficulties a crime play has to cross the Atlantic.

The next Christie novel to be adapted for the stage was *Peril at End House*. Like *The Murder of Roger Ackroyd*, *Peril at End House* is told by an unreliable first-person narrator, Poirot's usual "Watson" Captain Hastings. As pointed out by Hühn, the particular appeal of this narrative perspective resides in the fact that Hastings' view of the events he relates is frequently inaccurate – he fails to attribute the correct meaning to the incidents he witnesses – or distorted by prejudice, as for instance when he automatically considers Challenger "all right," because his behaviour corresponds to the conventions of his class. Arnold Ridley, who adapted *Peril at End House* for the stage, retained the character of Hastings and subjected him to much comic banter from Poirot. Since in the stage transfer, Hastings loses his function as narrator, he is instead given the new function of "straight man" in the traditional comedy couple of clown and stooge. For instance, at several points in the play, Poirot credits Hastings with improbable specialist knowledge or interests in order to effect an entrance, making Hastings suffer all the embarrassment of the victim of a practical joke.[8] After he has asked Hastings to introduce him, he bestows much praise on himself, remarking how much more agreeable it would have been had Hastings praised him:

Poirot:	One does not keep the dog and bark oneself.
Hastings:	(angry) I'll be damned!
Poirot:	I speak only your English proverb, mon ami. (He turns up C.)
Nick:	(L. of the settee) Anyway, what's the dog's name? Doctor Watson, I presume?[9]

Hastings serves as a target for Poirot's jokes and is humiliated before their client, Miss Nick Buckley. The same joke is found in the novel, but here it is Nick who is speaking: " 'One should not keep a dog and have to bark oneself', agreed Nick with mock sympathy. 'Who is the dog, by the way? Dr. Watson, I presume.'"[10] In the novel, the flippant conversational style serves to characterize Nick and is reserved for her. Ridley, by contrast, makes no attempt to create an individual diction for each of his characters, apart from Poirot's usual linguistic idiosyncrasies. Like later adaptors of Christie, Ridley is sometimes misguided in his attempt to inject additional humour into the play.

Poirot, however, is not merely given a few additional jokes in the stage version, but is also allowed to fire a gun. While in the novel version, it is left unclear who shoots the menacing intruder in the revelation scene (because the unreliable Hastings himself fails to notice it), Ridley delegates this task to Poirot. This points to the different concepts of the Great Detective in fiction and drama: while in a novel, it is sufficient that Poirot exercises his "little grey cells," he must become something of an action hero in the Bulldog-Drummond mould on stage: "The Man takes fresh aim, but Poirot is too quick for him. He too has drawn a pistol and fires first."[11] In later Christie adaptations, the detectives were also expected to show physical courage and a certain aptitude in the handling of firearms.

With respect to condensation and simplification of plot, Ridley's adaptation shows solid craftsmanship: he removes the superfluous subplot concerned with drug traffic and considerably simplifies the plot complication revolving around the poisoned chocolates. The menacing intruder, who makes a last-minute entrance in the climactic scene of the novel, is introduced in the very first scene of the play, and the essential fact for the solution of the mystery – that both "Nick" and her cousin bear the same first name – is made absolutely clear to the audience and repeatedly rubbed in. Arnold Ridley's experience as a crime playwright is clearly in evidence, yet *Peril at End House* ran for only three months at the Vaudeville. The part of Poirot was played by Christie's friend Francis L. Sullivan, who, however, did not conform to her idea of her detective. Osborne even claims that the experience of watching Sullivan play Poirot prompted Christie to remove her detective from her subsequent stage adaptations.

In 1943, Christie herself wrote her first adapted crime play, choosing *Ten Little Niggers* for the purpose. For reasons of political correctness, the play had to be retitled *Ten Little Indians* for its New York run, and both play and novel have since been produced and reprinted under the title of *And Then There Were None*. The plot is fairly well-known: ten people, who each committed murder at one time in their lives, are imprisoned on an island where they are killed off one by one in accordance with the nursery rhyme of the "Ten Little Niggers". The novel recommended itself for dramatization since it follows a formula already well-established on the English-speaking stage, the type christened "murder house mystery" by LaBorde. Earlier examples of this type had been *The Bat* by Rinehart/Hopwood (New York: Morosco, 1920 – London: St. James's, 1923) and John Willard's *The Cat and the Canary* (New York: National, 1922 – London: Shaftesbury, 1922). The simple basic principle of construction – getting the characters into a remote locale where they can be murdered one after the other – had thus already been proved sufficiently theatrically effective.

While its plot is fairly conventional, the narrative technique of *Ten Little Niggers* is somewhat unusual. The novel begins by introducing the individual characters as they approach the island by various means of transport. In this part of the novel, the reader has direct access to the characters' thoughts, which are mostly related in free indirect discourse. This technique was recognized as something new in the field of detective fiction, as the reviewer of the *New Statesman* pointed out: "Mrs Christie even allows us to know what every character present is thinking – and still we can't guess!"[12] Later in the novel, free indirect and direct discourse are employed again in order to establish each character's individual guilt. While she freely borrows the trappings of modern narrative technique at the beginning of the novel, Christie returns to an obvious nineteenth-century literary device at the denouement. In an epilogue, the police are shown to be baffled by the insoluble mystery of finding ten corpses on a lonely island, and apparently nobody responsible for their demise. The epilogue is followed and completed by "A Manuscript Sent to Scotland Yard by the Master of the 'Emma Jane' Fishing Trawler," a message in a bottle, which is a device clearly announcing its provenance in the tradition of the Victorian novel. Both the technique of free indirect discourse and the message in the bottle of course proved impossible to transfer to the stage.

Hence, in the play version, Christie introduced the characters by means of a dialogue between servants discussing their betters, a time-honoured device. While the initial chapter of the novel offers a direct insight into the characters' consciousness, the play concentrates on the portrayal of the behaviour of a group of people in a socially awkward situation: they are all strangers, and their host is absent. Each character's personal guilt is revealed in a confession scene that immediately precedes their decease. Whenever a character feels the sudden urge to own up to his or her crime, the spectator may be sure that (s)he will be killed soon. They address their confession speeches to their fellow prisoners rather than to the audience, since direct audience address violates a genre convention of the well-made play. While readers had apparently been perfectly ready to accept a largely unexplained direct access to the characters' minds, audiences were not subjected to a comparable breach of the illusionist aesthetics of the proscenium stage.

It is, however, in the handling of the denouement that the difference between novel and play becomes most obvious. The action presented in the epilogue is cut, and thus the concept of the "insoluble mystery" so central to much of classic detective fiction becomes unimportant. The self-revelation of the culprit that occurred in the manuscript is refashioned as a speech of confession the murderer makes to his last victim.[13] However, in the play, Christie does not allow her murderer to kill his last victim and afterwards himself. In her autobiography she wrote that this type of ending would have been "impossible, because no one would be left to tell the tale."[14] This remark at first sounds cryptic; on closer examination, it suggests that it may be necessary to rethink the implications of the "fourth wall" convention. For Christie evidently does not intend her audience to pretend that they are actually watching the events as they unfold, but that they are watching a re-enactment of a sequence of events that actually did occur earlier in history. That is, Christie asks her audience to suspend their disbelief in the same manner as when watching, for instance, a history play, a medieval mystery drama, or a Christmas pageant. As in the novel version, where she mixes nineteenth- and twentieth-century literary devices, Christie in the stage play does not clearly distinguish between the conventions of contemporary theatre and a more ancient form of drama.

The solution that Christie eventually found for the ending of the play was to have two characters turn out innocent after all – they survive the ordeal and get married. Although the final embrace seems a little facetious – Lombard puts his head into the noose together with Vera's and recites the alternative ending of the "Ten Little Niggers" rhyme – it conforms to the conventions of the crime play, where some kind of love interest is thought indispensable. The new ending of "We got married – and then there were none!" abolishes what the serial killer of the novel first found attractive about the nursery rhyme "Ten Little Niggers": "the inexorable diminishment – the sense of inevitability."[15] In the novel version, the murderer is terminally ill, and he relentlessly proceeds to dispense justice according to degree of guilt, finally killing himself, and branding his forehead with the mark of Cain. In the play, he is not ill but mad, and Vera's protestations of her innocence do not move him, since he has planned out the various deaths schematically and insists that he cannot let her spoil his "lovely little rhyme."[16] While the novel stresses the inexorable progress of justice, the drama version stresses the playful quality of plot mechanics.

As a murder house mystery, *Ten Little Niggers* cleverly exploits the confinement to a single set. In terms of set design, however, the play tries to distance itself from its predecessors by describing the living-room set as follows: "It is a very modern room, and luxuriously furnished. It is a bright sunlit evening."[17] The difference from the gloomy mansions of earlier murder house mysteries is pointed out by Lombard's remark when he proceeds to search the house for the mysterious U. N. Owen: "House ought to be easy. No sliding panels or secret doors."[18] When René Clair filmed *Ten Little Niggers* in 1945, however, he returned to an earlier style in set design and "substituted a conventionally sinister mansion for the streamlined contemporary house of the play."[19]

Ten Little Niggers, which opened at St. James's in 1943, was very successful both in London and in New York. The actor who played Philip Lombard, Terence de Marney, directed Christie's next stage adaptation, *Appointment with Death*, which ran at the Piccadilly from March to autumn, 1945.

The novel *Appointment with Death*, which had been published seven years previously, revolves around the character of a "vampire mother," an invalid matriarch preying on her stepchildren. When she fittingly gets murdered, the children are in danger of remaining forever in her thrall, unless the culprit is identified and punished. *Appointment with Death* is a Hercule-Poirot novel, but in the play, he is no longer present. While writing her novel, Christie had complained about being fed up with her most popular creation,[20] and hence she removed him when she adapted her novel for the stage. Even in the novel version, Poirot is absent for most of Part I, and the Boynton family is introduced mainly through descriptions drawn from the perspective of Dr Gerard, who later assumes the function of a surrogate detective and as such is largely exempt from suspicion.

As in the case of *Ten Little Niggers*, it is in the final section concerned with the solution of the mystery that the differences between novel and play become most obvious. In the novel, the solution is presented in a revelation scene, in which the Great Detective summons the various suspects, recapitulates the evidence and clears the innocent characters one by one. The murderess, who committed the murder because she had been blackmailed by the victim, kills herself on hearing the evidence. In the play version, there is no formally staged revelation scene; on the contrary, it is feared that the case may never be solved. In the absence of the Great Detective, it is one of the suspects who eventually hits on the correct solution, aided by a surprise witness. The fact that the functional detective doubles as a suspect effectively helps to reduce cast size. Marvin Carlson's observation on the blurring of actantial roles in the crime play may hence also be explained by the necessity of economy in cast size. The functional detective is a young doctor in love with one of the Boynton children and hence has a definite stake in the case; a notable departure from the ideal of the impartiality and detachment of the Great Detective.

The question of why Christie chose to remove Poirot from this play as well as from her other stage adaptations has been much discussed. It has been suggested that she thought him "too overwhelming a personality," or, alternatively, "but a collection of mannerisms and [...] all too clearly revealed as such when impersonated on stage."[21] The latter of the two arguments is dubious, since it is probably easier to get away with sketchy characterization in a play than in a novel, since in drama the author can still rely on the actor to flesh out the character so as to make him more convincing. Apparently, Christie was never really satisfied with the way Poirot had been cast in stage and film versions of her novels.[22] The main reason for the removal of Poirot in the plays, however, seems to be that in most of Christie's stage adaptations, the Great Detective has not really got much to do, since the proportion of the plays devoted to the actual investigation tends to be much smaller than in the novels. It is therefore not the type of detective that is at issue, but the function of "detective" in general.

Christie's next stage adaptation, *Murder on the Nile*, based on her novel *Death on the Nile*, opened at the Ambassadors in 1946, had disappointingly bad reviews and closed after a comparatively short run. The play's lack of popularity is also reflected in the fact that it has not been included in any of the published collections of Christie's plays.[23] What is most remarkable about the adaptation is the way in which Christie reduces plot complexity and cast size at the same time. However, the play's general lack of success suggests that Christie may have gone too far in removing more than 50 per cent of her personnel, or that the large-cast whodunit does not really recommend itself for stage adaptation.

The limited success of *Murder on the Nile*, however, did not deter Christie's fellow playwrights: in 1949, the team of Moie Charles and Barbara Toy approached Bertie Meyer, the producer of *Ten Little Niggers*, with their project of adapting the Miss Marple novel *The Murder at the Vicarage* for the stage. Since the novel was by then almost 20 years old, Charles/Toy updated the action considerably. *The Murder at the Vicarage* recommends itself for stage adaptation since it is based on the classic crime play configuration of the amorous triangle, it contains a Most Likely Victim, the "murderable squire" Colonel Protheroe, and the character of the functional detective Miss Marple is a very attractive part for a mature actress. Looking back at *The Murder at the Vicarage* several years later, Christie criticized that she had included too many different characters and too many subplots, but she still approved of her portrayal of village life in St Mary Mead. Charles/Toy reduced the number of subplots and the number of characters involved. The severe reduction of cast size, however, had an adverse effect on the portrayal of village life. In the first chapter of the novel, the vicar's young wife Griselda has to spend the afternoon in the company of four gossipy village ladies, Mrs Price Ridley, Miss Wetherby, Miss Hartnell and Miss Marple. In the play, Miss Wetherby and Miss Hartnell have been cut, and instead, Anne Protheroe is present. Since two of the characters have an extra function – Anne as the culprit and Miss Marple as the detective – only Griselda and Mrs Price Ridley are left to represent village life, and Mrs Price Ridley has been caricatured beyond recognition. She "still wears the tussore suits left over from her days in India"[24] and acidly remarks to Miss Marple: "All spinsters should be put away at 50!,"[25] a remark calculated to raise a laugh immediately before the curtain goes down on I.i.

While the adaptors hit the flippant tone of conversation between Griselda and her nephew Dennis admirably, they are surely misguided in extending that tone to other characters. Like Ridley, Charles/Toy insufficiently distinguish between the individual characters' conversational styles. The additional humour is not always particularly funny and, as Osborne complains, must mostly be provided by a dim-witted domestic. The snobbishness of which British detective fiction has often been accused is even more conspicuous in the drama version.

Murder at the Vicarage was first performed at the Playhouse in 1949 and ran for four months. When the play was revived at the Savoy in 1975, it played for two years, having obviously profited from the nostalgic appeal the intervening years had bestowed on it.

In 1951, the next Christie adaptation appeared on the London stage: *The Hollow*, dramatized once again by Christie herself. The source novel in this case does not rely on intricate plotting but on the creation of convincing characters. Charles Osborne comments: "The actual plot-line of *The Hollow* is not one of her best, [...] but the characterization is superb."[26] "The Hollow" is the name of a house, the family home of the Angkatells, but it "is doomed to take perpetual second place in the hearts of the Angkatell family to Ainswick, the immemorial family estate [...]; the Hollow is a hollow of Ainswick, existing only as its trace, its privation."[27] The characters' obsession with their family home would seem to be a theme particularly suitable for stage adaptation within the confines of the single set. However, in the eventual stage version, the limitations imposed by the single set sometimes lead to very awkward solutions. The sculptress Henrietta, for instance, must be moved from her London studio to the garden room of the Hollow with the somewhat unconvincing excuse that there has been a fire in her London flat. And the climactic scene of the murder develops a strong potential for unintentional humour.

An interesting choice with respect to set design is the fact that Henrietta's sculpture of "The Worshipper" is "supposed to stand in the alcove though not visible to the audience."[28] Instead of presenting the object itself on stage, Christie relied on the suggestive power of language – the characters' responses to the sculpture in the play are copied almost word by word from the novel, and their effect probably would not have been enhanced had a piece of statuary also been present on stage. This observation fits in with the loss of symbolic potential inherent in the use of stage props mentioned in Chapter 7. Accordingly, the motif of the matchbox as a symbol of illicit desire and secrecy has not been taken over from the novel to the stage version.

As has also been pointed out above, the crime play noticeably prefers the subgenre of the thriller to that of the puzzle, which suggests that in a transfer from page to stage, the amount of suspense must be increased. The most conspicuous technique is the insertion of a "cliffhanger," a device anticipating an imminent state of great danger. In order to increase the degree of suspense, a detective novelist may also revise his policy in dealing out information to the audience when adapting his work for the stage. While the surprise ending of the traditional whodunit detective novel requires that a substantial amount of information be withheld from the reader,[29] suspense is achieved only if the reader/viewer knows more than the characters, as Alfred Hitchcock has persuasively explained:

> There is a distinct difference between "suspense" and "surprise," and yet many pictures continually confuse the two. I'll explain what I mean. We are now having a very innocent little chat. Let us suppose that there is a bomb underneath this table between us. Nothing happens, and then all of a sudden, "Boom!" There is an explosion. The public is *surprised*, but prior to this surprise, it has seen an absolutely ordinary scene, of no special consequence. Now, let us take a *suspense* situation. The bomb is underneath the table and the public *knows* it, probably because they have seen the anarchist place it there. The public is *aware* that the bomb is going to explode at one o'clock and there is a clock in the decor. The public can see that it is a quarter to one. In these conditions this same innocuous conversation becomes fascinating because the public is participating in the scene. The audience is longing to warn the characters on the screen: "You shouldn't be talking about such trivial matters. There's a bomb beneath you and it's about to explode!" In the first case we have given the public 15 seconds of *surprise* at the moment of the explosion.

> In the second we have provided them with 15 minutes of *suspense*. The conclusion is that whenever possible the public must be informed.[30]

The ultimate scene of *The Hollow* might have been specifically designed to illustrate Hitchcock's point about the difference between surprise and suspense. The murderess Gerda tries to poison her accomplice Henrietta. In the novel version, this murder attempt comes as a complete surprise: Poirot arrives and prevents Henrietta from drinking her tea. Gerda returns, has her tea, collapses and dies. Henrietta is surprised by her suicide, but Poirot tells her that the poison was in Henrietta's cup. In the play, the surprise is replaced by a prolonged scene of suspense. The audience watches Gerda poisoning Henrietta's drink, but so does Henrietta, and in the ensuing scene Henrietta attempts to avoid drinking her tea, while Gerda tries to force it on her. Inspector Colquhoun, the play's Poirot replacement, intervenes and hands Gerda the poisoned drink by mistake. The resulting scene in the stage version is an almost perfect illustration of Swiss dramatist Friedrich Dürrenmatt's definition of suspenseful dramatic dialogue:

> If I merely show two people sitting together and drinking coffee while they talk about the weather, politics or the latest fashions, then I provide neither a dramatic situation nor dramatic dialogue, no matter how clever their talk. Some other ingredient must be added to their conversation, something to add pique, drama, double meaning. If the audience knows that there is some poison in one of the coffee cups, or perhaps even in both, so that the conversation is really one between two poisoners, then this little coffee-for-two idyl becomes through this artistic device a dramatic situation, out of which and on the basis of which dramatic dialogue can develop.[31]

Christie's refashioning of her material was successful with her audience, and *The Hollow* achieved a respectable run of 376 performances.

In her next stage adaptation of a novel, *Towards Zero*, Christie also introduced a sequence of heightened suspense immediately before the end. Pursued by the police, the culprit jumps out of a window overlooking a cliff, but returns immediately to strangle his ultimate victim, who is only saved by the timely entrance of the police. The suspense sequence is completed and finished by an instance of comic relief; the anti-climactic remark of one of the bystanders: "I say, are you all right?"[32] The return of a person presumed dead is a standard device of the stage thriller, but it is rare in Christie's works. It is therefore likely that the final scene was introduced on the initiative of Christie's co-author Gerald Verner. *Towards Zero* opened at the St. James's in 1956 and ran for six months, occasionally playing to half-empty houses. A newspaper critic who gave away the identity of the culprit has been blamed for the play's moderate success. However, its lack of popularity may also indicate that the age of the stage whodunit was gradually coming to an end – always excepting the apparently immortal *Mousetrap*.

Christie's last dramatization of one of her novels, *Go Back for Murder*, was even less successful and closed after one month. *Go Back for Murder* is based on the novel *Five Little Pigs*, in which Poirot investigates a murder committed 16 years ago. The novel is subdivided into three major parts, the first of which – Book I – introduces the murder case and the five suspects, the "little pigs" of the title. The murder victim is a painter, and in Book II, Christie inserts an experiment in narrative technique clearly indebted to modern painting, more precisely, to the theories of cubism. As cubism attempts to render a three-dimensional object on a two-dimensional canvas by showing the same object from different perspectives at once, Christie shows the same sequence of events from five different perspectives: each of the five suspects writes a report of the crucial afternoon of the murder. It is the combination of the various limited subjective accounts that allows the ordering intelligence of the detective to arrive at objective truth.[33]

Due to its retrospective plot interest and experimental narrative perspective, *Five Little Pigs* is particularly difficult to adapt for the stage. Christie eventually chose to present the material of Book II in an extended collective flashback. The flashback is indicated in the following manner: one of the witnesses, Meredith, begins to relate his version of the events of 16 years ago. As he is speaking, the lights go off. Meredith's voice is heard in the dark, and when the lights come on again, the events of 16 years ago are acted out, beginning with young Meredith's arrival at the scene of the crime. This manner of introducing a flashback suggests to the audience that the flashback is meant to be an illustration of old Meredith's speech. However, after a short conversation, young Meredith exits while the flashback continues. Accordingly, the audience must now suppose that the flashback does not illustrate the evidence given by Meredith but shows what really happened 16 years ago. It is only after the end of the flashback that Justin Fogg, the functional detective in charge of the case, points to certain inconsistencies in the flashback which only then is revealed to be a collated version of the various suspects' narratives. In contrast to the readers of the novel version, the audience is not in a position to disentangle who said what. Therefore, Justin alone can solve the case, which he does by proving that one of the witnesses consciously misrepresented a conversation she overheard. Another, shorter flashback follows, which shows the correct version of the conversation. This flashback seems intended as an illustration of Justin's speech. Thus, in the course of the play, the device of the flashback serves different functions; and it is hardly surprising that the audience failed to ascribe the correct status – illustration of speech, of individual memory or of collective memory – to the respective instances. *Go Back for Murder* does not even create the illusion that the spectators might have solved the mystery by themselves. This is a violation of the "fair play rule" applying to classic detective fiction – the drama critic of the *Daily Mail* accordingly complained that he "felt cheated" by the play.[34]

Probably it was the successful revival of *Murder at the Vicarage* in 1975 that prompted producer Peter Saunders and playwright Leslie Darbon to adapt the Miss Marple novel *A Murder Is Announced* (1950) for the stage. The basic plot motif of the novel – an elderly woman impersonates her rich, recently deceased sister – may have been inspired by R. C. Sherriff's crime play *Miss Mabel* that was first performed in 1948 and published by Gollancz in 1949. Osborne criticizes the novel, claiming that it contains "perhaps a little too much coincidence for comfort,"[35] and in particular doubts its suitability for stage adaptation:

> the complications of the plot were such that they must have proved too perplexing for the average audience. Revelations about the various family relationships, and surprises concerning people's identities come thick and fast in the play, and there is no going back to check that you have got it right![36]

In spite of the obvious nostalgic appeal of the material, Darbon uneasily hovers between an updating of the action and the attractions of a period piece. The directions in the published acting edition require the play to be set in "Agatha Christie time," which can hardly be considered an exact specification, since Christie's life spanned eight decades and her career more than 50 years.

As in *Murder at the Vicarage*, the portrayal of village life suffers from the removal of too many minor characters. While the novel *A Murder Is Announced* had presented a true "English microcosm" in the manner postulated by Grella – including an Indian Colonel, two elderly maiden ladies, a representative of the gentry, and a working-class girl – the play retains only those characters that may serve as potential suspects. Since the lesbian couple has been removed, the reconstruction of the crime is taken over by Miss Marple. It is also Miss Marple who finds out that the Locked Door is still in use, a discovery made by Inspector Craddock in the novel. Thus, the interest focuses entirely on the amateur detective: all activities that lead to the solution of the crime seem to gravitate towards the heroine.

While the reduction of cast size affects the play adversely, the confinement to a single set is rather fortunate, because the living-room set for once mirrors precisely the central theme of the plot. The set is described as follows:

> Two drawing rooms in an early Victorian house in Chipping Cleghorn which have been made into one room. [...] The wall which originally separated the two rooms was clearly a bearer-wall since part of it has had to be left intact. This is upstage and has a door almost directly next to it. This door is kept locked and will be referred to as the Locked Door.[37]

The central plot device of the surviving sister taking over the social role and part of the personality of the dead sister is reflected in the design of the set. The *doppelgänger* motif is also taken up by the props, two almost identical Dresden lamps representing a shepherd and a shepherdess. Hence the two lamps acquire a symbolic potential while at the same time figuring as clues in the solution of the mystery.

The additional humour injected into the play tends towards the facetious and is somewhat coarser than in the novel. One example may suffice: in the novel, Miss Marple uses the term "fall guy" she claims to have read in a novel by Dashiell Hammett. This sudden shift of register causes a mildly humorous effect. In the corresponding passage of the play, Dora Bunner confesses to a penchant for the more lurid type of sensational fiction, mentioning a "nice book" she has recently read, entitled "Death Wore a Bra and Panties."[38] The contrast created is even more striking, though not necessarily funnier. As usual, however, most of the humour has to be provided by an unfortunate domestic, the unspecified "mid-European" Mitzi. *A Murder Is Announced* opened at the Vaudeville in 1977 and saw over 400 performances.

In 1981, Darbon adapted yet another Christie novel for the stage of the Vaudeville, *Cards on the Table*. The novel had been published in 1936, and therefore the adaptor had to come to a clear decision about whether to update the play or aim for the appeal of a period piece. The play was eventually set in 1935. The action of the novel starts with a bridge party to which Mr Shaitana has invited four famous detectives – Poirot, Battle, Colonel Race and detective novelist Ariadne Oliver – to show them his rare and precious collection: a collection of murderers who got away with their crime. In the course of the evening, Shaitana is murdered and the four detectives must find out which of the four supposed murderers killed their host for fear of exposure. The principle of plot construction is the same as in *Ten Little Niggers*: the crimes in the past must be solved in order to reveal who committed the crime in the present. However, *Cards on the Table* lacks the suspense of *Ten Little Niggers*, since there is no clear and present danger to the characters involved. In the play version, Darbon refrained from using the device of flashback, with the result that "the murders committed in the past by the various suspects are mentioned too perfunctorily to be properly understood by an audience."[39] The subplot revolving around the character of Major Despard is particularly unsatisfactory, leaving the audience with the general impression that the man must have been innocent since he is to get the girl. Darbon reduces cast size by removing two of the four original detectives – Poirot and Race – thereby unfortunately destroying the symmetry of the character configuration. As a result, Mrs Oliver, who is conceived as a mildly comic character in the novel, becomes a far more efficient detective on stage. As in the case of Miss Marple in *A Murder Is Announced*, investigative activities undertaken by other characters in the novel gravitate to the female principal detective. Some of the play's attempts at humour were ill-judged, and its reviews accordingly were mixed, but it still managed a respectable run of over six months.

Reduction of plot complexity and cast size was a necessary consideration in the case of all the stage adaptations discussed so far. However, besides the 11 novels adapted for the stage, Agatha Christie produced three short stories that were later dramatized: "The Witness for the Prosecution" (1933), "Philomel Cottage" (1934) and "Three Blind Mice" (1950). "Three Blind Mice" is the story on which *The Mousetrap* is based, and although it might be interesting to examine how such a tremendously popular play came into being, *The Mousetrap* will be omitted from the discussion of the present chapter, since a comparison between the stage play and the short story does not promise to yield any definite insights into the adaptation process. For – as already pointed out in Chapter 4 – "Three Blind Mice" originally was a radio play, and the tale and the stage play still show traces of the original format.

Before comparing "The Witness of the Prosecution" and "Philomel Cottage" to their respective stage adaptations, it may be appropriate to consider briefly what may be termed the "poetics" of the detective short story. Ellery Queen even argued that the short story was the original form of detective fiction:

> Modern readers tend to think of "detective stories" as novels, and admittedly the novels are numberless. But the original, the "legitimate," form was the short story. The detective novel is a short story inflated by characterization and description and romantic nonsense, too often for purposes of padding, and adds only one innovation to the short-story form: the by-plot, or red herring, which when badly used serves only to irritate when it is meant to confuse.[40]

Thus Queen argues persuasively that the essentials of a detective narrative may be accommodated in the short format of some 6,000 words. Historically, the short story and the genre of detective fiction evolved simultaneously. Poe, the first recognized author of detective fiction, also was the first major theorist of the short story. Their common origin in the era of the mid-nineteenth century, and, more particularly, in the writings of Poe, may account for certain shared features of detective fiction and the short story. Both are concerned with a moment of discovery – the revelation of the culprit in detective fiction, and the more spiritual form of discovery in the short story dubbed "epiphany" by James Joyce. And both in detective fiction and in the short story, information is deliberately omitted – either in order to create a "mystery" or in a deliberately elliptic style of writing. While today criticism of the short story tends to focus on the impressionistic "slice-of-life" variety, it must not be overlooked that the "plotted" short detective stories, for instance of Conan Doyle, had an important part in establishing the newly evolving genre in the literary market.[41]

In stage adaptation, the short story may at first glance appear to present little difficulty. Critics like A. L. Bader and writers like Somerset Maugham have tended to define the short story largely in terms of plot construction borrowed from the Aristotelian poetics of drama.[42] While the appropriate length for a short story has been traditionally defined according to what may be consumed "in one sitting," drama literally *is* consumed in one sitting, if you disregard the interval. The short story's preference for an elliptic manner of representation, however, may cause problems in dramatization. Drama's constitutional inability to withhold, or, more precisely, *partly* withhold, information makes it difficult to render this specific technical device on stage. Ellipsis in short fiction is most frequently found at the beginning and the end of a story. Anton Chekhov, for instance, "actually advised writers to write a story and then delete the beginning and the end."[43] While a short story frequently begins *in medias res*, with essential information on the characters and their past life provided by a passage of narrative inserted later, a stage play needs a more orderly exposition. The stories discussed in the present chapter, for instance, "grow a head" in stage adaptation: the play sets in at a considerably earlier stage of the action than the short story. The "open" ending that denies the reader his accustomed pleasure of closure, though frequent in short fiction, is rare in detective stories, where it clashes with the conventions of detective fiction, which arouses "end-oriented desire" in the reader. A characteristic type of short story ending frequently to be found in detective short stories is the "final twist" popularized by O. Henry and mainly associated today with the stories of Roald Dahl. In stage adaptation, the convention of the twist collides with the competing dramatic convention of ending a play by dealing out punishment and reward according to desert.

"Philomel Cottage" (first published in the collection *The Listerdale Mystery*, 1934) admirably manages to reconcile poetic justice and the convention of the final twist. The story was dramatized soon after its first publication by actor and playwright Frank Vosper, author of *Murder on the Second Floor* and *People Like Us*. Vosper dyed his hair a crisp golden and played the character of Bruce, a serial killer whose last victim turns the tables on him. The short story "Philomel Cottage" is told in the third person from the perspective of Alix Martin, a young housewife who gradually discovers that the husband she has married after a whirlwind romance is a Bluebeard. She manages to defend herself by storytelling, inventing a past career of crime of her own and tricking her husband into believing that she has poisoned him. The killer, who has a weak heart, dies of fright.

The stage version has a typical early start. While the short story sets in after Alix has been married for some time and relates some necessary expository information in a flashback, the play shows the courtship between Bruce and Cecily (Alix' name in the drama version). Besides, Vosper adds a few minor characters who serve mainly "to deliver in dialogue on the stage information embedded in the expository prose of the story."[44]

While the narrative perspective of the short story focuses the interest on Alix, whose perceptions and reflections the reader is allowed to follow, in the stage version, the character of Bruce gains in stature. Since during his fits of "happy self-aggrandisement" he is given to soliloquizing, the audience is granted an inside view of the mental processes of a dangerous lunatic. Instead of merely watching him collapse in one of his fits, the spectators learn what the world looks like for a homicidal maniac:

> I'm afraid you've been wasting your sympathy over my little attacks, my dear. It may sound funny to you, but d'you know, they're almost enjoyable ... The sense of pace inside one, the pounding faster and faster, becomes almost music – wild, rushing music ... The outlines of things become blurred, but their colours brighter.[45]

The shift of interest from victim-cum-detective to villain (frequent in the crime play) is confirmed by a change of title: Vosper called his play "Love from a Stranger," the stranger of course being Bruce.[46] Bruce himself cunningly exploits the erotic appeal of foreignness, posing as an Englishman in order to attract American women and as an American towards his English victim Cecily. Following the same logic, one of the two film versions of Vosper's play turned Bruce into the Latin American Manuel.

Cecily in *Love from a Stranger* is less occupied with actual detection than her counterpart in the short story. She does not search her husband's desk drawers for clues, nor does she attempt to summon help via telephone. The searching of a suspect's room is apparently perceived as indecorous on the stage: it is for instance also omitted from the stage versions of *Death on the Nile* and *Towards Zero*. Since the wife no longer searches her husband's room, the Bluebeard motif is less strongly developed than in the story. The Philomel motif with its disturbing connotations of sexual violence and mutilation is also absent from the adaptation. Instead, two new motifs are introduced: the theme of Sheherazade, who saved her life by telling gripping stories, and the musical motif of the "Merry Widow" waltz Bruce hums to himself. The replacement of the Bluebeard and Philomel motifs by Sheherazade and the Merry Widow suggests a general lightening of the tone, but still the play was so exciting that reportedly spectators fainted during the third act. Nevertheless, it saw a run of merely 149 performances at the small St. Martin's Theatre and closed after only 31 New York performances. At most a qualified success, *Love from a Stranger* has yet been occasionally revived by repertory companies.

"The Witness for the Prosecution" was first published in *The Hound of Death*, a volume containing 12 short stories dealing with crime and detection or with supernatural forces. It became the title story of the American volume *The Witness for the Prosecution and Other Stories* (1948), in which Christie replaced the stories containing a supernatural element by detective stories. The plot revolves around a woman who goes to great lengths to save her husband from the gallows. In order to make his weak alibi more convincing, she feigns open hostility and appears in court as witness for the prosecution, giving incriminating evidence, which her husband's defence counsel mercilessly breaks down.

It was not until 1953 that Christie adapted her story for the stage, following a suggestion by Peter Saunders, who had actually produced a first draft. In order to get the courtroom scenes right, Christie read up on legal procedure and consulted lawyers. Although 20 years had passed between the short story's first publication and its stage adaptation, only one of the changes introduced by Christie can be accounted for as an attempt to update the material. While Romaine is an Austrian in the short story, Christie turned her into a German in the play version, making use of the specific living conditions in postwar Germany – Romaine's fictional lover Max, for instance, is turned into a Communist agent.

All the other changes Christie effected must be explained by the different generic conventions of short story and crime play. The short story characteristically begins *in medias res*: the solicitor goes to see his client Leonard Vole in prison. The play sets in at an earlier point of the action, shortly before Vole's arrest.[47] Since the play is considerably longer than the short story, which covers a mere 30 pages, some additional – albeit inconclusive – evidence is presented, which necessitates several intermittent plot summaries; in the play, every bit of information is presented at least twice. In spite of these additional clues, the whodunit question of who committed the original crime oddly moves into the background. The central question no longer is "Who murdered Emily French?" but "What are Romaine's intentions?" The short story at least volunteers one alternative suspect, Miss French's nephew, while the stage play does not bother to explain who committed the murder if Leonard is really innocent. The question of the identity of the culprit is superseded by the problem of the motivation of one of the central characters.

In the play's last scene, Christie actually changes her plot. The final twist of the short story is that the "innocent accused" Leonard Vole turns out to be guilty after all. In the play, Romaine finds out that he has betrayed her and stabs her unfaithful husband while still in the courtroom. Osborne explains the changed ending by the development of Christie's own persuasions: "By 1953, however, the author was no longer prepared to see murder (or at any rate this particular murderer) go unpunished."[48] And indeed a consultation of Wingate's list of Christie's endings proves that, after the irresponsible 1930s, a Christie culprit was highly unlikely to get away with murder.

The central plot motif of "The Witness for the Prosecution" is Romaine's disguise as "the other woman" who provides letters, written by Romaine, that eventually serve to dispel suspicion against Leonard. A woman's successful impersonation of another character is a favourite plot motif throughout Christie's fiction, occurring most prominently in *Lord Edgware Dies, After the Funeral, Dead Man's Folly* and *Third Girl*. On stage, this plot motif is much more effective, since actress, director and make-up artist can collaborate to prove to the audience that it can be done – instead of marvelling at the characters' credulity, the audience may be forced to admit that they would have been fooled, too. It is debatable, however, whether the audience is meant to recognize Romaine. Marvin Carlson argues that the audience is intended to be kept in the dark, just as much as Vole's defence counsel Sir Wilfrid Robarts:

> [T]he major surprise in the play (and the film) comes from the audience's not being aware that the wife and the "other woman," whose testimony provides a surprise defense for the hero, are in fact the same person, played by the same actress. In a conventional theatre program, this trick would be revealed, but the highly ingenious conclusion to the play not only gives an unexpected turn to the narrative but also solves this information problem. In the final moments, a real "other woman," who is a different person, played of course by a different actress, does appear. By the time the audience realizes that this is the "other woman" mentioned in the program, and not the false witness played by the wife, that more fundamental trick has already achieved its effect."[49]

Critics of Billy Wilder's film version, by contrast, have argued that every viewer recognizes Marlene Dietrich, in spite of her heavy make-up and Cockney accent, and that the whole point is Sir Wilfrid's unwillingness and consequent inability to penetrate her disguise – he only sees what he wants to see.[50] However, when the film was shot in 1957, visitors to the set were made to sign a "secrecy pledge," which suggests that the producers of the film thought mystery and surprise the main operative elements. From what has been stated above about the relative claims of surprise and suspense in the theatre, it would clearly be preferable to have the audience recognize Romaine at an early point of the scene, but still remain in suspense whether Sir Wilfrid will recognize her, too. If the audience does recognize Romaine in this scene, her initial remark to Sir Wilfrid: "Didn't recognize you without your wig"[51] gains an additional ironic slant, because the main issue of the scene is whether Sir Wilfrid will recognize Romaine in *her* wig. A perspicacious viewer is thus rewarded with an extra joke.

Witness for the Prosecution (without the definite article) opened at the Winter Garden in 1953 and had excellent reviews. It ran for 468 performances and was even more successful in New York. Christie herself considered it her best play, in spite of the greater financial success of *The Mousetrap*.

A number of the problems of stage adaptation discussed in this chapter are related to the specific type of detective fiction Agatha Christie wrote: the classic English whodunit with a complex plot, a large cast of characters and a surprising solution. Other features of the classic whodunit are the dominance of the detective and the importance of narrative perspective. All these characteristics had to be modified in stage adaptation. The following chapter will examine a number of adaptations of detective novels that do not (or not wholly) conform to the standard format of English detective fiction.

13
Other Types of Detective Fiction Adapted for the Stage

Due to the sheer number of her detective stories deemed suitable for stage adaptation, Agatha Christie presents a particularly interesting test case. However, since most of her novels conform to the same type of detective fiction – the whodunit or, as it is called in the present study, the "detective puzzle" – it is appropriate to consider at least a selection of novels and short stories by other writers that belong to other subgenres, most frequently varieties of the thriller. While Christie herself decided to eliminate her Great Detective from her stage adaptations, earlier writers like Arthur Conan Doyle and A. E. W. Mason experimented with the possibility of transferring their fictional detective heroes onto the stage.

In 1910, Doyle had rented the Adelphi for six months to put on his play on prize fighting *The House of Temperley*, which proved an immediate flop. In order to provide a replacement, Doyle dramatized his favourite short story "The Speckled Band" (*Strand Magazine*, 1890) in an emergency operation within a single week. It is a classic Locked Room Puzzle, in which the villainous Dr Roylott ("Rylott" in the play) kills his first stepdaughter by letting a poisonous snake into her bedroom through a ventilator shaft and later proceeds to kill his second stepdaughter in the same manner. The story has proved consistently popular and frequently been anthologized. Apart from its popularity, the story recommends itself for stage adaptation, because it creates considerable suspense by putting the heroine in immediate danger. The main mystery is not the identity of the culprit, which is obvious from the start, but the manner in which the murder was accomplished, the meaning of the enigmatic term "the speckled band." This permits the presence of a "strong villain," and – as Alfred Hitchcock has argued, "the more successful the villain, the more successful the picture."[1] Besides, "The Adventure of the Speckled Band" offers the pleasure of a particularly satisfying case of poetic justice when the snake turns on its master, demonstrating how violence recoils on its perpetrator.

Since the stage play lacks the context of the short story cycle, Doyle introduced a passage which is obviously intended to place the play firmly within the Sherlock Holmes saga. Watson arrives at Holmes's rooms in Baker Street and fondly remembers the times when he shared them with the Great Detective.

> Just the same as ever. There are the old chemicals! Heavens! what have I not endured from those chemicals in the old days? Pistol practice on the wall. Quite so. I wonder if he still keeps tobacco in that Persian slipper? Yes, here it is. And his pipes in the coal-scuttle – black clays. Full of them – the same as ever. (takes one out and smells it) Faugh! Bottle of cocaine.[2]

Although this scene in a sense serves to "introduce" Holmes (as if such an introduction were necessary), it is by no means the first scene of the play. The stage version begins with a newly invented scene showing the inquest on the heroine's dead sister, thus steering an uneasy middle course between showing the original crime and having it reported in retrospect at the beginning of the consultation scene, as in the short story. The introduction of the short story presents substantial problems for the adaptation, because it so strongly stresses the retrospective attitude of all narrative: all the main characters – the villain, the damsel in distress and (arguably) even the Great Detective himself – are already dead at the time of the telling of the tale. This effect obviously could not be reproduced on stage.

Watson is already present at the inquest, because Doyle has for the stage version turned him into a close personal friend of the two sisters and their deceased mother, following the crime play's general tendency to employ a detective personally involved in the case rather than an impartial investigator. Holmes, by contrast, is just as impartial as ever. Doyle does not follow William Gillette's example and fit the Great Detective with a love interest – on the contrary, he has Holmes declare categorically: "I am not for love, nor love for me!"[3] Doyle did imitate Gillette, however, in adopting the character of the office boy Billy, largely for purposes of comic relief. When Holmes disguises himself as Peters to enter Rylott's service, Billy must accompany him disguised as Peters' daughter Amelia. The disguise motif does not occur in the short story, and Doyle presumably borrowed it from another Holmes story, "Charles Augustus Milverton." The plot element of the detective in disguise has already been pointed out as a generic constant of the crime play.

The character of the villain has also been slightly modified for the stage version. While Dr Roylott of the short story merely has a violent temper, Dr Rylott of the play is positively insane – he is said to have been not in prison but in a madhouse. This change accords with the crime play's generic preference for the homicidal maniac. Rylott's insanity manifests itself in sudden changes of mood to which other symptoms were added in performance. Rylott was played by the celebrated Shakespeare actor Lyn Harding, who tried to turn the part into a psychological case study. Since unfortunately, however, the lines Doyle had written for him did not seem to offer sufficient material to build a character on, Harding resorted to physical theatre, inventing various mannerisms and nervous tics. Doyle was unhappy with Harding's interpretation and called in James Barrie as a referee. Barrie reportedly advised him to let Harding do as he pleased, and popular success proved him right: on the opening night, Harding received over a dozen calls.[4]

The fact that Rylott is keeping a snake in his room (which constitutes the main mystery of the short story), is revealed fairly early in the play, at the end of II.i. Rylott's native servant Ali takes a wickerwork basket out of a cupboard and starts playing an Eastern flute to it, in an attitude strongly suggesting a snake charmer. Possibly, by 1910, the meaning of "the speckled band" was no longer a mystery to anyone in England, but in all likelihood the early revelation of the central mystery is also related to the different attitude of narrative and play towards the respective claims of mystery and suspense.

It may also be related to drama's problem with ellipsis. In the short story, the snake is famously *not* seen to creep through the ventilator shaft by the narrator:

> Then suddenly another sound became audible – a very gentle, soothing sound, like that of a small jet of steam escaping continually from the kettle. The instant that we heard it, Holmes sprang from the bed, struck a match, and lashed furiously with his cane at the bell-pull. "You see it, Watson?" he yelled. "You see it?" But I saw nothing.[5]

It is several paragraphs later, after the death of the villain, that the mystery is explained. Watson and Holmes find Dr Roylott dead in the adjacent room:

> Round his brow he had a peculiar yellow band, with brownish speckles, which seemed to be bound tightly round his head. As we entered he made neither sound nor motion. "The band! The speckled band!" whispered Holmes. I took a step forward. In an instant, his strange head-gear began to move, and there reared itself from among his hair the squat diamond-shaped head and puffed neck of a loathsome serpent.[6]

Through the choice of restricted narrative perspective, the author can determine what the reader is allowed to "see," an option not open to the playwright. On stage, the appearance of the snake through the hole becomes the climax of the play. Originally, Doyle had intended to use a live snake and a snake man to handle it.[7] In the event, however, a property snake was used, "so ingeniously jointed, and worked on invisible packthreads, that it could move with hideous realism even away from the bell-rope."[8] While in the tale, the presence of the snake is first deduced and only perceived later, the stage play does not hint but show. Due to the fact that deduction is less important in the play than in the story, the band of gipsies serving as a false clue for the interpretation of the term "the speckled band" has been removed from the stage adaptation. A linguistic "red herring" hinging on the double meaning of "band" was obviously considered less appropriate on stage than in the purely verbal medium of the written text.

In the following year, A(lfred) E(dward) W(oodley) Mason presented his version of the Great Detective on the stage of the St. James's Theatre. Mason was an actor and author of historical romances and detective novels, of which the best known are *At the Villa Rose* (1910), *The House of the Arrow* (1924) and *The Prisoner in the Opal* (1928), all of them featuring Monsieur Hanaud of the Sûreté, an adaptation of Gaboriau's Monsieur Lecoq, and his Watson Mr Ricardo.

Although *At the Villa Rose* is narrated in the third person, the events are invariably described from the perspective of Mr Ricardo. The novel might be used as a model example to illustrate Hühn's remarks about the reticence of the Great Detective: much of the novel's suspense is created through the tension between Ricardo's limited awareness of the state of affairs, and Hanaud's superior awareness, the extent of which, however, remains concealed even from the reader. In spite of the conventional narrative perspective he employed, Mason professed scepticism towards the effectivity of the mere puzzle. In the novel itself, Hanaud remarks that the interest of the case does not reside in "the puzzle or its solution"[9] And in a radio interview, Mason rejected the mere puzzle in favour of a type of novel that manages "to enchant the interest of its readers on the different ground of the clash of its characters and the diversity of their interests."[10] Daniel P. King quotes Mason's declared intention to "combine the crime story which produces a shiver with the detective story which aims at a surprise."[11]

In the case of *At the Villa Rose*, the result of this endeavour is a distinct tripartite structure that clearly demonstrates the distinction between the "two stories" of detective fiction: the overt story of the investigation and the covert story of the crime. In *At the Villa Rose*, the story of the investigation is placed at the beginning, making up about two thirds of the entire novel. It is followed by the story of the crime (about one quarter) and a final chapter entitled "Hanaud Explains" revealing the Great Detective's reasoning process to his admiring Watson. It is Ricardo who writes down the story of the crime,[12] but it is Hanaud who makes sure that the tripartite structure is kept intact, when he refuses to anticipate information that properly belongs to second part, the story of the crime, imparted by the innocent accused and terrified witness Celia Harland: "I, too, Mr Ricardo, have artistic inclinations. I will not spoil the remarkable story which I think Mlle. Celie will be ready to tell us."[13] Critics have tended to disapprove of the novel's tripartite structure, complaining that the "puzzle is quickly solved when Mason begins to throw light upon it and the story loses its hold before the concluding chapter"[14] and that the "essential element of surprise is largely lost through the adoption of this technique [with the result that] a book which might have been a landmark trails off disappointingly."[15]

In order to keep up the reader's interest in the second part, Mason introduced a particularly gruesome murder committed in the course of a séance, a scene that also promised to be very effective on stage. For the stage adaptation, Mason disposed of his Watson character, whom he clearly recognized as a mere narrative device, and of the tripartite structure of "investigation – crime – explanation of reasoning process." Instead, the play shows the events in their chronological order including the crime, and hence without any mystery concerning the identity of the culprits or the manner in which the crime was committed. Here, Mason is obviously following the example of William Wilkie Collins, who had also dismissed the puzzle element as untheatrical in his stage adaptation of *The Moonstone*. While the novel *At the Villa Rose* is part puzzle and part thriller, the stage version is a thriller all through. A case in point is the treatment of Celia's kidnapping. Unless the reader is much quicker in the uptake than Ricardo, he will only find out that Celia is innocent and has been kidnapped shortly before she is rescued. In the play version, the audience is aware that Celia is in the hands of the criminals before the investigation even begins, and their opportunity to tremble for Celia's safety hence is prolonged considerably.

Although the puzzle element in the play has been diminished, the investigation process, which begins in Act III of the play, retains some of its interest, which is largely due to Mason's clever management of clues. The clue, of course, is an important element particularly in classic detective fiction, but the handling of clues on stage may present a problem, as LaBorde has pointed out:

> The play cannot totally copy the story in the parading of concrete clues such as fragments of glass, watches stopped at the time of the murder, and fingerprints on the murder weapon. Instead, mystery dramas need to include the scenes of drama and adventure that Wright said were not indigenous to the pure mystery story. The best method to introduce the activity necessary to the stage is to employ what novelist Erle Stanley Gardner termed "clues of action," in which deductions are drawn from what the detective sees people do rather than from objects gathered at the scene of the murder.[16]

In the stage adaptation of *At the Villa Rose*, Mason accordingly drops several concrete clues – the footprints on the gravel, Celia's dress, the murderer's road map – and introduces a "clue of action" eminently suitable for drama: Hanaud places a policeman in a position from where he can overlook the whole room to check if the murderer will attempt to signal to his accomplice. Thus, the audience is again invited to participate and given a clearly formulated task that is particularly appropriate to their position in the theatre: to observe the interactions of the characters on stage as closely as possible.

A remarkable feature of the play *At the Villa Rose* that distinguishes it from classic "cozy" English detective fiction is its extreme violence: apart from the murder of a helpless elderly woman during a séance, it features the use of vitriol in order to intimidate a witness. Yet, the play somehow passed the censor – possibly because the effects of vitriol are never demonstrated, and the murder itself takes place in the dark.

While Mason had attempted to combine the pleasures of the puzzle with that of the thriller, a number of English detective novelists produced pure thrillers, characterized by action, movement and violence. This strand in English detective fiction has been largely ignored in criticism, possibly because the style of writing is considerably inferior to that of the major American hardboiled writers Hammett and Chandler. However, the enormous sales figures of Sapper, Wallace and Cheyney suggest that the ideological climate in England was just as much in favour of the hardboiled variant of detective fiction as in the United States, which makes it necessary to qualify the traditional distinction between the English "cozy" puzzle and the American hardboiled thriller to some extent.

Sapper's Bulldog Drummond was the first action hero from a hardboiled English thriller to find his way onto the stage. "Sapper" was the pseudonym of Herman Cyril MacNeile, who had served in the Royal Engineers ("sappers") in World War I. MacNeile wrote nearly 30 books of crime fiction, including *Bulldog Drummond* (1920), *The Black Gang* (1922), *The Third Round* (1924) and *The Final Count* (1926). This series of novels relates the "four round fight" between Bulldog Drummond, a demobbed officer who finds peace "incredibly tedious," and his arch enemy, the cosmopolitan villain Carl Peterson. In contrast to Mason's Hanaud, Drummond does not distinguish himself by his superior deductive powers but by his ability to bear and inflict pain. His reasoning amounts to such basic formulations of strategy as: "First – They know where Potts is. Two – They will try to get Potts. Three – they will not get Potts."[17] The hero makes up for his lack of brains by immense physical prowess – he is for instance capable of strangling an adolescent gorilla with his bare hands. Moreover, the hero joyfully anticipates any opportunity to inflict violence: "his hands, apparently hanging listless by his sides, were tingling with the joy of what he knew was coming."[18]

The plot of *Bulldog Drummond* starts with a capitalist world conspiracy to bring about a General Strike and ruin England's economy. An American millionaire is kidnapped in order to force him to collaborate. Bulldog Drummond gets involved because he has advertised seeking a new exciting occupation, whereupon Phyllis, the heroine, comes to see him and reports sinister goings-on at the Elms, her neighbour's house in Godalming.

In the 1921 stage version, Gerald du Maurier played Bulldog Drummond and also assisted Sapper in the adaptation. The plot was reduced considerably: to begin with, the harebrained "world conspiracy" scheme was removed – the millionaire is now kidnapped for purely financial reasons. While the doped millionaire in the novel is kidnapped repeatedly by Peterson and Drummond in turn, one kidnapping suffices for the play, and a subplot concerned with a small-scale "cozy" crime, the theft of a string of pearls, is not shown but merely mentioned in dialogue.

Two further plot changes in the stage version are related to the crime play's fondness for role play and disguise. In the novel, Drummond takes Potts, the kidnapped millionaire, to his rooms in Half Moon Street. When Peterson's henchmen arrive to retrieve Potts, Drummond has already replaced him by a drunk demobbed soldier named Mullings. The criminals kidnap Mullings by mistake and only realize their error when they reach the Elms. In the play, the substitution of the millionaire (here named Travers) is effected in a both more complicated and distinctly funnier manner: Drummond dresses up in a greatcoat and impersonates Travers. Since he is supposedly still under the influence of drugs, he has to be helped down the stairs by his friend Algy, who in turn has to impersonate Drummond, merely mouthing the words while Drummond speaks under the cover of his coat. While in the novel, it is merely suggested that Peterson's gang has got the wrong man, and Mullings' identity is revealed only when Peterson himself discovers it, the theatre audience is aware of the exchange while the criminals are not, and is thus granted the greater pleasure of watching the villains being fooled.

The second instance of role play and disguise occurs right at the end of the play. Both in the novel and in the play, Peterson is allowed to escape in the end, but in the novel the precise manner of his escape is not divulged. In the play, Peterson phones his girlfriend Irma, telling her to use "the old circus gag," and is soon "arrested" by an accomplice disguised as a policeman. As has been variously noted above, the bogus policeman was soon to become one of the favourite characters in the crime play. Gerald du Maurier used a slightly modified version of this ending again in the final scene of *The Ringer*.

Although the play *Bulldog Drummond* has the unusually high number of five different stage sets, the numerous dynamic scene shifts of the novel – a typical feature of hardboiled fiction – proved impossible to reproduce on stage. In the novel, for instance, Drummond flies to Paris in an aeroplane for a single night – a more remarkable feat in 1920 than today. Even though the stage version is not limited to a single set in this case, it still exploits the dramatic effect of confinement: a telephone wire is cut to obstruct communication with the outside world (not in the novel), and the electrified staircase of the novel is turned into an electric door: "No one can come in or get out while the current is switched on."[19] Apart from the electrified staircase, several other gadgets and special effects have been removed from the stage version: the half-grown gorilla, the thumbscrews, and the acid bath in which Lakington dissolves his victims. Hence, the play is considerably less spectacular and less violent than the novel. In the novel, Drummond forces Lakington into an acid bath that eats his face away, Lakington escapes and is killed by the electrified staircase. In the stage version, Drummond fights Lakington in the dark, using a variety of weapons including guns, knives and bottles and finally strangles him. Although the result is the same, the element of deliberate torture inflicted by the hero is mercifully absent from the play. The reduction of spectacle and violence did not impair the success of the play, which opened at the Wyndham's in 1921 and saw almost 500 performances.

Although his novels rarely provide a comparable amount of graphic violence, the works of Edgar Wallace belong to the same subgenre of detective fiction, the variety of "hard-boiled British." Wallace was an extremely prolific writer who wrote over 200 books, including more than 80 detective novels and over 50 volumes of detective short stories. Critic Jack Adrian lists the following typical plot motifs one associates with a Wallace novel: "sinister cowled figures carrying swooning girls through shadowy corridors; bullets crashing through the windows of old manor houses; shrill cries of terror echoing through the murk of a dockside dawn."[20] As this quotation suggests, Wallace's novels have a definite theatrical (and cinematic) potential. However, it was only during the last six years of his life that Edgar Wallace took up crime writing for the stage in earnest, producing eight successful plays, three of which were adaptations of previously published novels: *The Ringer* (1926, based on his novel *The Gaunt Stranger*), *The Squeaker* and *The Flying Squad* (both 1928).

The novels' emphasis on suspense and thrill makes them particularly suitable for stage adaptation. Dramatization is also considerably facilitated by Wallace's specific technique of plotting. His method of plot management is derived from serial writing, with a basic plot construction scheme that consists of one overarching main plot interest and several local arches of suspense providing cliffhangers from instalment to instalment. Margaret Lane comments: "Unlike most writers of mystery stories, who have to devote the last couple of chapters to sorting out the tangles, he set and then solved new problems all the way through the book, keeping only his basic mystery unexplained until the end."[21]

In Wallace's novel *The Gaunt Stranger* (1925), the main mystery is the identity of the "Ringer," an arch criminal and master of disguise, possibly modelled on the French master criminal Fantômas but lacking the latter's anarchic strain, since the Ringer only kills from a perverted sense of justice. The title character "the gaunt stranger" seems to be identical with the Ringer for a major portion of the book, until he is finally revealed to be Police Sergeant Wills from Headquarters, the man to bring the Ringer to bay. The

Ringer himself has all the time been disguised as the slightly ridiculous elderly Scottish police surgeon Dr Lomond.

The minor mysteries are provided by a subplot revolving around the heroine Mary Lenley caught up in an amorous triangle between poor but honest Detective-Inspector Alan Wembury and the glamorous villain Meister, who combines the lucrative professions of solicitor, fence and informer. This character configuration of hero – heroine – villain is familiar from Victorian melodrama and recurs in all of Wallace's adapted plays. Usually the heroine is getting involved in the criminal milieu through a tainted family member – in *The Gaunt Stranger*, it is Mary's brother Johnny. Meister's attempts to incriminate Johnny and seduce Mary provide some of the minor plot interests. In Chapter 17, Meister is finally and satisfactorily killed, and the rest of the novel may now be devoted to chasing his murderer.

When Edgar Wallace adapted the novel for the stage under Gerald du Maurier's guidance, he made the Ringer the title character and removed the mysterious "gaunt stranger." His function in the plot is taken over by Central Detective-Inspector Bliss, a much more solid presence. The Ringer has been turned into a positive character – for instance, he no longer attempts to poison the heroine, a deed always likely to alienate an audience. Moreover, he only kills men who are definitely guilty, and he is motivated by a personal grudge against Meister, who is responsible for the death of Gwenda, the Ringer's sister.

Most conspicuously, *The Gaunt Stranger* and *The Ringer* differ with respect to their endings. In the novel, the Ringer is demasked and reminds Cora of a promise she once gave him. She obeys and shoots him, and as the Ringer escapes into death, Cora, too, manages to escape arrest in the general commotion. In the play, instead of killing her husband on demand, Cora shoots him with blanks, thus initiating a venerable tradition in the crime play genre. In the confusion, the Ringer escapes but returns disguised as a policeman, who is ordered to watch over Cora. They embrace and escape together. While in the novel, the master criminal suffers a qualified punishment, he is rewarded by a reunion with his loving wife in the stage version.

As Lane has noted, Wallace's cast of characters usually includes "a comic petty thief."[22] In *The Gaunt Stranger*, this position is filled by Sam Haggitt, a Cockney criminal who for some time serves as Meister's valet before he tries to get away with his master's old silver. On stage, the minor character of Haggitt, who is renamed Hackitt, evolves into a full-blown comedy part played by Gordon Harker, whose specific talents contributed much to the success of Wallace's plays. Lane comments: "Harker's Sam Hackitt in *The Ringer*, his Detective-Sergeant Totty in *The Case of the Frightened Lady* and his offensive criminal butler in *The Calendar* are the purest and funniest comedy that Wallace ever wrote."[23]

The Ringer opened at the Wyndham's in 1926 and ran for a year, earning Wallace about £7,000 pounds in royalties, half of which he made over to du

Maurier in recognition of his contribution to the shaping of the play. *The Ringer* marked the beginning of the period of Wallace's great financial successes, and he was so exhilarated by it that he contemplated giving up novel writing altogether and concentrating entirely on his work for the stage.

Apparently, he did not seriously consider giving up novel writing, for he immediately dashed off a novelization of the stage play *The Ringer*, dedicating the book to du Maurier:

> My dear Gerald, This book is "The Gaunt Stranger" practically in the form that you and I shaped it for the stage. Herein will you find all the improvements you suggested for "The Ringer" – which means that this is a better story than "The Gaunt Stranger." Yours, Edgar Wallace

Apart from Wallace's gratitude to du Maurier, this dedication shows that Wallace thought of stage adaptation not mainly in terms of a transfer to another medium but mostly as an opportunity of revising his works, which he did not normally have. That he did not consider the adaptation process particularly difficult may be related to his rather unadorned prose style, which often already resembles a stage direction, as in the following example from *The Squeaker* describing the murder of the "squeaker" Frank Sutton in Chapter 27:

> [Sutton] heard a noise and looked up. The door was opening slowly. A hand holding a pistol came through. [...] A little pause, and then the door of the board-room opened, and John Leslie stepped in, and in his hand was the smoking pistol.[24]

The Squeaker opened at the Apollo in 1928 and transferred to the Shaftesbury, clocking up a run of six months, which ensured a handsome profit. It had only been running for about a week when it was followed by another Wallace play that opened at the Lyceum, *The Flying Squad*, which turned out to be Wallace's last stage adaptation. As in *The Ringer* and *The Squeaker*, the heroine must choose between a policeman and a criminal after initially having been involved in the criminal milieu by a family member. The parallel main mystery revolves around the ambiguous character of Li Yoseph, an old man subject to mild hallucinations – or possibly gifted with psychic powers. He is involved in the drug traffic and gets killed by his accomplices, but even after his death, various witnesses claim that they have seen him. In this respect, *The Flying Squad* is clearly identifiable as a product of its era, when detective fiction and the crime play were still constantly flirting with the supernatural. The novel version of *The Flying Squad* ends with a qualified happy ending. The heroine Ann, who was once involved in the drug traffic, must patiently wait for her beloved policeman in Brazil, which has no extradition treaty. In the theatre, Wallace did not think it appropriate to deny his audience the conventional happy ending, and the last scene shows Ann and her policeman locked in an embrace.

While Wallace and Sapper were still firmly located in the British tradition, Peter Cheyney imitated the style of the American pulps. His first successful series of crime novels was launched in 1936 with the Lemmy Caution novel *This Man Is Dangerous*, set in the United States and written in "ludicrous pidgin-American."[25] For his Slim Callaghan series, Cheyney shifted to an English setting and to third-person narrative. The first Slim Callaghan novel, *The Urgent Hangman* (1938) contains obvious references to Hammett's *The Maltese Falcon*. Callaghan is an anglicized version of Sam Spade, and his secretary is named Effie Perkins, a deliberate echo of Spade's Effie Perine. However, Cheyney uses the references to *The Maltese Falcon* not only to establish the general mode of his novel, but also as a red herring: in contrast to Hammett's novel, the *femme fatale* client in *The Urgent Hangman* does not turn out to be the murderess. This use of an intertextual red herring is rare in detective fiction, but since Cheyney set out to create a new type of detective fiction, the "hard-boiled English private eye novel," it is not surprising that *The Urgent Hangman* should abound in self-conscious references to the genre of crime fiction.

One of Cheyney's main concerns was to establish London as a suitable hardboiled setting, a sordid metropolis on a level with American cities, as Hammett's San Francisco or Chandler's Los Angeles:

> For a few seconds he saw London objectively! Saw it as a place mainly consisting of two parts – a very thin upper crust and a damn thick lower one. The upper crust was the veneer of respectability, "niceness," cleanliness, which London showed the world; the lower crust – the thick one – all the rottenness, cheap crookery and general lousiness that existed in that Jungle in the heart of the metropolis whose boundaries are known to every intelligent police officer.[26]

At the same time, Cheyney expressly distances himself from the English tradition of cozy detective fiction, in which well-bred detectives in armchairs leisurely examine clues. The efficient Inspector Gringall is introduced as follows:

> He neither chewed gum, smoked curved pipes, had an income of his own, solved cases over a chessboard, asked crime reporters to give him a hand in solving cases, or looked under the edge of carpets for clues. Nobody had ever heard Gringall talk about "a clue," and he once said that if he met one he probably wouldn't know what it was.[27]

Slim Callaghan is no avid clue-seeker either. His investigation technique consists in tricking the suspect into betraying himself. Here the private detective has a definite advantage over the policeman, as Callaghan points out to Gringall: " not being a policeman, when I have suspicions I can create circumstances that either corroborate or break them down."[28] According to Hühn and Stowe, hardboiled fiction is characterized by a shift from the semiotic mode of detection – the "reading" of clues – to the hermeneutic mode in which neither the detective nor the crime he investigates remain a stable entity:

> Here the interpretation, as practiced by the private eye, is presented as an *interaction* between the reading subject and the object (the text) in which neither side remains a stable entity. The detective's reading activities affect the text (the mystery he investigates) and he in turn is affected by it.[29]

This kind of "interactive" investigation is evidently very suitable for the stage. The "Producer's Note" of *Meet Mr Callaghan*, Gerald Verner's adaptation of *The Urgent Hangman*, points out the dramatic effectivity of Callaghan's investigation technique:

> The chief interest of this play, apart from its value as a Murder Mystery, is its situational comedy. The fact that Callaghan is telling deliberate lies and faking evidence to gain the ends of his clients, must not only completely convince the character he is talking to, but must also leave the audience slightly worried as to his real intentions.[30]

The morally ambiguous hero makes an attractive protagonist in drama, since the audience can be kept guessing about his real intentions. Another feature of *The Urgent Hangman* that makes it attractive for stage adaptation is the pronounced courtship element evolving between the private detective and his beautiful lady client, who may be a murderess. The detective seducing (or being seduced by) his woman client of course is a stock motif in hardboiled fiction.

What is problematic when stage-adapting hardboiled fiction, however, is that it is a genre characterized by frequent dynamic scene shifts – Drummond's trip to Paris was mentioned above as an example. Slim Callaghan does not fly to Paris, but he is always on the move, though mostly within a clearly circumscribed area between Holborn, Soho and St. James's. Besides Slim's agency in Cursitor Street, the novel's settings include a mortuary, a hospital, Scotland Yard, several hotels, restaurants and nightclubs. The play, by contrast, limits itself to one single set – Callaghan's office – with the somewhat unfortunate result that the suspects must come to see Callaghan rather than the other way round, which seems occasionally rather unlikely. This departure from the conventional living-room set also indicates that the private eye adapted from American hardboiled fiction has no home, no background, and is hence difficult to "place" socially.

Apart from removing the essential hardboiled quality of dynamic movement, the stage adaptation also dispenses with a number of plot elements characteristic of hardboiled fiction. An indispensable plot element of a hardboiled detective novel is the beating up of the private detective, occurring in Chapter 10 of *The Urgent Hangman*. This scene has been removed from the stage version. Other typical hardboiled elements cut from the adaptation include the raiding of a nightclub, the attempt of the heroine's fiancé to frame the heroine, and the detective chloroforming and kidnapping the heroine. As a result, the stage play is not only less violent than the novel, but also considerably less pessimistic about human relationships.

The denouement of *The Urgent Hangman* reveals an interesting difference between hardboiled fiction and the whodunit. The last chapter of *At the Villa Rose*, as will be remembered, is entitled "Hanaud Explains" and reveals the reasoning process of the Great Detective to his admiring Watson. This chapter closes the gap between the discourse of the investigation provided by Ricardo and the story of the investigation provided by Hanaud. Callaghan explains his deduction process to Gringall in Chapter 14, even though Gringall does not serve as a mediator between detective and reader. By making Gringall his Watson, Callaghan asserts the superiority of the private eye over the institutionalized police. In the stage adaptation, even this modified remnant of the Detective-Watson interaction is absent. In *Meet Mr Callaghan*, the explanation of the detective's reasoning process becomes part of the showdown – the detective uses his knowledge to assert his dominance over the culprit. The intellectual battle of wits is then followed by a modest shoot-out, as befits the hardboiled genre.

The original production of *Meet Mr Callaghan* opened at the Garrick in 1952, directed by Derrick de Marney. The part of Slim Callaghan was taken by Derrick and Terence de Marney in turns, and Terence de Marney later also starred in another Cheyney adaptation, *Dangerous Curves*.

In stage adaptation, the various varieties of the thriller present problems that differ considerably from those involved in the dramatization of a classic whodunit. The plot of a thriller generally consists of a succession of more or less violent encounters between the detective and the criminals – the "rounds" between Drummond and Peterson – in which each of the opponents attempts to maintain the upper hand. These fights for dominance between a protagonist and an antagonist transfer rather well to the stage, although extreme physical violence may sometimes have to be moderated. The identity of the criminal (or, as in *The Ringer*, at least one of the criminals) is usually obvious from an early point in the action, which helps the audience to focus their aggression. Since the identity of the culprit is hardly ever in doubt, there is no necessity for the meticulous gathering and interpretation of clues. While the whodunit tends to revolve around a private crime, thrillers are more frequently concerned with professional crime. On stage, details concerning the criminal organization are usually dropped – this applies to all stage adaptations of thrillers discussed above. Organized crime, however, is a theme suitable for stage adaptation since it allows the use of several villains. The minor villain(s) may be defeated in the course of the action, while the main villain is left for the showdown. The defeat of the various villains structures the action and replaces the tedious tidying up of minor mysteries in classic detective fiction. As the plays *The Ringer* and *Meet Mr Callaghan* show, a morally ambiguous hero makes a particularly good crime play protagonist. In the characters of the Ringer and Slim Callaghan, features of the criminal and the detective have been combined; another instance of the crime play's tendency to blur the actantial roles.

The Golden Age of detective fiction produced yet another subgenre that turns the villain into the hero and tells the story of a crime from the point of view of its perpetrator, the subgenre dubbed "inverted" detective fiction and traditionally traced back to Francis Iles' two novels *Malice Aforethought* (1931) and *Before the Fact* (1932). However, even before Iles, C. S. Forester had written two "inverted" detective novels, *Payment Deferred* and *Plain Murder* (1930), two naturalistic studies of crime and punishment set in a sordid lower-class environment. In comparison to Iles' novels, planning and plotting is of secondary importance. William Marble in *Payment Deferred*, for instance, commits murder on the spur of the moment in the first chapter, and the rest of the novel is then devoted to his attempts to evade discovery, and the eventual "payment" of his moral debt. His actual crime remains undetected, but he ends up being hanged for a murder he did not commit. Although Forester himself was also interested in the theatre, *Payment Deferred* was adapted for the stage by Jeffrey Dell. The play opened at the St. James's in 1931 and saw 80 performances.

The novel has a pitiless heterodiegetic narrator, who provides occasional glimpses of the inner life of the characters but remains generally unsympathetic towards them. An air of fatality pervades the text, beginning with the title, which promises that, though justice may be delayed for some time, Marble will not escape punishment. On stage, the impression of the events leading logically and inevitably to their predestined conclusion is created by revealing the outcome first. An estate agent's clerk shows the Marbles' house at 53, Malcolm Road, Dulwich, to a prospective tenant, giving a correct account of Marble's trial but a mistaken version of the events leading up to it. An extended flashback then shows the audience what really happened. The play ends with an epilogue in which the prospective tenant agrees to take the house but mentions in passing that he will have to dig the garden right through from end to end in order to get rid of the weeds. Since Marble has buried his victim in the garden, the tenant's remark foreshadows that the real crime will eventually be brought to light. This ironic twist corresponds to a similarly ironic ending in the novel; the narrator dryly comments that Marble went down in history as an unusually clumsy murderer.

In the novel, Marble becomes increasingly obsessed with having to watch over the grave in his garden until he is almost unable to leave his house. This allows the adaptor to exploit the confinement of the single living-room set. The Marbles cannot move house, in spite of their newly acquired riches, nor can they have their present house redecorated, since Marble lives in mortal fear of snooping workmen. In order to beautify the home, Marble furnishes it with hideous gilded Empire-style furniture. The stage direction prescribes: "The old paintwork and wallpaper look shabbier than ever by contrast with the over-gilded rococo and pseudo-Empire monstrosities which replace the old furniture."[31] The set design reveals Marble as a social climber and also visualizes his imprisonment in a gilded cage.

The central event in both novel and play is Annie Marble's discovery that her husband is a murderer. In the novel, Marble makes a desperate attempt to convince his wife that he loves her because he fears for his safety: "It was of the first importance that she should continue to think he loved her."[32] In the play, by contrast, Annie's discovery brings the couple closer together, as the initial stage direction of III.i prescribes: "In their conversation there is now a more intimate warmth noticeable, her tone being one of maternal solicitude and protectiveness, and his rather wistful and childlike timidity."[33] The development of Marble's character has been moved into the stage directions and hence largely becomes the responsibility of the actor. Scenes like III.i humanize Marble to some extent. While the narrator of the novel clearly distances himself from its central character, it is possible to pity Marble when one watches him on stage. This is also due to the fact that Marble on stage shows a variety of strong emotions and is seen sobbing violently, laughing hysterically and suffering a heart attack, whereas in the novel his only display of emotion is to beat up his wife. While the reader is supposed to take an attitude of emotionally detached "clinical" interest, the spectator is allowed to indulge in the feeling of pity. This change from novel to play may be explained once again by the absence of narrative mediation – in the novel, the distinct narrative voice functions as a distancing device.

A particularly successful adaptation of an inverted detective story was the stage version of Roy Vickers' "Double Image" that had won the first prize in the short story contest of *Ellery Queen's Mystery Magazine* in 1953. It was subsequently published in Vickers' short story collection *Double Image and Other Stories* (London: Jenkins, 1955). Vickers had been writing detective fiction since the 1930s under a variety of pseudonyms and in a variety of styles. Most of his short stories belong to the type of "inverted" detective fiction, and usually the culprit is finally brought to bay neither by any brilliant act of detection nor by a fatal mistake of his own but merely by accident, frequently by a single significant concrete clue. In "Double Image" the protagonist Julian Fanshaw invents a long-lost twin brother in order to commit the perfect murder of his uncle Ernest. The motif of the fictional twin brother goes back to Milne's *The Red House Mystery*, and its main appeal resides in the fact that one of the "brothers" may disappear without a trace.

"Double Image" was adapted by the team of Roger MacDougall and Ted Allan for the Savoy stage in 1956. In accordance with tradition, MacDougall/Allan moved most of the play's action into the Fanshaws' living-room. Julian, a slow-transport agent in the story, is an unsuccessful composer in the stage version, conforming to the stereotype of the criminal as (failed) artist. He establishes his alibi by playing a record in an upstairs room on which he is heard to play the piano and curse intermittently. He then forces his way into his own home in the character of the evil twin brother David and stages a fight between the two brothers upstairs which is presented as an audiodrama to the audience and to his mother-in-law, a highly welcome earwitness downstairs.

The move from Julian's office to the family home indicates a general shift in emphasis from public to private life. While in the short story, Julian's difficulties are mainly professional and financial, his problems in the drama version are mostly personal and marital. While Vickers is interested mainly in the technical aspects of the ingenious murder plan, MacDougall/Allan venture a psychological explanation for Julian's actions. For instance, Julian accuses his uncle of always secretly having wished that Julian had a brother gifted with a sound business instinct. While Julian's wife Elsa increasingly believes that Julian is suffering from a split personality, her mother believes that he consciously enacts the parts of both Julian and David: "It's Julian acting Jekyll and Hyde. [...] Perhaps he doesn't like himself as he is."[34] Elsa, however, maintains that "[a]ll those things he was doing without realizing it. As if he *had* to create a twin to do the things, say the things, that he'd like to do."[35] Even at the end, when it is proved beyond any doubt that Julian killed his uncle, she still insists that he is not responsible for his actions: "My husband needs a doctor, not the police."[36] As if to prove her diagnosis correct, Julian begins to mix up his two identities. When he is arrested, his last words to his wife are: "Oh, Elsa, just in case there are any complications, will you ring our solicitor? After all, I am your brother-in-law."[37] It remains doubtful whether Julian really is insane or merely feigning madness to escape capital punishment.

While in the short story, Julian stumbles over the concrete clue of the newspaper clipping, in the play version his wife is instrumental to establishing his guilt.[38] Taunting her husband with "David's" superior faculties as a lover, Elsa wrings a full confession from him. The stage direction prescribes: "Elsa follows him with her eyes as a cat might follow the movements of a mouse."[39] The plot motif of the reading of clues is replaced by the battle of wits of the sexes.

From the perspective of audience reception, *Double Image* raises a similar question as *Witness for the Prosecution*, namely, whether the spectators are supposed to realize that David is Julian in disguise. Once again, the theatre programme is used as an instrument for audience mystification. On the cast list,[40] both Julian and David are listed as played by Richard Attenborough, which suggests that Julian and David exist as separate characters that are merely played by the same actor. The play's plot mechanics immediately made sense to theatre managers abroad, and *Double Image* was translated for instance into French and German.

Besides the thriller and "inverted" detective fiction, one further subgenre of crime fiction shall be considered in the present chapter: the genre of "courtroom drama" the very name of which already suggests a certain suitability for stage adaptation. In the 1960s, lawyer Henry Cecil dramatized three of his novels, which all revolve around a specific legal practice and are triggered off by a miscarriage of justice. These plays are: *Settled Out of Court* (with William Saroyan, Strand, 1960 – based on Cecil's novel of 1959), *Alibi for a Judge* (with Felicity Douglas and Basil Dawson, Savoy, 1965 – based on Cecil's novel of 1960) and *According to the Evidence* (also with Felicity Douglas and Basil Dawson, Savoy, 1967 – based on Cecil's novel of 1954).

The earliest of these plays, *Settled Out of Court*, is a courtroom drama outside the courtroom, using courtroom procedure for investigating a crime in a private setting. A short plot synopsis will clarify what this means: Lonsdale Walsh, a man "allergic to lies," is sentenced to life imprisonment for the murder of a business rival. Eager to prove that he has been framed, he escapes from prison and enforces a retrial by taking the entire personnel of his original trial hostage at the judge's house, including the presiding judge himself, the prosecuting counsel and the witnesses. The kidnapping is organized by Walsh's daughter Angela, who is assisted by a professional criminal who commits crimes on commission.

As this summary suggests, the plot of the novel recommends itself for stage adaptation, because the central idea not only allows but positively requires confinement to a single set. Accordingly, the play opens with the scene in which the judge Sir George realizes that he is a prisoner in his own house. In the novel version, this incident is preceded by a character sketch of Lonsdale Walsh and his dominant trait of compulsive honesty, and by Angela's attempts to find an intelligent and open-minded lawyer to assist in her experiment. In order to do so, she attends various criminal trials, which means that the novel begins by presenting actual courtroom practice before showing Lonsdale's mock trial, which substantially departs from customary legal practice.

While the novel is concerned with the mechanics of the legal apparatus, the play is more interested in the techniques of theatrical performance. The criminal who arranges the kidnapping, for instance, somewhat unoriginally addressed as "the boss" in the novel, is called "Mr I" in the stage version, in order to draw attention to his function as a surrogate author. Fittingly, he speaks of his crimes as theatrical productions and comments upon the proper way of employing graphic violence: "The trick is to keep violence both concealed and unexploited, but always a very real possibility."[41] Like Mr I, Sir George himself is interested in the mechanics of the crime play; he has written a book entitled *The Criminal Act in Art and Reality*, which contains a chapter on "The Art of Tragedy." This plot motif occurs in the stage version only. The book is used in Act II in order to test the veracity of a witness, Mrs Meadowes, who has aroused suspicion because she recited her late husband's evidence at the original trial by heart, word for word. She claims to have a remarkable memory but turns out to be unable to repeat a passage from Sir George's book that Walsh reads to her. Eventually, Mrs Meadowes admits that her husband learnt his testimony by heart and adds that he always used to rehearse his evidence before appearing in court: "so he wouldn't stand there like a dummy the way so many people do."[42] The difficulty of producing spontaneous "natural" dialogue unrehearsed is demonstrated in Act III. The police are at the front door, about to liberate the hostages, but by now a major reversal has taken place, and Sir George is inclined to believe in Walsh's innocence. Accordingly, the judge tries to get rid of the police at the front door while the other hostages are asked to talk naturally, as if at an ordinary luncheon party. Of course, they cannot think of anything to say, a phenomenon familiar from theatrical production, where actors are said to resort to mumbling "rhubarb" when asked to provide background noises. This notable concern with theatre and drama is not a characteristic of *Settled Out of Court* alone but may be observed in English courtroom drama in general.[43]

Apart from shifting the focus from legal practice to (meta)drama, the confinement to a single set also brings about a reduction of plot complexity. Two parallel subplots from the novel are removed, and the number of actors is reduced by having Walsh plead his own case at the mock trial, a solution that is also theatrically more effective.[44] This is yet another instance of the blurring of actantial roles typical of the crime play; at the same time it demonstrates the difficulty of allotting a specific actantial role to the character of the "innocent suspect" – is he "criminal" or is he "detective"?

As in *Settled Out of Court*, the action of Cecil's next crime play, *Alibi for a Judge* takes a legal practice and a legal problem as its starting point. The legal practice criticized is the English way of appointing judges from the Bench, which means that a judge looks back not on a legal training specifically designed for the position of judge but on a previous career as a barrister, which requires a completely different set of abilities and personal qualities. Mr Justice Carstairs' blatant unfairness in court thus is a result not merely of his personal shortcomings but also a result of the shortcomings of the legal system. The legal problem that forms the basis of the novel is the fact that once an obviously unfair verdict has been passed, nobody – not even the presiding judge himself – can have it revised. A judge like Carstairs, prone to worrying and second thoughts, is therefore in an unpleasant position. The play version, by contrast, concentrates on the question of whether Lesley

Burford, the wife of the convicted man, is trying to clear her husband or attempting to manoeuvre the judge into a compromising position. Hence, the focus of interest is moved from legal practice to amorous intrigue, a topic in which drama traditionally is very interested.

The action of the novel *According to the Evidence* revolves around the question of whether it may be legitimate to take the law into one's own hands. Alec Morland kills a homicidal maniac before he can strike again. When the police are on his track, Alec enlists the help of the reformed criminal Ambrose Low (a character appearing in a number of Cecil's novels), who decides to have Alec tried for murder and acquitted in order to relieve him from suspicion once and for all. In the stage version the relationship between Low and Morland becomes more complex, since Morland was once engaged to Low's wife Liz. Thus, the dramatization introduces an amorous triangle absent from the novel. The new plot motif of sexual jealousy arouses doubts as to Low's intentions: is he really trying to get Alec acquitted, or is he attempting to have him sentenced to life imprisonment? As in *Alibi for a Judge*, the central mystery revolves around the true intentions of one of the main characters. The shift of interest from the public realm of the courtroom to the private realm of personal relations is also indicated by the set: while about 50 per cent of the novel is set in a courtroom, the play abandons the courtroom setting altogether, and the entire action unfolds in Low's house.

The fact that in each of these adaptations Cecil moved away from the legal themes of his novels raises the question of why he chose to dramatize his books in the first place. Cecil's interest in drama is most plausibly explained by his fondness for writing dialogue – each of his novels contains long passages consisting entirely of dialogue, which frequently are not even taken over into the stage versions. The reluctance to employ courtroom scenes on stage is more difficult to account for. At first glance, courtroom practices like the questioning and cross examination of witnesses would seem to offer abundant opportunities for dramatic dialogue and hence have definite theatrical potential. It is remarkable, however, that even playwrights who choose to write in the format of courtroom drama usually include a number of scenes set outside the courtroom. Legal procedure seems too rigid and too little varied to offer thrilling entertainment if simply reproduced on stage. In particular, the strict rules of court procedure do not sufficiently permit the blurring of actantial roles that is such an important device in the crime play. Successful courtroom plays always include a reversal of actantial roles, as is best demonstrated by Bayard Veiller's *The Trial of Mary Dugan* (1917). Not only does the accused – the character assigned the actantial role of culprit – turn out to be innocent, the real culprit is her defence attorney, who in the course of the trial is replaced by another lawyer who eventually turns out to be the heroine's long-lost brother. The example of *The Trial of Mary Dugan* may also serve as a timely reminder that the various formulas of the crime play had vastly different degrees of success in Britain and the United States – while courtroom drama in Britain never acquired a particular popularity, in the United States, the subgenre was even seminal for the genesis of the genre of the crime play in the first place. Therefore, even though the results of this chapter and the previous one suggest that the differences between the conventions of detective fiction and those of the crime play may largely be accounted for by the differences between narrative and drama, cultural factors must by no means be ignored.

Coda: The Twentieth-Century English Crime Play – A View from 2015

This study has sketched the development of the English crime play between 1900 and 2000; the present chapter, however, will consider to what extent the general observations made above may be validated by the present state of English crime drama. I will also discuss whether the crime play is an exclusively twentieth-century phenomenon or whether it is still alive today. I do not propose to discuss the last 15 years in detail, however, but will present a snapshot view from 2015. At the time of writing, four crime plays are running in the London West End, on which my study has focused: *The Mousetrap* at the St. Martin's, *The Play That Goes Wrong* at the Duchess, *The 39 Steps* at the Criterion and *The Nether* at the Duke of York's. *The Mousetrap* represents the twentieth-century crime play in a fossilized state, having turned into a tourist attraction, almost a London landmark. In this case, the characteristic ephemeral quality of live theatre has become extinguished and *The Mousetrap* arguably has proved John Russell Taylor's dictum quoted in Chapter 8 wrong according to which an architectural metaphor for drama is misleading since theatre is a temporal art form. *The Play That Goes Wrong* by Mischief Theatre Company started off on the fringe and consecutively won best comedy in the whatsonstage awards. It is a play based on the principle of Michael Frayn's *Noises Off* and the mechanicals' play in Shakespeare's *A Midsummer Night's Dream*: an inept amateur dramatic company performs a 1920s' murder mystery. Hence the resulting play satirizes both amateur dramatic endeavour and a recognizable dramatic formula. In order for the parody to work, the play that is being deconstructed by the well-meaning cast needs to belong to a recognizable genre with clearly defined conventions. Thus, *The Play That Goes Wrong* answers the question posed in Chapter 11 of the present study, affirming that the crime play indeed *is* a genre with clear-cut rules. This play, as well as *The 39 Steps*, also exploits the nostalgic appeal of the crime play whose supposed heyday were the 1920s and 1930s. *The 39 Steps* by Patrick Barlow also started on the fringe but has by now been running for several years in the West End. It is a stage version of Alfred Hitchcock's celebrated 1935 film thriller (not of Buchan's novel), on which I have written more extensively elsewhere.[1] For the purposes of the present chapter, suffice it to say that *The 39 Steps* was one of the earliest representatives of a new trend in theatre, that of transferring successful films onto the stage. This is in tune with the crime play's general aptitude for innovation and experiment already noted in the previous chapters. *The 39 Steps* expressly negotiates the relationship between the different media, which is another feature of the crime play in general. As the historical overview in Chapters 2 through 6 of the present study has demonstrated, the crime play as a genre experienced a special blossoming in 1929 and 1955. Previously, I have argued that these peaks in productivity always take place 10 years after the end of an armed conflict, which suggests that the crime play assists audiences in coping with nationwide experiences of violence and aggression. However, these especially productive phases also coincide with substantial shifts in the media landscape of Britain: 1929 with the advent of sound film, 1955 with the approaching saturation of the British Isles with TV sets. With plays like Ben Elton's *Popcorn* or Alan Ayckbourn's *Man of the Moment*, the crime play has indulged in a critique of the mass media, which are considered more corrupt than the traditional medium of theatre. The crime play's eagerness to engage with new media is also in evidence in Jennifer Haley's *The Nether*, a US import that transferred from Los Angeles to the Royal Court. This play uses the format of police interrogation to discuss the moral implications of virtual reality. This issue may be related to the crime play's fascination with mediated communication in general that was already in evidence in the preference for the telephone as a crime play prop discussed in Chapter 7.

An exclusive focus on the London West End, however – even though it was assumed in the present study for practical reasons – fails to present the full picture of the crime play in Britain. As suggested by *The Play That Goes Wrong*, crime plays have become a mainstay of amateur drama. The single living-room set, whose ubiquity was pointed out in Chapter 7, may be considered one central aspect of the plays' "well-madeness," since it facilitates the creation of verisimilitude even in amateur productions or performances by touring repertory companies. The website amdram lists over 30 productions of crime plays recently or currently performed or prepared by British amateur dramatic societies, with (perhaps unsurprisingly) Agatha Christie as the most popular individual author and Patrick Hamilton's *Gas Light* the most popular individual play. As pointed out in the present study, the crime play has always been remarkably sensitive to cultural shifts, including shifts in gender relations. The rising divorce rates of the period after World War II, for instance, were mirrored in the increasing popularity of plays revolving around murders of spouses. The current popularity of *Gas Light* in turn may be related to the current post-feminist fascination with the theme of sadomasochism in relationships between the sexes, a trend also in evidence in the phenomenal success of *50 Shades of Grey*. Other plays popular with amateur theatre companies include Barlow's *The 39 Steps* and Leslie Sands' *Something to Hide*, a 1950s' battle-of-wits murder-of-spouse play. Apart from Christie, Anthony Shaffer and Francis Durbridge are also popular authors on amdram, each represented by several plays from the 1970s conforming largely to Marvin Carlson's concept of the postmodern comedy thriller. Among crime plays performed by touring repertory companies, the stage adaptations of Peter James's bestselling novels *The Perfect Murder* (produced in 2014) and *Dead Simple* (produced in 2015), adapted by Shaun McKenna, ought to be mentioned. As observed above, from the 1960s onwards, the crime play increasingly became divorced from contemporary detective fiction, a fact also in evidence in the selection of crime plays adapted from novels discussed in Chapters 12 and 13. Peter James may serve as an interesting counterexample to this trend, and he appropriately mentions Agatha Christie as his model in stage adaptation. *The Perfect Murder* is an instance of a murder-of-spouse comedy thriller while *Dead Simple* revolves around the primal fear of being buried alive. As the actors' credits reveal, the cast of both plays mostly looked back on previous experience in West End crime plays and TV soap opera.

This reference to TV is a useful cue, since even more than through the offerings of amateur dramatic societies and touring repertory companies, audiences today consume crime plays widely in the form of TV drama. The crime play is the dominant genre in TV drama in Britain, making up about 50 per cent of all serial formats, coming in over 200 different shows (compared to a humble 40 in my native Germany). However, while the stage crime play tends to favour the criminal as its protagonist, the TV crime play prefers the detective as central character and consequently employs whodunit or other puzzle formulas, in this respect imitating detective fiction rather than the traditional crime play. This preference for the character of the detective, however, may be due to the fact that British crime fiction (as well as the crime play) has always favoured private crime over organized professional crime. US TV, by contrast, has recently produced serials revolving around the criminal as central character, for instance *The Sopranos*, *Dexter*, *Breaking Bad* and – most recently – *Orange Is the New Black*. *Dexter* may also serve as an interesting instance of the blurring of actantial roles (here of criminal and detective) that is a characteristic feature of the crime play, as already discussed in Chapters 9 and 11. Moreover, he is also a doctor, one of the crime play's favourite criminal characters (as also pointed out in Chapter 9). A good contrasting example of a British TV serial might be *Midsomer Murders* (running since 1997), a police procedural revolving around a "plodder" rather than a Great Detective, creating little or no suspense and certainly deserving the nickname of "cozy." *Midsomer Murders* deliberately exploits the nostalgic appeal of British crime fiction and projects an idyllic image of Englishness that has rendered it an attractive product for export. A similar approach may be discerned in the recently completed filming of the entire Poirot saga and constantly revisioned and rejuvenated versions of Miss Marple, as well as in the new Sherlock Holmes programme for which the material has been very rigorously updated indeed. Even though the TV crime play is largely modelled on crime fiction, it has also learnt from the stage crime play. The idea of the dismemberment of the dead body as spectacle memorably presented in the opening scene of Anthony Shaffer's *Murderer* arguably prompted an entire new genre, including the British TV drama *Silent Witness* and numerous other forensic TV serials created in its wake.[2] The idea of employing continuity, a temporal organization of the plot in which performance time equals theatre time (discussed in Chapter 8 of the present study), for the sake of creation of suspense, has been imitated by yet another US TV programme, *24*. The crime play's innovative use of non-chronological time also discussed above has been adopted mainly by post-2000 crime film, particularly *Memento* (dir. Christopher Nolan, 2000) and the recent *Before I Go to Sleep* (dir. Rowan Joffé, 2014). As the above observations prove, the crime play is not extinct, nor is it altogether a museum piece, its influence is still discernable in present-day entertainment.

Notes

1 The Body in the Library and the Body on Stage

1. The term "whodunit" for instance suggests that the main concern of a crime play must be the mystery concerning the identity of the criminal, while "detective drama" presupposes that the detective must be the central character.
2. See Michael Jefferson, *Criminal Law* (Harlow: Pearson Education, 2001), 10.
3. The selection of theatres that may be said to belong to the London West End largely follows J. P. Wearing's useful compendium *The London Stage – A Calendar of Plays and Players* (Metuchen, NJ: Scarecrow, 1976–1993), which lists the performances in all West End theatres between 1890 and 1959.
4. See John Russell Taylor, *The Rise and Fall of the Well-Made Play* (London: Methuen, 1967), 101.
5. Stanley Richards (ed.), *Best Mystery and Suspense Plays of the Modern Theatre* (New York: Dodd, 1971) and *10 Classic Mystery and Suspense Plays of the Modern Theatre* (New York: Dodd, 1973).
6. Charles LaBorde, *Form and Formula in Detective Drama* (PhD dissertation: Ohio State, 1976).
7. Generally, critics tend to consider 200 performances or more a "long run."
8. LaBorde, 4.
9. LaBorde's technique of classification and categorization will be examined in greater detail in Chapter 11.
10. In: Robin W. Winks (ed.), *Detective Fiction – A Collection of Critical Essays* (Woodstock: Foul Play Press, 1988), 25–34.
11. LaBorde, 51.
12. Ibid., 64, italics mine.
13. It must be added, however, that in the case of *Ten Little Niggers*, the solution in the shape of a voluntary confession by the culprit appears already in the novel version. A more detailed comparison between the novel and its stage adaptation is presented in Chapter 12.
14. LaBorde, 77.
15. Ibid., 94.
16. Ibid., 97.
17. Ibid., 98.
18. Marvin Carlson, *Deathtraps: The Postmodern Comedy Thriller* (Bloomington: Indiana UP, 1993), 2.
19. Ibid., 9.
20. Ibid., ibid.
21. Ibid., 13.
22. Ibid., 10.
23. Ibid., 16.
24. Ibid., 18.
25. Ibid., 3.
26. Ibid., 7.
27. Ibid., 8.

28. In his penultimate Chapter 8, "Pigs and Angels: The Postmodern Private Eye," Carlson inserts a slightly misplaced analysis of a few American musicals of the 1980s.
29. This strikingly corresponds to Michael Holquist's decision to exclude the American hardboiled thriller as an "impure exception" from his discussion of the development from detective story to postmodern novel. Cf. Carlson, 158.
30. Carlson, 22.
31. Cf. George Grella, "Murder and Manners: The Formal Detective Novel," in: Winks, 84–102.
32. Carlson, 43f.
33. Ibid., 49.
34. Ibid. The term "guarantor of meaning" was originally coined by Slavoj Žižek in *Looking Awry: An Introduction to Jacques Lacan through Popular Culture* (Cambridge: MIT Press, 1991), 58.
35. Carlson, 51.
36. In: *Who Done It? A Guide to Detective, Mystery and Suspense Fiction* (New York and London: Bowker, 1969)
37. Carlson, 67.
38. This plot motif actually already appears in a crime play of 1931, Michael Barringer's *Inquest.*
39. Carlson, 93.
40. Ibid., 135.
41. Ibid., 183.
42. Cambridge UP, 2000.
43. John Stokes, "Body Parts: The Success of the Thriller in the Inter-War Years," in: Clive Barker and Maggie B. Gale (eds), *British Theatre between the Wars, 1918–1939* (Cambridge: Cambridge UP, 2000), 38–62, 46.
44. In this respect, Stokes is obviously indebted to the distinction between "curiosity" and "suspense" established by Tzvetan Todorov in his "Typologie du roman policier."
45. Stokes, 59.
46. *Journal of American Drama and Theatre*, 13 (Winter 2001), 1–35.
47. Thomson, 3.
48. Ibid., 20. She does not explain what these universal subjects may be. Presumably, she means that in detective plays, crime is not conceived as a result of social conditions but of some kind of primal evil.
49. In: Winks, 15–24.
50. This confusion is pervasive and, in all likelihood, deliberate: "Auden, in 'The Guilty Vicarage' comments about detective plays that ..." (24); "W. H. Auden, in 'The Guilty Vicarage', affirms that in detective plays, the archetypal parlors and libraries are Edens infected by and then purged of corruption" (30f.) and "The structure does not expose a dialectic between innocence and guilt, as Auden avers is the essential mission of the detective play ..." (33).
51. In this respect, Thomson imitates Raymond Chandler, who claimed the superiority of American over English detective fiction also partly on grounds of the social criticism inherent in the former.
52. Thomson, 23.
53. Frank Rahill, *The World of Melodrama*, (University Park and London: Pennsylvania State UP, 1967), 291.
54. Ibid., 295.
55. Ibid., ibid.
56. Ibid., 296.

57. James L. Smith, *Melodrama* (The Critical Idiom 28, London: Methuen, 1973), 46.
58. *English Melodrama* (London: Jenkins, 1965).
59. Taylor, 101.
60. This list was based on the results of an opinion poll carried out by the National Theatre, with the single restriction that each author was included only once, represented by their most significant work.
61. Amnon Kabatchnik, *Blood on the Stage, 1900–1925, Blood on the Stage, 1925–1950; Blood on the Stage, 1950–1975; Blood on the Stage, 1975–2000* (Lanham, MD: Scarecrow, 2008–2012).
62. See Martin Priestman, *Detective Fiction and Literature: The Figure on the Carpet* (London: Macmillan, 1990).
63. Ulrich Suerbaum, *Krimi: Eine Analyse der Gattung* (Stuttgart: Reclam, 1984), 16–29.
64. Columbia, SC: Camden House, 1994.
65. Geraldine Pederson-Krag, "Detective Fiction and the Primal Scene," *Psychoanalytic Quarterly* 18 (1949), 207–214.
66. Jacques Lacan, "Seminar on 'The Purloined Letter'," *Yale French Studies* 48 (1973), 39–72; originally published in *Écrits* (Paris: Seuil, 1966).
67. George Orwell, "Raffles and Miss Blandish," in: *The Collected Essays* (London: Secker and Warburg, 1961), 249–263; first published in 1944.
68. William Aydelotte, "The Detective Story as a Historical Source," in: Larry N. Landrum, Pat Browne and Ray B. Browne (eds), *Dimensions of Detective Fiction* (Bowling Green: Bowling Green University Popular Press, 1976), 68–82, first published in *Yale Review* 39 (1949/50).
69. For instance Umberto Eco, "Narrative Structures in Fleming," in: *The Role of the Reader: Explorations in the Semiotics of Texts* (Bloomington and London: Indiana UP, 1979), 144–163; first published in *Il caso Bond*, 1965; Tzvetan Todorov, "The Typology of Detective Fiction," in: *The Poetics of Prose* (Oxford: Blackwell, 1977), 42–52; Peter Hühn, "The Detective as Reader: Narrativity and Reading Concepts in Detective Fiction," *Modern Fiction Studies* 33:3 (Autumn 1987), 451–466.
70. For instance Patricia Merivale, "The Flaunting of Artifice in Vladimir Nabokov and Jorge Luis Borges," in: L. S. Dembo (ed.), *Nabokov: The Man and His Work* (Madison and London: University of Wisconsin Press, 1967), 209–224; Stefano Tani, *The Doomed Detective: The Contribution of the Detective Novel to Postmodern American and Italian Fiction* (Carbondale and Edwardsville: Southern Illinois UP), 1984; Martin Priestman (1990).
71. For instance John P. Muller and William J. Richardson (eds), *The Purloined Poe: Lacan, Derrida and Psychoanalytic Reading* (Baltimore and London: Johns Hopkins UP, 1988) and Slavoj Žižek's *Looking Awry*.
72. See William W. Stowe, "From Semiotics to Hermeneutics: Modes of Detection in Doyle and Chandler," in: Glenn W. Most and William W. Stowe, *The Poetics of Murder: Detective Fiction and Literary Theory* (San Diego, New York and London: Harcourt Brace Jovanovich, 1983), 366–383; Umberto Eco and Thomas A. Sebeok (eds), *The Sign of Three: Dupin, Holmes, Peirce* (Bloomington and Indianapolis: Indiana UP, 1983).
73. Cf. George Grella's essays "Murder and Manners: The Formal Detective Novel" and "Murder and the Mean Streets: The Hard-Boiled Detective Novel," reprinted in Winks, 84–102 and 103–120.
74. E.g. Kathleen Gregory Klein, *The Woman Detective: Gender and Genre* (Urbana and Chicago: University of Illinois Press, 1988), and Maureen Reddy, *Sisters in Crime: Feminism and the Crime Novel* (New York: Frederick Ungar, 1988).

75. Peter Rabinowitz, "Chandler Comes to Harlem: Racial Politics in the Thriller of Chester Himes," in: Barbara A. Rader and Howard G. Zettler (eds), *The Sleuth and the Scholar: Origins, Evolutions, and Current Trends in Detective Fiction* (New York: Greenwood Press, 1988, 19–29).

2 From Victorian Melodrama to the End of the Twenties

1. Owen Davis, *I'd Like to Do It Again* (New York: Farrar and Rinehart, 1931), 85.
2. Denis Salter, "Henry Irving, the 'Dr Freud' of Melodrama," in: James Redmond (ed.), *Melodrama* (Themes in Drama 14, Cambridge UP, 1992), 161–182, 167.
3. Peter Brooks, *The Melodramatic Imagination* (New Haven: Yale UP, 1976), 27.
4. Michael R. Booth, *English Melodrama* (London: Jenkins, 1965), 159.
5. Rahill, 286.
6. Maurice Willson Disher, *Melodrama – Plots That Thrilled* (London: Rockliff, 1954), 53.
7. Arnold Ridley and Bernard Merivale, *The Wrecker,* New, 1927. Incidentally, *The Wrecker* was the first crime play to introduce a female professional detective on stage.
8. LaBorde, 43.
9. Smith, 75.
10. Taylor, 115.
11. Ibid., 116.
12. James Gindin, "John Galsworthy," in: Stanley Weintraub (ed.), *Modern British Dramatists, 1900–1945* (Dictionary of Literary Biography 10, Detroit: Gale, 1982), vol. 1, 194–206, 203. Subsequently abbreviated to *DLB* 10.
13. Anthony Stephenson, "Frederick Lonsdale," *DLB* 10:1, 294–300, 289.
14. Barbara J. Small, "Ben Travers," *DLB* 10:2, 184–188, 186, italics mine.
15. Later crime plays completely overshadowed by their film versions include Agatha Christie's *Witness for the Prosecution* and Frederick Knott's *Wait until Dark.*
16. J. C. Trewin, "The Great War and the Theater, 1914–1918," in *DLB* 10,2, 277–284, 281.
17. LaBorde, 124.
18. Ibid., 178.
19. Rahill, 295.
20. A. A. Milne, *The Fourth Wall – A Detective Story in Three Acts* (London: French, 1929), n.p.
21. What was later to be termed "metadrama" had already been a major preoccupation of Milne's earlier playlet *The Man in the Bowler Hat,* which is based on the principle of *Seven Keys to Baldpate.* The sensational action of the hero pursued by the villain who wants to obtain the Rajah's Ruby is finally revealed to be but the rehearsal of a Victorian-style melodrama, which also explains the enigmatic presence of the play's title character, who is sitting on stage throughout the play, entirely unmoved: he is the director. From the point of view of the history of the genre, *The Man in the Bowler Hat* is mainly interesting as a possible model for a much later short play, Tom Stoppard's *The Real Inspector Hound.*
22. Ordean A. Hagen, *Who Done It? A Guide to Detection, Mystery and Suspense Fiction* (New York and London: Bowker, 1969), 471.
23. Ibid.

3 From the 1930s to the End of World War II

1. Cf. Dorothy Sayery and Muriel St. Clare Byrne, *Busman's Honeymoon – A Detective Comedy in Three Acts* (London: Gollancz, 1937), 7–9, 7.

2. Ibid.
3. Ibid., 9.
4. Ingram d'Abbes and Fenn Sherie, *Murder in Motley – A Comedy-Thriller* (London: French, 1953), iii.
5. This plot element is a standard motif of the crime play, as LaBorde has pointed out, who dubbed the plot motif of the disappearing corpse the "busy body" (cf. LaBorde, 224, n. 11).
6. A parallel example in detective fiction is Agatha Christie's *Murder on the Orient Express* of 1934.
7. Richard Courtney, *Outline History of British Drama* (Littlefield: Adams, 1982), 226.
8. Auden, 15.
9. *The Plays of J. B. Priestley*, vol. II (London: Heinemann, 1949), ix.
10. *Laburnum Grove* was followed by a minor experiment of Priestley's in the field of the crime play: *Mystery at Greenfingers* was written for an amateur dramatic society competition organized by the *News Chronicle* in 1938. The most intriguing aspect of the play is that it is a murder house mystery without a murder: it eventually turns out that the murder was staged by the functional detective in order to frighten a criminal gang.
11. Simon Trussler, *The Cambridge Illustrated History of British Theatre* (Cambridge: Cambridge UP, 1994), 284.
12. Cf. blurb of Roland Pertwee, *Pink String and Sealing Wax – A Play in Three Acts* (London: English Theatre Guild, 1945).
13. Martin Vale is the pen name of Marguerite Veiller, the wife of Bayard Veiller, author of *The Thirteenth Chair.*
14. Andrew Cramp, "Patrick Hamilton," in *DLB* 10:2, 216ff., 217.
15. As in the case of *Rope,* Hamilton's other successful crime play, *Gas Light,* the play has been almost completely overlaid in the collective memory by George Cukor's famous film version. It may therefore be appropriate to point out a few major differences between the play and the film version. In Hamilton, the heroine is not related to the murder victim, and the murderer's marriage to her therefore is not motivated by his desire to inherit the house in order to continue his search. In Cukor, the villain's antagonist is a young man attracted (and, presumably, attractive) to the heroine; Hamilton's heroine is rescued by an elderly policeman and her sole perspective consists of returning to her family and relatives in Devon. The film starts with the murder (though the camera remains outside the house), while Hamilton, keeping intact the three unities, shows a typical domestic scene, which – but for the undercurrent of deliberate humiliation and sadism – might be straight from *A Doll's House,* and is only later revealed to be the result of downright criminal behaviour.
16. Cf. Annegret Maack "Dramatiker der Vorkriegszeit im englischen Theater nach 1945," in: Heinz Kosok (ed.), *Drama und Theater im England des 20. Jahrhunderts* (Düsseldorf: Bagel, 1980), 105–114, 105.
17. Courtney, 224.
18. Trussler, 301.
19. This clearly demonstrates the importance of the personage of Winston Churchill as a figurehead in the British war effort.

4 From the End of World War II to 1955

1. Agatha Christie, *A Murder is Announced* (London: Collins, 1979), 127.
2. The play's continuing popularity has been explained variously by clever marketing strategies, by its essential "Englishness," which renders it appropriate

as a tourist attraction, and by a kind of "avalanche effect" by which immense popularity engenders more of the same.

3. Arthur Nicholas Athanason, "The Mousetrap Phenomenon," *Armchair Detective* 12:2 (Spring 1979), 152–157, 154f.
4. Michael Hutton, *Power without Glory – A Play in Three Acts* (London: French, 1948), 5.
5. Gordon Glennon, *Gathering Storm – A Play in Three Acts Dramatised from Reyner Barton's Novel "Envy My Simplicity"* (London: Rylee, 1948), n. p. From a present-day perspective, the move from the drawing-room to the kitchen may not seem a very radical shift, but in the drama of the time, it was obviously something new, since the plays written in the wake of Osborne's *Look Back in Anger* and Wesker's *Chicken Soup with Barley* were later dubbed "kitchen sink drama."
6. Ludovic Kennedy, *Murder Story – A Play in Three Acts with an Epilogue on Legal Killing* (London: Gollancz, 1956), 149.
7. Ibid.
8. In contrast to Galsworthy's *Justice*, *Murder Story* did not bring about an immediate judicial reform – it took another 11 years until the death penalty for murder was abolished in the United Kingdom.
9. The recent discovery of the German concentration camps had evidently injected new life blood into the stock character of the "evil German."
10. Cf. Christopher Innes, *Modern British Drama, 1890–1990* (Cambridge: Cambridge UP, 1992), 95.
11. *An Inspector Calls* premiered in Moscow in 1945 and saw its first London production at the New Theatre in 1946.
12. J. B. Priestley, *An Inspector Calls – A Play in Three Acts* (London: Heinemann, 1947), 57.
13. Stanley Lourdeaux, "J. B. Priestley," *DLB* 10:2, 110–116, 114.
14. John Gassner, *The Theatre in Our Times* (New York: Crown, 1963), 409.
15. Simon Shepherd and Peter Womack, *English Drama – A Cultural History* (Oxford: Blackwell, 1996), 272.
16. Ibid.
17. Christopher Innes, *Modern British Drama 1890–1990* (Cambridge: Cambridge UP, 1992), 377.
18. Priestley himself had of course already used a similar device in *Dangerous Corner*.
19. J. B. Priestley, *Bright Shadow – A Play of Detection* (London: French, 1950), 11.
20. As Ulrich Suerbaum has convincingly argued in *Krimi*, the structuring of the text by a succession of guiding questions is *the* distinct characteristic of the classic detective novel (cf. 22).
21. Priestley, 19.
22. A subplot revolving around a black market racketeer is rigorously cut down from the novella version.
23. This assortment of suspects admirably fits Grella's specifications of the "English microcosm" represented by the cast of the traditional English whodunit.
24. Agatha Christie, *The Mousetrap*, in: *The Mousetrap and Selected Plays* (London: HarperCollins, 1994), 285–366, 304.
25. *Kinematograph Weekly*, 17 November 1960. Quoted from Agatha Christie, *Spider's Web – A Play* (London: French, 1956), n.p.
26. Arthur Marwick, *British Society Since 1945* (Harmondsworth: Penguin, 1982), 40.
27. Similarly, of the three plays in which couples conspire to commit a murder, two are dramatizations of novels.
28. The familiar surname is not a coincidence: Campbell Christie was the brother of Agatha's first husband Archibald.
29. Trussler, 280.

30. Lesley Storm, *Black Chiffon – A Play in Three Acts* (London: English Theatre Guild, 1950), 78.

5 From 1956 to the End of the Sixties

1. Cf. Trussler, 321.
2. Martin Esslin, *The Theatre of the Absurd* (Harmondsworth: Penguin, 1991), 238f.
3. Ibid., 433–434.
4. Tom Stoppard, *Jumpers* (London: Faber and Faber, 1972), 50.
5. Eric Bentley, *The Life of the Drama* (New York: Atheneum, 1975), 227.
6. M. D. Allen, "Joe Orton," in: Stanley Weintraub (ed.), *British Dramatists Since World War II* (Dictionary of Literary Biography 13, Detroit: Gale, 1982) vol. 2, 364–370, 366. Subsequently abbreviated to *DLB* 13.
7. Ibid., 367.
8. Ibid., 268.
9. Allen, 365.
10. Anthony Stephenson, "Simon Gray," in *DLB* 13:2, 199–208, 201.
11. His ambition to become a famous murderer is shared by the protagonist of Anthony Shaffer's later crime play *Murderer* (1975).
12. Stephenson, 202.
13. However, the audience never learns with complete certainty whether Baxter is guilty.
14. Barry England, *Conduct Unbecoming* (London: Heinemann, 1971), 100.
15. Roger Cornish, "Howard Brenton," *DLB* 13:1, 100–108, 102.
16. Peter Paul Schnierer, "Crime and Detective Story Structures in Modern British Drama," in: Stephen Kaplan and Will Wright (eds.), *The Image of Crime in Literature, the Media, and Society* (Pueblo, CO: Society for the Interdisciplinary Study of Social Imagery, 1991), 53–64, 58.
17. Howard Brenton, *Christie in Love, and Other Plays* (London: Methuen, 1970), 22f.
18. J. C. Trewin, *Plays of the Year 1959* (London: Elek, 1960), vol. 20, 14.
19. Story and play will be discussed in greater detail in Chapter 13.
20. John Chapman, *Simple Spymen* (London: English Theatre Guild, 1960), 20.
21. Ordean A. Hagen, *Who Done It? A Guide to Detection, Mystery and Suspense Fiction* (New York and London: Bowker, 1969), 468.
22. Original stage plays broadcast by TV in the period included *Murder Mistaken* (1956), *Laburnum Grove* (1957), *The Amazing Dr. Clitterhouse* (1957), *Plunder* (1957), *Ladies in Retirement* (1958), *Suspect* (1958), *An Inspector Calls* (1961), *The Winslow Boy* (1961), *Dial "M" for Murder* (1962) and *Something to Hide* (1962).
23. Jack Popplewell, *Busybody* (London: French, 1965), 28.
24. For this doubling of parts, Chapman/Cooney might have taken their cue from the Christies' *Grand National Night*, in which the same actress (Hermione Baddeley) had also played both the murder victim and her sister, who arrives to solve the murder mystery.
25. Roat disguises himself when he first goes to see Susy, even though he knows that she is blind.
26. Frederick Knott, *Wait Until Dark* (London: French, 1967), 69.

6 From *Sleuth* to In-Yer-Face Theatre

1. Carlson, 15.
2. Ibid., 18.

3. Ibid., 13.
4. Francis Gillen, "Anthony Shaffer," *DLB* 13:2, 445–450, 445.
5. Carlson, 12.
6. *Murder by the Book* is one of Carlson's prime examples; however, the play never made it to the London West End.
7. Carlson, 107.
8. Melvyn Barnes, "Francis Durbridge," in: John M. Reilly (ed.), *Twentieth-Century Crime and Mystery Writers* (London: St. James Press, 1985), 293–295, 294f.
9. Aleks Sierz, *In-Yer-Face Threatre – British Drama Today* (London: Faber and Faber, 2001).
10. *The Amateur Cracksman* by E. W. Hornung was a collection of short stories first published in 1899.
11. This idea may have been inspired by Stoppard's *Jumpers*.
12. Schnierer, 62.
13. Roger Cornish, "David Hare," *DLB* 13:1, 234–243, 239.
14. Sierz, 4–5.
15. See Ibid., 72.
16. Ibid., 39.
17. Ibid., 164ff.
18. Interview with Bennett, in: Simon Bennett, *Drummers* (London: Nick Hern, 1999), n.p.
19. Apparently, an in-yer-face play is incomplete without at least one anal-rape scene.
20. Sierz, 207.
21. Ibid.
22. Carlson, 17.
23. *Copenhagen*, like *Hapgood*, deals with the act of betraying secret scientific discoveries to an enemy agent. However, while *Hapgood* treats the theme in the manner of a spy thriller, *Copenhagen* stresses the aspect of political responsibility. Hence, Frayn's play has not been considered in depth in the present study.
24. Hence the title "Murmuring Judges," which means "speaking ill of the judiciary," still an offence in Scottish law.
25. Cf. the example of Galsworthy, who had the courage to use a guilty protagonist in *Justice* and yet succeeded in convincing the public that the judicial system was at fault.

7 Settings and Stage Sets

1. Heinz Kosok, "Traditionen und Konventionen des 19. Jahrhunderts im englischen Drama und Theater des 20. Jahrhunderts", in: Heinz Kosok (ed.), *Drama und Theater im England des 20. Jahrhunderts* (Düsseldorf: Bagel; Bern and München: Francke, 1980), 9–24.
2. James Cawelti, *Adventure, Mystery, and Romance: Formula Stories as Art and Popular Culture* (Chicago: University of Chicago Press, 1976).
3. Booth, 120.
4. Percy Fitzgerald, *Principles of Comedy and Dramatic Effect* (London: Tinsley, 1870), 12.
5. Carlson, 22.
6. LaBorde, 106.
7. Tom Stoppard, *The Real Inspector Hound* (London: Faber and Faber), 1968, 15.
8. Stoppard, 17.
9. Auden, 19.
10. Quoted from Edward Stirling, *Old Drury Lane* (London: Chatto and Windus, 1881), vol. 2, 239. London melodrama, a subgenre thriving in the 1880s,

conceived of the great city as a place of poverty, danger and moral decay akin to the American cityscapes of the twentieth-century hardboiled thriller.

11. London: French, 1976, n. p.
12. Frank Vosper, *Love from a Stranger* (London: French, 1937), 36.
13. Cf. Suerbaum, 91.
14. LaBorde, 141, n. 30.
15. Disher, 37.
16. Leslie Sands, *Something to Hide* (London: English Theatre Guild, 1959), 7.
17. Anthony Armstrong, *Mile-Away Murder* (London: Play Company, 1937), 5.
18. Armstrong, 62.
19. Cf. Barré Lyndon, *The Man in Half Moon Street* (London: Hamish Hamilton, 1939), 1.
20. Frederick Knott, *Write Me a Murder* (London: French, 1962), 1.
21. Lesley Storm, *The Day's Mischief* (London: French, 1952), 1.
22. John Hopkins, *This Story of Yours* (Harmondsworth: Penguin, 1969), 7.
23. Frank Vosper, *People Like Us* (London: French, 1951), 7.
24. Vosper, 28.
25. Arnold Ridley and Mary Cathcart Borer, *Tabitha* (London: French, 1956), 1.
26. Ludovic Kennedy, *Murder Story* (London: Gollancz, 1956), 15.
27. Max Catto, *They Walk Alone* (London: Secker and Warburg, 1939), 13.
28. Joan Morgan, *This Was a Woman* (London: Fortune, 1946), 5.
29. Morgan, 7.
30. Gerald Anstruther, *The Third Visitor* (London: English Theatre Guild, 1950), 7.
31. James Sexton, "Aldous Huxley's Three Plays, 1931–1948," in: C. C. Barefoot (ed.), *Aldous Huxley between East and West* (Amsterdam: Rodopi, 2001), 59–74, 73.
32. Lesley Storm, *Black Chiffon* (London: English Theatre Guild, 1950), 7.
33. Anthony Armstrong and Harold Simpson, *Without Witness* (London: Gollancz, 1934), 7.
34. Percy Robinson and Terence de Marney, *The Crime of Margaret Foley* (London: French, 1946), 8.
35. Anthony Armstrong, *Ten-Minute Alibi* (London: Gollancz, 1933), 5.
36. Charles Bennett, *Blackmail* (London: Rich and Cowan, 1934), 5.
37. Patrick Hamilton, *Rope* (London: Constable, 1929), 13.
38. Edward Percy and Reginald Denham, *Trunk Crime* (New York: Dramatists' Play Service, 1940), 13.
39. James Parish, *Distinguished Gathering* (London: French, 1936), 5.
40. Priestman, 48.
41. Keir Elam, *The Semiotics of Theatre and Drama* (London: Methuen, 1980), 22f.
42. Although spectators are asked to imagine that there is another voice speaking at the other end of the line, a telephone conversation in a stage play tends to be represented technically as a monologue: a speech performed by a single actor.
43. Eckart Voigts-Virchow, "Your Voice Is Around My Neck – The Media History and Poetics of Theatre Telephony," *Forum Modernes Theater* 19:2 (2004), 131–151, 135f. The final quotation is from John Brooks, *Telephone: The First Hundred Years* (New York: Harper and Row, 1976), 174. The quote from Marshall McLuhan is from *Understanding Media: The Extension of Man* (London: Routledge and Kegan Paul, 1964).
44. Ian Hutchby, *Conversation and Technology – From the Telephone to the Internet* (Oxford: Blackwell, 2001), 84f.
45. Voigts-Virchow, 132.
46. Ibid.
47. Tom Stoppard, *Jumpers* (London: Faber and Faber, 1972), 10.

48. Voigts-Virchow, 148, n. 35.
49. Ibid., 133.
50. In a further application of this technique, a recent cinematic remake of *Dial "M" for Murder* entitled *A Perfect Murder* fuses the parts of the lover ("Max") and the paid assassin ("Lesgate") into one.
51. This is the term employed by Knott himself.
52. The "MacGuffin" is a term employed by Alfred Hitchcock, referring to the secret information or object around which the suspense plot ostensibly revolves, but which merely serves to trigger the action. An example is the wine bottle filled with uranium in *Notorious*.
53. *Wait until Dark*, 7.
54. *Thrills and Regressions* was first published by Hogarth in 1959. This book will be consulted in greater depth in Chapter 11.
55. Cf. for instance Mladen Dolar, "Hitchcock's Objekte," in: Slavoj Žižek (ed.), *Ein Triumph des Blickes über das Auge: Psychoanalyse bei Alfred Hitchcock* (Wien: Turia & Kant, 1992), 25–42.
56. Thomson has noted that the object of theft in the "crook plays" of the early twentieth century frequently is a pearl necklace.
57. Discussion following Richard Alewyn's paper "Die Literarische Angst," in: Hoimar von Ditfurth (ed.), *Aspekte der Angst – Starnberger Gespräche 1964* (Stuttgart: Thieme, 1965), 43.
58. LaBorde, 108.
59. Ibid., 107.
60. Hagen, 468.
61. Frederick Knott, *Wait until Dark* (London: French, 1967), 67.

8 Timing and Plot "Construction"

1. Dennis Porter, *The Pursuit of Crime: Art and Ideology in Detective Fiction* (New Haven and London: Yale UP, 1981), 29.
2. Charles Derry, *The Suspense Thriller – Films in the Shadow of Alfred Hitchcock* (Jefferson, NC, and London: McFarland, 1988), 32.
3. François Truffaut with the collaboration of Helen G. Scott, *Hitchcock – Revised Edition* (London: Truffaut/Hitchcock, 1986), 89.
4. Ibid., 88.
5. Francis Durbridge, *Deadly Nightcap* (London: French, 1986), 35.
6. Ibid., 36.
7. Anthony Shaffer, *Murderer* (London: Boyars, 1979), 50.
8. Ibid., 53.
9. Porter, 25.
10. Ibid., 30.
11. Erdmut August, *Dramaturgie des Kriminalstücks – Untersuchung zur Intellektualisierung theatralischer Vorgänge am Beispiel einer neu ausgebildeten Gattung* (PhD dissertation, München, 1961) 65f. (my translation). Since August's dissertation is available only in German, it has not influenced criticism of detective fiction and the crime play to any noticeable extent.
12. In Winks, 25–34.
13. Ibid., 30.
14. Ibid., 27.
15. *Murder Mistaken* is an exception to this rule.

16. In the use of the terms "theatre time" and "dramatic time" I am following Terry Hodgson, *The Batsford Dictionary of Drama* (London: Batsford, 1988), 399. Hodgson further subdivides "dramatic time" into "story time" and "plot time," a distinction derived from Russian formalism, which will not be employed here, although the central difference between chronological representation and flashback will of course be considered.
17. Dorothy Sayers, "Aristotle on Detective Fiction," in: Winks, 25–34, 27.
18. J. B. Priestley, *Man and Time* (London: Star Books, 1978), 89.
19. Ibid., 99.
20. Holger Klein, *J. B. Priestley's Plays* (Basingstoke: Macmillan, 1988), 206f.
21. Aldous Huxley, *The Gioconda Smile* (London: Chatto and Windus, 1948), 113.
22. Carlson, 75.
23. Booth mentions the examples of C. Z. Barnett's *Victorine; or, The Orphan of Paris* of 1831 and James MacClosky's *The Fatal Glass; or, The Curse of Drink* (1872).
24. Gene D. Phillips, *Alfred Hitchcock* (Boston: Twayne, 1984), 130.
25. However, this convention seems to be about to be broken: spectators largely accepted the final twist of the film *The Usual Suspects* (1995), which reveals the entire succession of flashbacks as a clever invention by the culprit, made up on the spur of the moment during the interrogation.
26. Charles Osborne, *The Life and Crimes of Agatha Christie* (London: Collins, 1982), 34.
27. Preface to *The Letter* in Richards (1971), 267. Gladys Cooper, who produced *The Letter* and played Leslie recalls the matter differently – she remembers having had doubts about the dramatic value of a blackout in the third act, upon which Maugham kindly agreed to change the script. Maugham's account probably is the more reliable of the two, since he wrote it immediately after the production, for the 1927 publication of *The Letter*, whereas Cooper's is from her autobiography.
28. W. Somerset Maugham, "The Letter", in: Stanley Richards (ed.), *Best Mystery and Suspense Plays of the Modern Theatre* (New York: Dodd, 1971), pp. 269–343, 335.
29. Guy Barefoot, "*East Lynne* to *Gas Light* – Hollywood, Melodrama and Twentieth-century Notions of the Victorian," in: Jacky Bratton, Jim Cook, Christine Gledhill (eds.), *Melodrama – Stage, Picture, Screen* (London: British Film Institute, 1994), pp. 49–105; pp. 95–101.
30. Graham Greene, *The Return of A. J. Raffles* (London: Bodley Head, 1975), p. 9.

9 Dramatic Characters

1. Sayers, 25.
2. See Booth, 13–39.
3. Ibid., 24–25.
4. Grella, 94–95.
5. Ibid., 98.
6. Disher, 43.
7. Derry, 57.
8. LaBorde, 64.
9. Carlson, 49.
10. Sayers, 30.
11. Grella, 96.
12. Ibid.
13. Ibid.
14. Ibid., 96–97.

15. Ibid., 97.
16. Ibid.
17. Auden, 19.
18. Ibid.
19. Michael Gilbert, *The Shot in Question* (London: Constable, 1963), 9.
20. Anthony Armstrong, *Ten-Minute Alibi* (London: Gollancz, 1933), 10.
21. Agatha Christie, *Appointment with Death*, in: *The Mousetrap and Selected Plays*, 79–171, 83.
22. *The Return of A. J. Raffles*, 43.
23. It must be noted, however, that technically, *Distinguished Gathering* does not deal with a case of blackmail.
24. Robert Bloomfield, *Portrait of Murder* (London: French, 1964), 3–4.
25. Michael Gilbert, *The Bargain* (London: Constable, 1961), 12.
26. Cf. Truffaut/Scott, 145.
27. Georg Seeßlen, *Kino der Angst: Geschichte und Mythologie des Film-Thrillers* (Reinbek: Rowohlt, 1980), 101 (my translation).
28. Sayers, 33.
29. Arthur Watkyn, *Not in the Book* (London: French, 1958), 6.
30. Emlyn Williams, *Someone Waiting* (London: Heinemann, 1954), 12.
31. J. B. Priestley, *Laburnum Grove* (London: Heinemann, 1934), 11.
32. *A Clean Kill*, 6.
33. Emlyn Williams, *Night Must Fall*, in: *Famous Plays of 1935* (London: Gollancz, 1935), 9–128, 42.
34. N. J. Crisp, *Dangerous Obsession* (London: French, 1989), 1–2.
35. Howard Brenton, *Christie in Love, and Other Plays* (London: Methuen, 1970), n. p.
36. Grella, 98.
37. Francis Durbridge, *House Guest* (London: French, 1982), 9.
38. S. S. Van Dine, "Twenty Rules for Writing Detective Stories," in: Haycraft, 189–193, 191.
39. G. K. Chesterton, *The Innocence of Father Brown* (Harmondsworth: Penguin, 1950), 12.
40. S. E. Sweeney, "Locked Rooms: Detective Fiction, Narrative Theory, and Self-Reflexivity," in: Ronald G. Walker and June M. Frazer (eds.), *The Cunning Craft – Original Essays on Detective Fiction and Contemporary Literary Theory* (Macomb, IL: Western Illinois UP, 1990), 1–14, 8.
41. Carlson, 55.
42. Ibid., 115.
43. Frank Vosper, *Love from a Stranger* (London: Collins, 1936), 111.
44. Brenton, n. p.
45. Michael Clayton Hutton, *Power without Glory* (London: French, 1948), 52.
46. Barry England, *Conduct Unbecoming* (London: Heinemann, 1971), 113.
47. Edward Percy and Reginald Denham, *Trunk Crime* (New York: Dramatists' Play Service, 1940), 9.
48. Cf. Stowe in: Most/Stowe, 366–383.
49. *A Clean Kill*, 21.
50. Williams, 20.
51. *Rope*, 31.
52. LaBorde, 158ff.
53. It is perhaps interesting to ponder for an instant on the notable hostility towards the police in both traditional types of detective fiction, British and American.

In British fiction, the policeman is unacceptable because he is not a gentleman. In American hardboiled fiction, the policeman is an objectionable character, because he has let himself be corrupted by the governing classes. In both traditions, the policeman is thus implicitly accused of having betrayed his class – this observation fits in perfectly with crime fiction's general hostility towards social climbers.

54. Carlson therefore considers *Deadly Nightcap* a typical example of the postmodern comedy thriller; however, the policeman corrupted by sentiment already appears in Bennett's *Blackmail* of 1928.
55. Julian Symons, *Bloody Murder – From the Detective Story to the Crime Novel: A History*, revised edition (Harmondsworth: Penguin, 1974), 207.
56. Brenton, n. p.
57. *Sleuth*, 56.

10 Violence, Crime and Punishment

1. Dorothy Sayers, "The Omnibus of Crime," in: Winks, 53–83, 77.
2. David Hare, *Knuckle*, in: *Plays 1* (London: Faber and Faber, 1996), 221–317, 270.
3. In David Rudkin's *Afore Night Comes* of 1962, the censor for instance objected to the presentation of a severed head.
4. Mel Gordon, *The Grand Guignol – Theatre of Fear and Terror* (New York: Amok Press, 1988), 38f. It should be noted that Patrick Hamilton explicitly declared his aversion to the graphic violence of the Grand Guignol in his preface to *Rope*.
5. This definition is quoted above, in Chapter 6.
6. See Claude Schumacher (ed.), *Artaud on Theatre* (London: Methuen, 1989), 108.
7. Schumacher, 108.
8. Ibid., 109.
9. Evadne Price and Kenneth Attiwell, *Once a Crook* (London: French, 1943), 66.
10. Auden, 17.
11. Van Dine, 190.
12. Bentley, 305.
13. Raymond Chandler, "The Simple Art of Murder," in: Haycraft, 222–237, 236.
14. Van Dine, 192.
15. Alec Coppel, *The Gazebo* (London: English Theatre Guild, 1962), 46.
16. William Saroyan and Henry Cecil, *Settled Out of Court* (London: French, 1962), 10.
17. *Murderer*, n. p.
18. Ibid., 12–13.
19. P. D. James, *Original Sin* (London: Faber and Faber, 1994), 24.
20. LaBorde, 224, n. 11.
21. LaBorde, 226, n. 12.
22. Hence, the motif of the busy body is closely related to one of the crime play's favourite plot motifs: the return of a person presumed dead. This tradition goes back to the conventions of Victorian melodrama, which, as Booth has pointed out, is "full of corpses who revive at crucial moments" (Booth, 113.).
23. Percy Robinson, *To What Red Hell* (London: Putnam's, 1928), 9f.
24. LaBorde, 224, n. 11.
25. The entire cinematic genre of the "big caper movie," represented for instance by Jules Dassin's *Du rififi chez les hommes* (1955), is based on this mechanism.
26. Roy Vickers, "Crime on the Stage – The Criminological Illusion," in: Michael Gilbert (ed.), *Crime in Good Company: Essays on Criminals and Crime-writing* (London: Constable, 1959), 179–191, 190.

27. London: Heinemann, 1924.
28. Thomson, 17. Stokes, by contrast, believes that the crime of murder also illuminates the socio-political conditions of the era in which it occurs.
29. Jack Popplewell, *Dear Delinquent* (London: Evans, 1958), 79.
30. Cf. for instance Auden and Sayers in Winks, or the articles by Chesterton, Wrong, Nicholson, Thomson and Krutch collected in Haycraft.
31. *Journal of Popular Culture,* 12:4 (Spring 1979), 581–603.
32. Wingate, 581.
33. See Auden, 20: "Execution [...] is the act of atonement, by which the murderer is forgiven by society."
34. Wingate, 597.
35. Ibid., 600.
36. Van Dine, 192.
37. Cf. Cawelti, 96 and 107.
38. Although Wingate speaks of the "immanence" of justice, the "imminence" of justice (i.e. punishment) seems the more important aspect.
39. Stokes, 47.
40. See *Plays of the Year* 16, 8.
41. J. C. Trewin, *Plays of the Year 1956–1957* (London: Elek, 1958), vol. 16, 8.
42. Vickers, 189.
43. Derry, 72.

11 Genre, Generic Development and Subgenres

1. The term is used by E. D. Hirsch in *Validity in Interpretation* (New Haven and London: Yale UP, 1967), 114. "The members are grouped together as a class for conceptual convenience, but there is no single specific difference common to them all." Although he starts off with a quotation from Wittgenstein, Hirsch does not advocate the use of this concept of "genre."
2. Ross Murfin and Supryia M. Ray, *The Bedford Glossary of Critical and Literary Terms,* 2nd ed. (Boston and New York: Bedford/St. Martin's, 2003), 190.
3. This poster is in Peter Haining, *Agatha Christie: Murder in Four Acts* (London: Virgin, 1990), 25.
4. Thomson, 31.
5. Tom Stoppard, *Rosencrantz and Guildenstern Are Dead* (London: Faber and Faber, 1967), 52.
6. The term "formula" is somewhat unfortunate, since it is generally perceived as derogatory – however, it is the term employed by Cawelti and hence has a certain history in the genre study of popular writing.
7. In fairness, it should be added that LaBorde is not the only critic to base his classification system on varying features of the plays in question. Critics of Victorian melodrama for instance categorize their plays according to setting and mode (for instance "Military and Nautical" and "Sensation Drama" in Booth) or, additionally, according to theme and plot, author, manager or house style (for example "The Sins of Society" and "William Terris and the Adelphi" in Disher). This approach produces an abundance of subcategories, and it can most easily be explained by the authors' interest being mainly historical rather than systematic.
8. *Rope,* viii.
9. Paul Hernadi, "Entertaining Commitments: A Reception Theory of Literary Genres," *Poetics* 10 (1981), 195–211, 198–207.

10. The only crime playwright who ever claimed the label of "tragedy" for one of his works was Galsworthy in *Justice*.
11. Tzvetan Todorov, "The Origin of Genres," *New Literary History* 8 (1976/77), 159–170, 161.
12. Karl Beckson and Arthur Ganz, *Literary Terms: A Dictionary* (London: Deutsch, 1990), 300f.
13. Cf. G. C. Ramsey, *Agatha Christie: Mistress of Mystery* (New York: Dodd, Mead & Company, 1967), 6.
14. To tragedy, Bentley relates the "higher" emotions of awe and compassion. See Bentley, 281–285.
15. Bentley, 200.
16. Robert B. Heilman, *Tragedy and Melodrama – Versions of Experience* (Seattle and London: University of Washington Press, 1968), 85.
17. Taylor, 120.
18. Derry, 58.
19. Cawelti, 91.
20. The term "mode" is used as defined by Rodway/Lee: "Basic class-name defining the general nature *of* the constituent kinds in terms of their final effect" (Allan Rodway and Brian Lee, "Coming to Terms," *Essays in Criticism* 14:2 (April 1964), 109–125, 116, italics by Rodway/Lee). Examples given by Rodway/Lee include comedy, farce and thriller.
21. Thomson, 3.
22. Bentley, 200.
23. Aristotle, *Poetics*, transl. Leon Golden (Englewood Cliffs, NJ: Prentice Hall), 1968, 21f., italics mine.
24. Bentley, 345. It is noteworthy that Bentley employs the same terminology as the theorists of theatre of cruelty and in-yer-face theatre: the image of shaking the spectator in order to wake him up to a fuller experience of life. In the context of the present study, we will assume – as a behaviorist psychologist would – that people enjoy an experience if they repeatedly and of their own accord subject themselves to it, which certainly applies to the watching of thrillers.
25. Derry, 22.
26. Ibid.
27. Ibid.
28. The early Ellery Queen novels contain an explicit "Challenge to the Reader" (cf. Symons, 126), but all classic detective fiction implicitly challenges the reader to a contest of wits with its detective protagonist.
29. Patrick Hamilton, *Gas Light* (London: Constable, 1939), 83.
30. Bentley, 283.
31. Cf. Johann N. Schmidt, *Ästhetik des Melodramas* (Heidelberg: Winter, 1986), 122.
32. Bentley, 283, italics mine.
33. Salter, 167.
34. Vickers, 182.
35. Cawelti, 261.
36. Ibid., 263.
37. Ibid., 266.
38. *The Letter*, 340.
39. Bentley, 304, Bentley's italics.
40. Carlson, 156.
41. Quoted in the "Introductory Note" by Richards, 1973, ixf.

42. Lionel Trilling, *Sincerity and Authenticity* (Cambridge, MA: Harvard University Press, 1972), 14.
43. Pyrhönen identifies Sayers, "The Omnibus of Crime," as the source of this idea, but see also Porter, Woods and Stephen Knight, *Form and Ideology in Crime Fiction* (London: Macmillan, 1980).

12 Stage Adaptations of Agatha Christie's Detective Stories

1. Edward Groff, "Point of View in Modern Drama" *Modern Drama* 2:3 (December 1959), 268–282, 270f.
2. See for instance Robert Barnard on Christie: "The main characteristic of Agatha Christie's writing is that one does not notice it. And that, perhaps, is about the highest praise one could give to a writer of popular literature" (Robert Barnard, *A Talent to Deceive – An Appreciation of Agatha Christie* (New York: Dodd, 1987), 123) and S. S. Van Dine, writing under his real name of Willard Huntington Wright: "The style of a detective story must be direct, simple, smooth, and unencumbered" (Willard Huntington Wright, "The Great Detective Stories," in: Howard Haycraft (ed.), *The Art of the Mystery Story* (New York: Biblo and Tannen, 1976), 33–70, 39).
3. Peter Hühn, "The Detective as Reader: Narrativity and Reading Concepts in Detective Fiction," *Modern Fiction Studies* 33.3 (Autumn 1987), 451- 466, 452.
4. Barnard, 37.
5. Hühn, 459.
6. Ronald Knox, "A Detective Story Decalogue," in: Robin W. Winks (ed.), *Detective Fiction: A Collection of Critical Essays* (Woodstock: Foul Play Press, 1988), 200–202, 200.
7. Osborne, 37.
8. This device in all likelihood was borrowed from the comic P. G. Wodehouse short story "Uncle Fred Flits By" of 1936, in which Uncle Fred subjects his nephew Pongo to a similar treatment.
9. Arnold Ridley, *Peril at End House* (New York and London: French), 1945, 26.
10. Agatha Christie, *Peril at End House* (London: Collins), 1932, 21.
11. Ridley, 79.
12. See Osborne, 183.
13. A prolonged confessional speech in drama is always in danger of becoming unintentionally funny; for instance if the killer raises his knife and then breaks out into a lengthy narrative covering events of twenty years ago – see for instance the ending of the Miss-Marple film *Murder Most Foul* (1964), based on the Poirot novel *Mrs McGinty's Dead*.
14. Agatha Christie, *An Autobiography* (London: Collins, 1977), 471.
15. Agatha Christie, *And Then There Were None* (London: HarperCollins, 1993), 213.
16. Agatha Christie, *And Then There Were None*, in: *The Mousetrap and Selected Plays* (London: HarperCollins, 1994), 1–78, 78.
17. Ibid., 3.
18. Ibid., 44.
19. Osborne, 113.
20. See the *Daily Mail* interview of 1938.
21. See Osborne, 89–90.
22. See Ibid.
23. *The Mousetrap and Other Plays*, *The Mousetrap and Selected Plays*, *Witness for the Prosecution and Other Plays*, all published by HarperCollins.

24. Moie Charles and Barbara Toy, *Murder at the Vicarage* (London: French, 1950), 9.
25. Ibid., 17.
26. Osborne, 141.
27. Nicholas Birns and Margaret Boe Birns, "Agatha Christie: Modern and Modernist," in: Ronald G. Walker and June M. Frazer (eds.), *The Cunning Craft: Original Essays on Detective Fiction and Contemporary Literary Theory* (Macomb, IL: Western Illinois UP, 1990, 120–134, 127).
28. Agatha Christie, "The Hollow," in: *The Mousetrap and Selected Plays*, 173–283, 177.
29. The withholding of information, however, must not interfere with the "fair play rule." Ideally, a reader of detective fiction should be in possession of all the relevant data, but unable to reach a satisfactory explanation.
30. Truffaut 91.
31. Friedrich Dürrenmatt, "Problems of the Theatre," in: James L. Calderwood and Harold E. Toliver (eds.), *Perspectives on Drama* (New York: Oxford University Press, 1968, 25–48, 35f).
32. Agatha Christie and Gerald Verner, "Towards Zero," in: *The Mousetrap and Other Plays*, ed. Ira Levin, New York: Harper Paperbacks, 1978, 477–559, 559.
33. The use of several different narrators is nothing new in detective fiction – Wilkie Collins used it in *The Moonstone* – but the innovation in *Five Little Pigs* is that the same events are re-narrated five times, albeit from different perspectives.
34. See Osborne, 130.
35. Ibid., 155.
36. Ibid., 157.
37. Leslie Darbon, *A Murder is Announced* (London: French, 1987), 1.
38. See Ibid., 14.
39. Osborne, 94.
40. Ellery Queen, "The Detective Short Story: The First Hundred Years," in: Haycraft, 476–491, 477.
41. See for instance Martin Kayman, "The Short Story from Poe to Chesterton," in: Martin Priestman (ed.), *The Cambridge Companion to Crime Fiction* (Cambridge: Cambridge UP, 2003), 41–58, 41.
42. See Ian Reid, *The Short Story* (London: Methuen, 1977), 59.
43. Frank Myszor, *The Modern Short Story* (Cambridge: Cambridge UP, 2001), 60.
44. Osborne, 95.
45. Frank Vosper, *Love from a Stranger* (London: Collins, 1936) 112.
46. For this phenomenon, Seeßlen/Weil have coined the phrase: "Everything strange is erotic and (hence) dangerous." Georg Seeßlen and Claudius Weil *Kino des Phantastischen – Geschichte und Mythologie des Horror-Films* (Reinbek: Rowohlt, 1980), 78.
47. The 1982 TV version starring Ralph Richardson and Diana Rigg uses an even earlier start and shows the discovery of the body before the main title.
48. Osborne, 174.
49. Carlson, 122.
50. See for instance Neil Sinyard and Adrian Turner, *Journey Down Sunset Boulevard: The Films of Billy Wilder* (Ryde: BCW Publishing, 1979), 100.
51. *Witness for the Prosecution*, 458.

13 Other Types of Detective Fiction Adapted for the Stage

1. Truffaut, 279.
2. Arthur Conan Doyle, *The Speckled Band* (London: French, 1912), 54.

3. Ibid., 57.
4. See Hesketh Parsons, *Conan Doyle* (London: Unwin, 1987), 161.
5. Arthur Conan Doyle, "The Speckled Band," in: William S. Baring-Gould (ed.), *The Annotated Sherlock Holmes* (London: Murray, 1979), 243–262, 259.
6. Ibid., 261.
7. Both are still included in the "Properties Plot" of the French acting edition.
8. John Dickson Carr, *The Life of Sir Arthur Conan Doyle* (New York: Harper, 1949), 205.
9. A. E. W. Mason, *At the Villa Rose* (London: Hodder and Stoughton, 1950), 142.
10. Cf. Symons, 121.
11. Daniel P. King, "A.E.W. Mason," in: John M. Reilly (ed.), *Twentieth Century Crime and Mystery Writers* (London: St. James Press, 1985), 608–609, 609.
12. His authorship, however, remains largely unnoticeable. There is no evidence that he edits the story, nor does he write in a distinct narrative voice that conspicuously differs from the style of the rest of the text.
13. Mason, 142.
14. King, 609.
15. Symons, 100.
16. LaBorde, 59. The references are to Willard Huntington Wright, "The Great Detective Stories" in Haycraft, 33–70, 42, and to Erle Stanley Gardner, "The Case of the Early Beginning," in Haycraft, 203–207, 206.
17. Sapper, *Bulldog Drummond* (London: Dent, 1983), 52.
18. Ibid., 83.
19. Sapper and Gerald du Maurier, *Bulldog Drummond* (London: French, 1925), 67.
20. Jack Adrian, "Edgar Wallace," in: John M. Reilly (ed.), *Twentieth Century Crime and Mystery Writers* (London: St. James Press, 1985), 877.
21. Margaret Lane, *Edgar Wallace: The Biography of a Phenomenon* (London: Hamish Hamilton, 1964), 233.
22. Cf. Ibid., 233.
23. Ibid., 253–254.
24. Edgar Wallace, *The Squeaker* (Leipzig: Tauchnitz, 1929), 226.
25. Symons, 215.
26. Peter Cheyney, *The Urgent Hangman* (London: Collins, 1952), 111.
27. Ibid., 78.
28. Ibid., 217.
29. Hühn, 461.
30. Gerald Verner, *Meet Mr Callaghan* (London: French, 1953), n.p.
31. Jeffrey Dell, "Payment Deferred," in: Van H. Cartmell and Bennett Cerf (eds.), *Famous Plays of Crime and Detection: From Sherlock Holmes to Angel Street* (Freeport, NY: Books for Libraries Press, 1971), 649–724, 684.
32. C. S. Forester, *Payment Deferred* (London: Michael Joseph, 1952), 173.
33. Dell, 705.
34. Roger MacDougall and Ted Allan, *Double Image* (London: French, 1957), 19.
35. Ibid., 31–32.
36. Ibid., 67.
37. Ibid.
38. However, even the short story ends with the Inspector's closing remark that it was not the newspaper clipping that broke Julian's neck but his neglect of his wife's feelings.
39. MacDougall and Allan, 65.

40. Included in the French acting edition.
41. Henry Cecil and William Saroyan, *Settled Out of Court* (London: French, 1962), 10.
42. Ibid., 54.
43. See my article "Representations of the British Legal System in Twentieth-Century Popular Courtroom Drama," in: Monika Fludernik and Greta Olson (eds.), *In the Grip of the Law – Trials, Prisons and the Space Between* (Frankfurt: Peter Lang, 2004), 65–81.
44. Jack Roffey for instance also has the innocent accused plead his own case in *Hostile Witness* (1964).

Coda: The Twentieth-Century English Crime Play – A View from 2015

1. Cf. my following two articles: "Sister Arts? Alfred Hitchcock und das Londoner Theater der 1920er und 1930er Jahre," in: Henry Keazor (ed.), *Hitchcock und die Künste* (Marburg: Schüren, 2013), 67–82, and "From Screen to Stage: The Case of *The 39 Steps*," in: Monika Pietrzak-Fragner and Eckart Voigts-Virchow (eds.), *Adaptations – Performing Across Media and Genres* (Trier: WVT, 2009), 143–158.
2. I have explored this aspect more fully elsewhere, see: "Dead Bodies on Stage," *Journal of Contemporary Drama in English* 1:1 (2013), 43–53.

Bibliography: The Plays

Ackland, Rodney, *Before the Party – A Comedy of Expediency Based on a Short Story by W. Somerset Maugham* (London: French, 1950), St. Martin's, 26 October 1949.

Ackland, Rodney, *A Dead Secret – A Play* (London: French, 1958), Piccadilly, 30 May 1957.

Anstruther, Gerald, *The Third Visitor – A Play in Three Acts* (London: English Theatre Guild, 1950), Duke of York's, 13 June 1949.

Archer, Jeffrey, *Beyond Reasonable Doubt – A Play* (London: French, 1989), Queen's, 22 September 1987.

Armstrong, Anthony, *Ten-Minute Alibi – A Play in Three Acts* (London: Gollancz, 1933), Haymarket, 8 February 1933.

Armstrong, Anthony, and Harold Simpson, *Without Witness – A Suspense Play* (London: Gollanz, 1934), Duke of York's, 12 March 1934.

Armstrong, Anthony, *Mile-Away Murder – A Detective Play in Three Acts* (London: Play Company, 1937), Duchess, 2 April 1937.

Atkinson, Alex, *Four Winds – A Play in Three Acts* (London: French, 1954), Phoenix, 29 September 1953.

Ayckbourn, Alan, *Man of the Moment* (London: Faber and Faber, 1990), Globe, 14 February 1990.

Ayckbourn, Alan, *The Revengers' Comedies* (London: Faber and Faber, 1991), Strand, 16–17 October 1991.

Ayckbourn, Alan, *Communicating Doors* (London: Faber and Faber, 1995), Gielgud, 7 August 1995.

Barringer, Michael, *Inquest – A Play in Three Acts* (London: Deane, 1935), Windmill, 22 June 1931.

Beckwith, Reginald, *Boys in Brown – A Play in Three Acts* (London: Marshall, 1948), Arts, 4 July 1947.

Bennett, Alan, *Single Spies* (contains *An Englishman Abroad* and *A Question of Attribution*) (London: Faber and Faber, 1989), Lyttleton, 1 December 1988.

Bennett, Charles, *Blackmail – A Play in Three Acts* (London: Rich and Cowan, 1934), Globe, 28 February 1928.

Bennett, Charles, *The Last Hour – A Melodrama in Three Acts* (London: Rich and Cowan, 1934), Comedy, 20 December 1928.

Bennett, Simon, *Drummers* (London: Nick Hern, 1999), New Ambassadors, 1 September 1999.

Black, Ian Stuart, *We Must Kill Toni – A Comedy in Three Acts* (London: Evans, 1953), Westminster, 29 June 1954.

Bloomfield, Robert, *Portrait of Murder – A Drama* (London: French, 1964), Savoy, 24 October 1963.

Bradley, Lilian Trimble, *What Happened Then? – A Melodrama in Three Acts* (London: Gollancz, 1934), Fortune, 11 September 1933.

Brenton, Howard, *Christie in Love, and Other Plays* (London: Methuen, 1970), Royal Court Upstairs, 12 March 1970.

Brenton, Howard, *Revenge* (London: Methuen, 1970), Royal Court Upstairs, 2 September 1969.

Butterworth, Jez, *Mojo* (London: Nick Hern, 1995), Royal Court, 18 July 1995.

Campion, Cyril, *Ladies in Waiting – A Play for Women in Three Acts* (London: French, 1934), Westminster, 9 May 1934.

Cary, Falkland, and Philip Weathers, *Madam Tic-Tac – A Drama in Three Acts* (London: French, 1951), Winter Garden, 11 March 1950.

Catto, Max, *They Walk Alone – A Play in Three Acts* (London: Secker and Warburg, 1939), Shaftesbury, 19 January 1939.

Chapman, John, *Simple Spymen – A Farce in Three Acts* (London: English Theatre Guild, 1960), Whitehall, 19 March 1958.

Chapman, John, and Michael Pertwee, *Look, No Hans! – A Comedy* (London: French, 1986), Strand, 4 September 1985.

Charles, Moie, and Barbara Toy, *Murder at the Vicarage* (London: French, 1950), Playhouse, 14 December 1949.

Christie, Agatha, *Black Coffee – A Mystery Play in Three Acts* (New York: French, 1934), St. Martin's, 9 April 1931.

Christie, Agatha, *Ten Little Niggers – A Play in Three Acts* (London: French, 1944), St. James's, 17 November 1943.

Christie, Agatha, *Murder on the Nile – A Play in Three Acts* (London: French, 1948), Ambassadors, 19 March 1946.

Christie, Agatha, *The Hollow – A Play in Three Acts* (London: French, 1952), Fortune, 7 June 1951.

Christie, Agatha, *The Mousetrap – A Play in Two Acts* (London: French, 1954), Ambassadors, 25 November 1952.

Christie, Agatha, *Witness for the Prosecution – A Play in Three Acts* (New York: French, 1954), Winter Garden, 28 October 1953.

Christie, Agatha, *Appointment with Death – A Play in Three Acts* (London: French, 1956), Piccadilly, 31 March 1945.

Christie, Agatha, *Spider's Web – A Play* (London: French, 1956), Savoy, 14 December 1954.

Christie, Agatha, and Gerald Verner, *Towards Zero* (New York: Dramatists' Play Service, 1958, and London: French, 1957), St. James's, 4 September 1956.

Christie, Agatha, *The Unexpected Guest – A Play* (London: French, 1958), Duchess, 12 August 1958.

Christie, Agatha, *Verdict – A Play in Two Acts* (London: French, 1958), Strand, 22 May 1958.

Christie, Agatha, *Go Back for Murder – A Play* (London: French, 1960), Duchess, 23 March 1960.

Christie, Dorothy and Campbell, *Someone at the Door – A Comedy Thriller in Three Acts* (London: French, 1936), Aldwych, 3 March 1935.

Christie, Dorothy and Campbell, *Grand National Night – A Play in Three Acts* (London: French, 1947), Apollo, 12 June 1946.

Christie, Dorothy and Campbell, *Carrington, V. C. – A Play in Three Acts* (London: French, 1954, and Heinemann, 1954), Westminster, 28 July 1953.

Christie, Dorothy and Campbell, *The Touch of Fear* (London. French, 1957), Aldwych, 5 December 1956.

Clowes, St. John L., *Dear Murderer – A Play in Three Acts* (London: Evans, 1954), Aldwych, 1 August 1946.

Coke, Peter, *Breath of Spring – A Comedy in Three Acts* (London: French, 1959), Cambridge, 26 March 1958.

Cooney, Ray, and John (Roy) Chapman, *My Giddy Aunt – A Comedy-Thriller* (London: English Theatre Guild, 1970; rev. ed. London: Chappell, 1987), Savoy, 20 June 1968.

Coppel, Alec, *I Killed the Count! – A Play in Three Acts* (London: Heinemann, 1938, and French, 1938), Whitehall, 10 December 1937.
Crimp, Martin, *Getting Attention – Two Plays and a Fiction* (London: Nick Hern, 1991), Royal Court Upstairs, 16 May 1991.
Crisp, N. J., *Dangerous Obsession – A Play* (London: French, 1989), Apollo, 9 November 1987.
D'Abbes, Ingram, and Fenn Sherie, *Murder in Motley – A Comedy-Thriller in Three Acts* (London: French, 1953), Royalty, 15 April 1934.
Darbon, Leslie, *Agatha Christie's Cards on the Table – A Play* (London: French, 1982), Vaudeville, 9 December 1981.
Darbon, Leslie, *Agatha Christie's A Murder is Announced* (London: French, 1987), Vaudeville, 21 September 1977.
Dell, Jeffrey, *Payment Deferred – A Play in a Prologue, Three Acts, and an Epilogue* (London and New York: French, 1934), St. James's, 4 May 1931.
Denham, Reginald, and Mary Orr, *Minor Murder – A Crime Play in Two Acts* (London: French, 1967), Savoy, 25 July 1967.
Dinner, William, and William Morum, *The Late Edwina Black – A Play in Three Acts* (London: French, 1950), Ambassadors, 12 July 1949.
Donald, Simon, "The Life of Stuff", in: Ian Brown and Mark Fisher (eds.), *Made in Scotland* (London: Methuen, 1995), Donmar Warehouse, 23 September 1993.
Douglas, Felicity, and Henry Cecil, with Basil Dawson, *Alibi for a Judge – A Comedy from the Book by Henry Cecil* (London: French, 1967), Savoy, 5 August 1965.
Douglas, Felicity, and Henry Cecil, with Basil Dawson, *According to the Evidence – A Comedy Adapted from the Book of the Same Name by Henry Cecil* (London: French, 1968), Savoy, 14 September 1967.
Doyle, Arthur Conan, *The Speckled Band – An Adventure of Sherlock Holmes* (London: French, 1912), Adelphi, 4 June 1910.
Doyle, Monte, *Signpost to Murder – A Thriller in Three Acts* (London: French, 1963) Cambridge, 9 February 1962.
Du Maurier, Daphne, *Rebecca – A Play in Three Acts* (London: French, 1939), Queen's, 5 April 1940.
Durbridge, Francis, *Suddenly at Home – A Play* (London: French, 1973), Fortune, 30 September 1971.
Durbridge, Francis, *The Gentle Hook – A Play* (London: French, 1975), Piccadilly, 21 December 1974.
Durbridge, Francis, *House Guest – A Thriller* (London: French, 1982), Savoy, 29 April 1981.
Durbridge, Francis, *Deadly Nightcap – A Play* (London: French, 1986), Westminster, 19 July 1986.
Durbridge, Francis, *A Touch of Danger – A Play* (London: French, 1989), Whitehall, 12 September 1988.
Elice, Eric, and Roger Rees, *Double, Double – A Play* (London: French, 1986), Fortune, 12 June 1986.
Ellis, David, *Make Me a Widow – A Play in Three Acts* (London: French, 1965), Comedy, 16 July 1964.
Elton, Ben, *Popcorn – A Play* (London: French, 1998), Apollo, 2 April 1997.
England, Barry, *Conduct Unbecoming* (London: Heinemann, 1971, and French, 1971), Queen's, 10 July 1969.
Fairchild, William, *Breaking Point – A Play* (London: French, 1963), Lyric, 5 September 1962.
Fairchild, William, *The Sound of Murder – A Play in Three Acts* (London: French, 1960), Aldwych, 5 August 1959.

Feely, Terence John, *Murder in Mind – A Play* (London: French, 1982), Strand, 3 February 1982.

Frisby, Terence, *Rough Justice – A Play* (London: French, 1995), Apollo, 5 May 1994.

Galsworthy, John, *The Silver Box – A Comedy in Three Acts* (London: Duckworth, 1910), Royal Court, 25 September 1906.

Galsworthy, John, *Justice – A Tragedy* (London: Duckworth, 1910), Duke of York's, 21 February 1910.

Galsworthy, John, *Loyalties – A Drama in Three Acts* (London: Duckworth, 1922), St. Martin's, 8 March 1922.

Galsworthy, John, *Escape – An Episodic Play in a Prologue and Two Parts* (London: Duckworth, 1926), Ambassadors, 12 August 1926.

Garner, Julian, "The Awakening", in: Matthew Lloyd (ed.), *First Run 3 – New Plays by New Writers* (London: Nick Hern, 1991), Hampstead, 19 April 1990.

Gates, Tudor, *Who Saw Him Die? – A Play* (London, French, 1975), Haymarket, 8 May 1974.

Gates, Tudor, *Who Killed Agatha Christie? – A Play* (London, French, 1979), Ambassadors, 18 October 1978.

Gilbert, Michael, *A Clean Kill – A Murder Mystery in Three Acts* (London: Constable, 1960), Criterion, 15 December 1959.

Gilbert, Michael, *The Bargain – A Play in Three Acts* (London: Constable, 1961), St. Martin's, 19 January 1961.

Gilbert, Michael, *The Shot in Question* (London: Constable, 1963), Duchess, 7 May 1963.

Glennon, Gordon, *Gathering Storm – A Play in Three Acts Dramatised from Reyner Barton's Novel "Envy My Simplicity"* (London: Rylee, 1948), St. Martin's, 18 February 1948.

Grant, Neil, *Dusty Ermine – A Play in Three Acts* (London: French, 1937), Comedy, 6 March 1936.

Gray, Simon, *Wise Child* (London: Faber and Faber, 1968), Wyndham's, 10 October 1967.

Gray, Simon, *Dutch Uncle* (London: Faber and Faber, 1969), Aldwych, 26 March 1969.

Gray, Simon, *Molly – A Play* (London: French, 1979), Comedy, 25 October 1978.

Gray, Simon, *Stage Struck – A Play* (London: French, 1979), Vaudeville, 21 November 1979.

Gray, Simon, *Cell Mates* (London: Faber and Faber, 1995), Albery, 16 February 1995.

Green, Janet, *Murder Mistaken – A Thriller* (London: French, 1953), Ambassadors, 4 November 1952.

Greene, Graham, *The Return of A. J. Raffles – An Edwardian Comedy in Three Acts Based Somewhat Loosely on E. W. Hornung's Characters in The Amateur Cracksman* (London: Bodley Head, 1975), Aldwych, 4 December 1975.

Hackett, Walter, *The Barton Mystery – A Play in Four Acts* (London and New York: French, 1930), Savoy, 22 March 1916.

Hamilton, Patrick, *Rope – A Play* (London: Constable, 1929), Strand, 3 March 1929.

Hamilton, Patrick, *Gas Light – A Victorian Thriller in Three Acts* (London: Constable, 1939), Apollo, 31 January 1939.

Hare, David, *Knuckle* (London: Faber and Faber, 1974), Comedy, 4 March 1974.

Hare, David, *Murmuring Judges* (London: Faber and Faber, 1991), Olivier, 10 October 1991.

Harris, Richard, *The Business of Murder – A Play* (London: French, 1981), Duchess, 2 April 1981.

Harris, Richard, *Dead Guilty – A Play* (London: French, 1996), Apollo, 17 July 1995.

Harvey, Frank, *Saloon Bar – A Play in Three Acts* (London: Deane, 1942), Wyndham's, 15 November 1939.

Hastings, Charlotte, *Bonaventure – A Play in Three Acts* (London: French, 1950), Vaudeville, 6 December 1949.

Hastings, Patrick, *The Blind Goddess – A Play in Three Acts* (London: French, 1948), Apollo, 2 December 1947.

Hodges, Horace, and T. Wigney Percyval, *Grumpy – A Play in Four Acts* (London: French, 1921), New, 13 May 1914.

Home, William Douglas, *"Now Barrabas..." – A Play* (London: Longmans, 1947), Vaudeville, 7 March 1947.

Hopkins, John, *This Story of Yours* (Harmondsworth: Penguin, 1969), Royal Court, 11 December 1968.

Horne, Kenneth, *A Lady Mislaid – A Comedy in Two Acts* (London: English Theatre Guild, 1950), St. Martin's, 25 January 1950.

Horne, Kenneth, "Trial and Error", in: J. C. Trewin (ed.), *Plays of the Year, 1953* (vol. 9, London and New York: Elek, 1954), 15–154, Vaudeville, 17 September 1953.

Horowitz, Anthony, *Mindgame* (London: Oberon, 2000), Vaudeville, 5 June 2000.

Hutton, Michael Clayton, *Power without Glory – A Play in Three Acts* (London: French, 1950), Fortune, 5 April 1947.

Huxley, Aldous, *The Gioconda Smile – A Play* (London: Chatto and Windus, 1948), New, 3 June 1948.

Isitt, Debbie, *The Woman Who Cooked Her Husband* (London: Warner Chappell, 1993), Royal Court, 25 November 1991.

Judd, Rex (i.e. Edward Percy Smith and Reginald Denham), *Suspect – A Play in Three Acts* (London: French, 1937), St. Martin's, 16 February 1937.

Keeffe, Barrie, *Sus* (London: Methuen, 1979), Soho Poly, 18 June 1979.

Kennedy, Ludovic, *Murder Story – A Play in Three Acts With an Epilogue on Legal Killing* (London: Gollancz, 1956 and Evans, 1955), Cambridge, 22 July 1954.

Kerr, Geoffrey, *Cottage to Let – A Play in Three Acts* (London: French, 1941), Wyndham's, 31 July 1940.

King, Philip, *"How Are You, Johnnie?" – A Murder Play in Three Acts* (London: French, 1963), Vaudeville, 19 February 1963.

Knott, Frederick, *Dial "M" for Murder* (London: French, 1955 [rev. ed.] and New York: Dramatists' Play Service, 1954), Westminster, 19 June 1952.

Knott, Frederick, *Write Me a Murder – A Play in Three Acts* (London: French, 1962), Lyric, 28 March 1962.

Knott, Frederick, *Wait Until Dark – A Play* (London: French, 1967), Strand, 27 July 1966.

Launder, Frank, and Sidney Gilliat, *Meet a Body – An Improbable Adventure in Three Acts* (London: French, 1954), Duke of York's, 21 July 1954.

Leach, Karoline, *The Mysterious Mr Love – A Play* (London: French, 1997), Comedy, 7 August 1997.

Levene, Philip, *Kill Two Birds – A Play in Three Acts* (London: French, 1963), St. Martin's, 25 October 1962.

Levy, Benn W., *A Man with Red Hair* (London: Macmillan, 1928), Little, 27 February 1928.

Llewellyn, Richard, *Poison Pen – A Play in Three Acts* (London: French, 1938), Shaftesbury, 9 April 1938.

Lonsdale, Frederick, *The Last of Mrs Cheyney – A Comedy in Three Acts* (London: Collins, 1925), St. James's, 22 September 1925.

Lucie, Doug, "Gaucho", in: Doug Lucie (ed.), *Plays I* (London: Methuen, 1998), Hampstead, 24 October 1994.

Lyndon, Barré, *They Came by Night – A Play in Five Scenes* (London: Hamish Hamilton, 1937), Globe, 7 July 1937.

Lyndon, Barré, *The Amazing Dr. Clitterhouse – A Play in Three Acts* (London: Hamish Hamilton, 1936 and French, 1938), Haymarket, 6 August 1936.

Lyndon, Barré, *The Man in Half Moon Street* (London: Hamish Hamilton, 1939), New, 22 March 1939.

MacDougall, Roger, and Ted Allan, *Double Image – A Play in Three Acts* (London: French, 1957), Savoy, 14 November 1956.

Mackie, Philip, *The Whole Truth – A Play in Three Acts* (London: Evans, 1956), Aldwych, 11 October 1955.

Mackie, Philip, *The Key of the Door – A Play in Three Acts* (London: Evans, 1959), Lyric, Hammersmith, 27 May 1958.

Mackie, Philip, *The Big Killing – A Drama in Three Acts* (London: Evans, 1962), Princes, 1 February 1962.

McLoughlin, Maurice, *Brush with a Body – A Play* (London: French, 1963), St. Martin's, 19 July 1962.

McNally, John, *Paper Chase – A Comedy in Three Acts* (London: French, 1926), Little, 10 April 1926.

Maltby, H. F., *For the Love of Mike – A Farcical Comedy in Three Acts* (London: French, 1934), Saville, 8 October 1931.

Mason, A. E. W., *The Witness for the Defence – A Play in Four Acts* (London: French, 1913), St. James's, 1 February 1911.

Mason, A. E. W., *At the Villa Rose* (London: Hodder, 1928), Strand, 10 July 1920.

Maugham, W. Somerset, *The Letter – A Play in Three Acts* (London: Heinemann, 1927), Playhouse, 24 July 1927.

Maugham, W. Somerset, *The Sacred Flame – A Play in Three Acts* (London: Heinemann, 1929), Playhouse, 8 February 1929.

Millar, Ronald, *A Coat of Varnish – A Play of Murder Suggested by the Novel by C. P. Snow* (London: French, 1983), Haymarket, 1 April 1982.

Mills, Hugh, *The House by the Lake – A Play in Three Acts* (London: Evans, 1956 and French, 1956), Duke of York's, 9 May 1956.

Milne, A. A., *The Fourth Wall – A Detective Story in Three Acts* (London: French, 1929), Haymarket, 29 February 1928.

Monkhouse, Allan, *The Grand Cham's Diamond – A Play in One Act* (London and Glasgow: Gowan and Gray, 1924), Arts, 15 March 1933.

Moon, Gerald, *Corpse! – A Comedy Thriller* (London: French, 1985), Apollo, 24 July 1984.

Morgan, Elaine, *License to Murder – A Play in Two Acts* (London: French, 1963), Vaudeville, 26 March 1963.

Morgan, Joan, *This Was a Woman – A Play in Three Acts* (London: Fortune, 1946), Comedy, 15 March 1945.

Morton, Michael, *Alibi – A Play in Three Acts from a Story by Agatha Christie* (London: French, 1929), Prince of Wales's, 15 May 1928.

Nagy, Phyllis, *Butterfly Kiss* (London: Nick Hern, 1994), Almeida, 13 April 1994.

Neilson, Anthony, "Normal", in: Anthony Neilson (ed.), *Plays I* (London: Methuen, 1998), Finborough, 1 October 1991.

Norton-Taylor, Richard, *The Colour of Justice – Based on the Transcripts of the Stephen Lawrence Inquiry* (London: Oberon, 1999), Victoria Palace, 3 March 1999.

O'Donnell, Peter, *Mr Fothergill's Murder – A Play in Two Acts* (London: English Theatre Guild, 1983), Duke of York's, 25 October 1982.

Orton, Joe, *Entertaining Mister Sloane – A Comedy* (London: Hamish Hamilton, 1964), New Arts, 6 May 1964.

Orton, Joe, *Crimes of Passion* (contains *The Ruffian on the Stair* and *The Erpingham Camp*) (London: Methuen, 1967), Royal Court, 6 June 1967.

Orton, Joe, *Loot* (London: Methuen, 1967), Jeanette Cochrane, 29 October 1966.

Parish, James, *Distinguished Gathering – A Play in Three Acts* (London: French, 1936), St. Martin's, 26 November 1935.

Parish, James, *Mrs Inspector Jones – A Comedy in Three Acts* (London: French, 1951), Savoy, 2 November 1950.

Paul, Jeremy, *The Secret of Sherlock Holmes – A Stage Play* (Romford: Ian Henry, 1989), Wyndham's, 22 September 1988.

Peach, L. Du Garde, and Ian Hay, *The White Sheep of the Family – A Felonious Comedy* (London: French, 1953), Piccadilly, 11 October 1951.

Percy, Edward, and Reginald Denham, *Trunk Crime – A Play in an Epilogue and Three Acts* (New York: Dramatists Play Service, 1940), Comedy, 29 September 1937 (as *The Last Straw*).

Percy, Edward, and Reginald Denham, *Ladies in Retirement – A Play in Three Acts* (London and New York: English Theatre Guild, 1940), St. James's, 12 December 1939.

Percy, Edward, *Play with Fire – A Play in Three Acts* (also published as *The Shop at Sly Corner*) (London: English Theatre Guild, 1942), St. Martin's, 11 April 1945.

Percy, Edward, and Lilian Denham, *The Man with Expensive Tastes* (London: English Theatre Guild, 1955), Vaudeville, 23 July 1953.

Pertwee, Roland, *Interference – A Play in Three Acts* (London: French, 1929), St. James's, 29 January 1927.

Pertwee, Roland, *Honours Easy – A Play in Three Acts* (London: French, 1932), St. Martin's, 7 February 1930.

Pertwee, Roland, and Harold Dearden, *To Kill a Cat – A Play in Three Acts* (London: English Theatre Guild, 1939), Aldwych, 7 June 1939.

Pertwee, Roland, *Pink String and Sealing Wax – A Play in Three Acts* (London: English Theatre Guild, 1945), Duke of York's, 1 September 1943.

Pinter, Harold, *The Dumb Waiter* (London: Methuen, 1960), Royal Court, 8 March 1960.

Popplewell, Jack, *Dead on Nine – A Play in Three Acts* (London: French, 1956), Westminster, 24 August 1955.

Popplewell, Jack, *Dear Delinquent – A Play in Three Acts* (London: Evans, 1958), Westminster, 5 June 1957.

Popplewell, Jack, *Policy for Murder – A Play in Three Acts* (London: French, 1963), Duke of York's, 7 November 1962.

Popplewell, Jack, *Busybody – A Comedy* (London: French, 1965), Duke of York's, 9 December 1964.

Popplewell, Jack, *Dead Easy – A Play* (London: French, 1975), St. Martin's, 24 January 1974.

Price, Evadne, and Kenneth Attiwell, *Once a Crook – A Play in a Prologue and Three Acts* (London: French, 1943), Aldwych, 22 August 1940.

Price, Evadne, *The Phantom Light – A Comedy Thriller in Three Acts* (London: French, 1949), Haymarket, 7 September 1937.

Priestley, J. B., *Dangerous Corner – A Play in Three Acts* (London: Heinemann, 1932), Lyric, 17 May 1932.

Priestley, J. B., *Laburnum Grove – An Immoral Comedy in Three Acts* (London: Heinemann, 1934), Duchess, 28 November 1933.
Priestley, J. B., *An Inspector Calls – A Play in Three Acts* (London: Heinemann, 1947), New, 1 October 1946.
Rattigan, Terence, *The Winslow Boy – A Play* (London: Hamish Hamilton, 1946), Lyric, 23 May 1946.
Rattigan, Terence, *Cause Célèbre – A Play in Two Acts* (London: Hamish Hamilton, 1978), Her Majesty's, 4 July 1977.
Ridley, Arnold, *The Ghost Train – A Drama in Three Acts* (New York and London: French, 1930), St. Martin's, 23 November 1925.
Ridley, Arnold, and Bernard Merivale, *The Wrecker – A Play in Three Acts* (New York and London: French, 1930), New, 6 December 1927.
Ridley, Arnold, *Recipe for Murder – A Play in Three Acts* (London: Deane, 1936), Duke of York's, 21 December 1932.
Ridley, Arnold, *Peril at End House – A Play in Three Acts Adapted from the Novel by Agatha Christie* (London: French, 1945), Vaudeville, 1 May 1940.
Ridley, Arnold, and Mary Cathcart Borer, *Tabitha – A Play in Three Acts* (London: French, 1956), Duchess, 8 March 1956.
Robinson, Percy, *To What Red Hell – A Drama of Two Families* (London: Putnam's, 1928), Wyndham's, 30 October 1928.
Robinson, Percy, and Terence de Marney, *The Crime of Margaret Foley – A Drama in a Prologue and Three Acts* (London: French, 1946), Comedy, 11 July 1947.
Roffey, Jack, *Hostile Witness – A Play in Two Acts* (London: Evans, 1965), Haymarket, 4 November 1964.
Ross, Charles, *Dead Ringer – A Play Adapted from* The Prime Pretender *by Logan Gourlay* (London: French, 1984), Duke of York's, 17 May 1983.
Ross, George, and Campbell Singer, *Guilty Party – A Play in Three Acts* (London: French, 1962), St. Martin's, 17August 1961.
Ryton, Royce, *The Unvarnished Truth – A Comedy* (London: French, 1978), Phoenix, 13 April 1978.
Sands, Leslie, *Something to Hide – A Thriller in Three Acts* (London: English Theatre Guild, 1959), St. Martin's, 29 April 1958.
Sapper and Gerald Du Maurier, *Bulldog Drummond – A Play in Four Acts* (London: French, 1925), Wyndham's, 29 March 1921.
Saroyan, William, and Henry Cecil, *Settled Out of Court – A Play in Three Acts* (London: French, 1962), Strand, 19 October 1960.
Sayers, Dorothy L., and Muriel St. Clare Byrne, *Busman's Honeymoon – A Detective Comedy in Three Acts* (London: Gollancz, 1937), Comedy, 16 December 1936.
Shaffer, Anthony, *Sleuth* (London: Calder and Boyars, 1971), St. Martin's, 12 February 1970.
Shaffer, Anthony, *Whodunnit – A Comedy Thriller* (New York: French, 1983) Her Majesty's, 13 September 1979 (as *The Case of the Oily Levantine*).
Shaffer, Anthony, *Murderer – A Play in Two Acts* (London: Boyars, 1979) Garrick, 12 March 1975.
Shairp, Mordaunt, *The Crime at Blossoms – A Play in Three Acts* (London: Allen and Unwin, 1932 and 1933), Playhouse, 21 May 1931.
Sherriff, R. C., *Miss Mabel – A Play in Three Acts* (London: Gollancz, 1949), Duchess, 23 November 1948.
Sherriff, R. C., *Home at Seven – A Play in Three Acts* (London: French, 1950), Wyndham's, 7 March 1950.

Sherriff, R. C., *A Shred of Evidence – A Play in Three Acts* (London: French, 1961), Duchess, 27 April 1960.
Simpson, Reginald, and Frank Gregory, *Living Dangerously – A Play in Three Acts* (London: Cassell, 1934), Strand, 7 June 1934.
Spencer, David, *Killing the Cat – A Play* (London: Methuen, 1990), Royal Court Upstairs, 29 August 1990.
Stoppard, Tom, *The Real Inspector Hound* (London: Faber and Faber, 1968), Criterion, 17 June 1968.
Stoppard, Tom, *Jumpers* (London: Faber and Faber, 1972), Old Vic, 2 February 1972.
Stoppard, Tom, *Hapgood* (London: Faber and Faber, 1988), Aldwych, 8 March 1988.
Storm, Lesley, *Black Chiffon – A Play in Three Acts* (London: English Theatre Guild, 1950 and New York: French, 1951), Westminster, 3 May 1949.
Storm, Lesley, *The Day's Mischief – A Play in Two Acts* (London: French, 1952), Duke of York's, 11 December 1951.
Stuart, Aimee, and Philip, *Nine Till Six – A Play in Three Acts* (London: Benn, 1930 and London and New York: French, 1931), Arts, 22 January 1930.
Sylvaine, Vernon, *A Spot of Bother – A Farce in Three Acts* (London: French, 1938), Strand, 6 July 1937.
Sylvaine, Vernon, *Warn That Man! – A Play in Three Acts* (London: French, 1943), Garrick, 23 December 1941.
Taylor, Edward, *Murder by Misadventure – A Thriller* (London: French, 1992), Vaudeville, 13 July 1992.
Terry, J. E. Harold, and Lechmere Worrall, *The Man Who Stayed at Home – A Play in Three Acts* (London: French, 1916), Royalty, 10 December 1914.
Thompson, J. Lee, *Murder without Crime – A Play in Three Acts* (London: French, 1943), Comedy, 31 July 1942.
Titheradge, Dion, *The Crooked Billet – A Play in Three Acts* (London and New York: French, 1930), Royalty, 13 October 1927.
Travers, Ben, *Plunder – A Farce in Three Acts* (London: Bickers, 1931), Aldwych, 26 June 1928.
Vachell, Horace Annesley, *The Case of Lady Camber – A Play in Four Acts* (London: Murray, 1916), Savoy, 16 October 1915.
Vale, Martin, *The Two Mrs Carrolls – A Play in Three Acts* (London: Allen, 1936), St. Martin's, 12 June 1935.
Van Druten, John, *Somebody Knows – A Play* (London: Gollancz, 1932), St. Martin's, 12 May 1932.
Verner, Gerald, *Meet Mr Callaghan – A Play in Two Acts* (London: French, 1953), Garrick, 27 May 1952.
Vosper, Frank, *Murder on the Second Floor – A Play* (London and New York: Putnam's, 1929, London and New York: French, 1930), Lyric, 21 June 1929.
Vosper, Frank, *People Like Us – A Play* (London and New York: Putnam's, 1929; also London: French, 1951), Strand, 3 November 1929.
Vosper, Frank, *Love from a Stranger – A Play in Three Acts* (London: Collins, 1936), Wyndham's, 2 February 1936.
Wallace, Edgar, *The Flying Squad* (London: Hodder and Stoughton, 1929), Lyceum, 7 June 1928.
Wallace, Edgar, *The Ringer – A Play in Four Acts* (London: French, 1929 and London: Hodder and Stoughton, 1929), Wyndham's, 1 May 1926.
Wallace, Edgar, *The Squeaker* (London: Hodder and Stoughton, 1929), Apollo, 29 May 1928.

Wallace, Edgar, *The Terror – A Play* (London: Hodder and Stoughton, 1929), Lyceum, 11 May 1927.

Wallace, Edgar, *The Calendar – A Racing Play in Three Acts* (London: French, 1932), Wyndham's, 18 September 1929.

Wallace, Edgar, *The Case of the Frightened Lady – A Play in Three Acts* (London: French, 1932), Wyndham's, 18 August 1931.

Wallace, Edgar, *The Green Pack – A Play in Three Acts* (London: French, 1933), Wyndham's, 9 February 1932.

Watkyn, Arthur, *Not in the Book – A Comedy* (London: French, 1958), Criterion, 2 April 1958.

Watkyn, Arthur, *Out of Bounds – A Comedy in Two Acts* (London: French, 1963), Wyndham's, 8 November 1962.

Watkyn, Arthur, *Amber for Anna – A Play in Three Acts* (London: French, 1965), Vaudeville, 28 April 1964.

Welsh, Irvine, *You'll Have Had Your Hole – A Play* (London: Methuen, 1998), Astoria 2, 2 February 1999.

Wilkes, Allene Tupper, *The Creaking Chair – A Farcical Mystery Play* (London: French, 1926), Comedy, 22 July 1924.

Williams, Emlyn, *A Murder Has Been Arranged – A Ghost Story in Three Acts* (London: Collins, 1930), Strand, 9 November 1930.

Williams, Emlyn, "Night Must Fall – A Play in Three Acts", in: *Famous Plays of 1935* (London: Gollancz, 1935), Duchess, 31 May 1935.

Williams, Emlyn, *Someone Waiting – A Play in Three Acts* (London: Heinemann, 1954), Globe, 25 November 1953.

Wooll, Edward, *Libel! – A Play in Three Acts* (London: Heinemann, 1934), Playhouse, 2 April 1934.

Bibliography: Works Cited

Primary texts cited

Burke, Stewart, *Key to Murder – A Play* (London: French, 1978).

Carr, John Dickson, *The Hollow Man* (London: Hamish Hamilton, 1935).

Cecil, Henry, *According to the Evidence* (London: Chapman and Hall, and New York: Harper, 1954).

Cecil, Henry, *Alibi for a Judge* (London: Joseph, 1960).

Cecil, Henry, *Settled Out of Court* (London: Joseph, and New York: Harper, 1959).

Cheyney, Peter, *The Urgent Hangman* (London: Collins, 1952 [[1]1938]).

Christie, Agatha, *A Murder Is Announced* (London: Collins, 1979 [[1]London: Collins, 1950]).

Christie, Agatha, *And Then There Were None* (London: HarperCollins, 1993 [[1]London: Collins, 1939 as *Ten Little Niggers*]).

Christie, Agatha, *Appointment with Death* (London: Collins, 1938).

Christie, Agatha, *Cards on the Table* (London: Collins, 1936).

Christie, Agatha, *Death on the Nile* (London: Collins, 1937).

Christie, Agatha, *Five Little Pigs* (London: Collins, 1942).

Christie, Agatha, *Peril at End House* (London: Collins, 1932).

Christie, Agatha, "Philomel Cottage", in: Agatha Christie (ed.), *The Witness for the Prosecution and Other Stories* (New York: Dodd, Mead and Company, 1985), 217–248 ([1]in: *The Listerdale Mystery*, London: Collins, 1934).

Christie, Agatha, *The Hollow* (London: Collins, 1946).

Christie, Agatha, *The Murder at the Vicarage* (London: Collins, 1930).

Christie, Agatha, *The Murder of Roger Ackroyd* (London: HarperCollins, 1993 [[1]London: Collins, 1926]).

Christie, Agatha, "The Witness for the Prosecution", in: Agatha Christie (ed.), *The Witness for the Prosecution and Other Stories* (New York: Dodd, Mead and Company, 1985), 3–32 ([1]in: *The Hound of Death*, London: Collins, 1933).

Christie, Agatha, "Three Blind Mice", in: Agatha Christie (ed.), *Three Blind Mice and Other Stories*, (New York: Dell, 1977 [[1]Dodd, Mead and Company, 1950]), 7–82.

Christie, Agatha, *Towards Zero* (London: Collins, 1944).

Cohan, George M., *Seven Keys to Baldpate* (New York: French, 1914).

Coke, Peter, *Midsummer Mink – A Comedy* (London: French, 1965).

Coppel, Alec, *The Gazebo – From a Story by Myra and Alec Coppel* (London: English Theatre Guild, 1962 [[1]1957]).

Cox, Constance, *The Murder Game* (London: French, 1976).

Doyle, Arthur Conan, "The Speckled Band", in: William S. Baring-Gould (ed.), *The Annotated Sherlock Holmes* (London: Murray, 1979), 243–262 ([1]1890).

DuMaurier, Daphne, *Rebecca* (London: Gollancz, 1938).

Forester, C(ecil) S(cott), *Payment Deferred* (London: Michael Joseph, 1952 [[1]1926]).

Frayn, Michael, *Copenhagen* (London: Methuen, 1998).

Gillette, William, *Sherlock Holmes* (New York: French, 1922).

Green, Janet, *Matilda Shouted Fire – A Play in Two Acts* (London: Evans, 1961).

Greenwood, Duncan, and Robert King, *Murder by the Book – A Thriller* (London: French, 1982).

Hopwood, Avery, and Mary Roberts Rinehart, *The Bat* (New York: French, 1932).
Huxley, Aldous, "The Gioconda Smile", in: Aldous Huxley (ed.), *Mortal Coils and Other Stories* (London: Chatto and Windus, 1971), 1–62 ([1]1922).
James, P. D., *Original Sin* (London: Faber and Faber, 1994).
Kesselring, Joseph, *Arsenic and Old Lace* (London: English Theatre Guild, 1948).
Mason, A. E. W., *At the Villa Rose* (London: Hodder and Stoughton, 1950 [[1]1910]).
Maugham, W(illiam) Somerset, "The Letter", in: W. Somerset Maugham (ed.), *The Complete Short Works of W. Somerset Maugham*, (London: Heinemann, 1976), Vol. III, 1414–1445 ([1]1924).
Milne, A. A., *The Man in the Bowler Hat – A Terribly Exciting Affair* (London: French, 1923).
Priestley, J. B., *Bright Shadow – A Play of Detection* (London: French, 1950).
Priestley, J. B., *Mystery at Greenfingers* (London: French, 1937).
Sapper, *Bulldog Drummond* (London: Dent, 1983 [[1]London: Hodder, New York: Doran, 1920 as *Bull-Dog Drummond: The Adventures of a Demobilized Officer Who Found Peace Dull*]).
Stoppard, Tom, *Rosencrantz and Guildenstern are Dead* (London: Faber and Faber, 1967).
Taylor, Tom, *The Ticket-of-Leave Man* (London: Lacy, 1863).
Veiller, Bayard, *The Thirteenth Chair* (New York: French, 1922).
Veiller, Bayard, *The Trial of Mary Dugan* (New York: French, 1928).
Vickers, Roy, "Double Image", in: Roy Vickers (ed.), *Double Image and Other Stories* (London: Jenkins, 1955), 7–45.
Wallace, Edgar, *The Flying Squad* (London: Hodder and Stoughton, 1928).
Wallace, Edgar, *The Gaunt Stranger* (London: Hodder and Stoughton, 1925).
Wallace, Edgar, *The Ringer* (London: Hodder and Stoughton, 1927).
Wallace, Edgar, *The Squeaker* (Leipzig: Tauchnitz, 1929 [1London: Hodder and Stoughton, 1927]).
Wodehouse, P. G., "Uncle Fred Flits By", in: P. G. Wodehouse (ed.), *Young Men in Spats* (London etc.: Hutchinson, 1980 [[1]London: Jenkins, 1936]), 120–137.

Secondary sources on detective fiction, drama, and detective drama

Adrian, Jack, "Edgar Wallace", in: John M. Reilly (ed.), *Twentieth Century Crime and Mystery Writers*, 2nd ed. (London: St. James, 1985), 873–877.
Alewyn, Richard, "Die literarische Angst", in: Hoimar von Ditfurth (ed.), *Aspekte der Angst – Starnberger Gespräche 1964* (Stuttgart: Thieme, 1965), 24–43.
Allen, M. D., "Joe Orton", in: Stanley Weintraub (ed.), *British Dramatists since World War II* (Dictionary of Literary Biography 13, Detroit: Gale, 1982), vol. 2, 364–370.
Aristotle, *Poetics*, transl. Leon Golden (Englewood Cliffs, NJ: Prentice Hall, 1968).
Athanason, Arthur Nicholas, "The Mousetrap Phenomenon", *Armchair Detective* 12:2 (Spring 1979), 152–157.
Auden, Wystan Hugh, "The Guilty Vicarage", in: Robin W. Winks (ed.), *Detective Fiction* (Woodstock: Foul Play Press), 1988, 15–24.
August, Erdmut, *Dramaturgie des Kriminalstücks – Untersuchung zur Intellektualisierung theatralischer Vorgänge am Beispiel einer neu ausgebildeten Gattung* (PhD Dissertation: München, 1967).
Aydelotte, William, "The Detective Story as a Historical Source", in: Larry N. Landrum, Pat Browne and Ray B. Browne (eds.), *Dimensions of Detective Fiction* (Bowling Green: Popular Press, 1976), 68–82.

Balint, Michael, *Thrills and Regressions* (London: Hogarth Press, 1959).

Barefoot, Guy, "*East Lynne* to *Gas Light* – Hollywood, Melodrama and Twentieth-century Notions of the Victorian", in: Jacky Bratton, Jim Cook, Christine Gledhill (eds.), *Melodrama – Stage, Picture, Screen* (London: British Film Institute, 1994), 49–105.

Barnard, Robert, *A Talent to Deceive: An Appreciation of Agatha Christie* (New York: Dodd, 1980).

Barnes, Melvyn, "Francis Durbridge", in: John M. Reilly (ed.), *Twentieth-Century Crime and Mystery Writers* (London: St. James Press, 1985), 293–295.

Barzun, Jacques, and Wendell Hertig Taylor, *A Catalogue of Crime* (New York: Harper and Row, 1971).

Beckson, Karl, and Arthur Ganz, *Literary Terms: A Dictionary* (London: Deutsch, 1990).

Bentley, Eric, *The Life of the Drama* (London: Methuen, 1964).

Birns, Nicholas, and Margaret Boe Birns, "Agatha Christie: Modern and Modernist", in: Ronald G. Walker and June M. Frazer (eds.), *The Cunning Craft: Original Essays on Detective Fiction and Contemporary Literary Theory* (Macomb, IL: Western Illinois UP, 1990), 120–134.

Boileau, Pierre, and Thomas Narcejac, *Le roman policier* (Paris: Payot, 1964).

Booth, Michael R., *English Melodrama* (London: Herbert Jenkins, 1965).

Brooks, Peter, *The Melodramatic Imagination* (New Haven: Yale UP, 1976).

Brooks, John, *Telephone – The First Hundred Years* (New York: Harper and Row, 1976).

Carlson, Marvin, *Deathtraps: The Postmodern Comedy Thriller* (Bloomington: Indiana UP, 1993).

Carr, John Dickson, *The Life of Sir Arthur Conan Doyle* (New York: Harper, 1949).

Cawelti, James, *Adventure, Mystery, and Romance: Formula Stories as Art and Popular Culture* (Chicago: U of Chicago P, 1976).

Chandler, Raymond, "The Simple Art of Murder", in: Howard Haycraft (ed.), *The Art of the Mystery Story* (New York: Biblo and Tannen, 1976 [[1]1946]), 222–237.

Christie, Agatha, *An Autobiography* (London: Collins, 1977).

Cornish, Roger N., "David Hare", in: Stanley Weintraub (ed.), *British Dramatists Since World War II* (Detroit: Gale, 1982, Dictionary of Literary Biography 13), vol. 1, 234–243.

Cornish, Roger N., "Howard Brenton", in: Stanley Weintraub (ed.), *British Dramatists since World War II* (Detroit: Gale, 1982, Dictionary of Literary Biography 13), vol. 1, 100–108.

Courtney, Richard, *Outline History of British Drama* (Totowa, NJ: Littlefield, Adams and Co., 1982).

Cramp, Andrew, "Patrick Hamilton", in: Stanley Weintraub (ed.), *Modern British Dramatists, 1900–1945* (Detroit: Gale, 1982, Dictionary of Literary Biography 10), vol. 1, 216ff.

Davis, Owen, *I'd Like to Do It Again* (New York: Farrar and Rinehart, 1931).

Derry, Charles, *The Suspense Thriller – Films in the Shadow of Alfred Hitchcock* (Jefferson, NC, and London: McFarland, 1988).

Disher, Maurice Willson, *Melodrama – Plots That Thrilled* (London: Rockliff, 1954).

Dolar, Mladen, "Hitchcocks Objekte", in: Slavoj Žižek (ed.), *Ein Triumph des Blickes über das Auge: Psychoanalyse bei Alfred Hitchcock* (Wien: Turia and Kant, 1992), 25–42.

Dürrenmatt, Friedrich, "Problems of the Theatre", in: James L. Calderwood and Harold E. Toliver (eds.), *Perspectives on Drama* (New York: Oxford UP, 1968), 25–48.

Eco, Umberto, "Narrative Structures in Fleming", in: Umberto Eco (ed.), *The Role of the Reader: Explorations in the Semiotics of Texts* (Bloomington and London: Indiana UP, 1979).

Eco, Umberto, and Thomas A. Sebeok (eds.), *The Sign of Three: Dupin, Holmes, Peirce* (Bloomington and Indianapolis: Indiana UP, 1983).

Elam, Keir, *The Semiotics of Theatre and Drama* (London: Methuen, 1980).

Esslin, Martin, *The Theatre of the Absurd*, 3rd, rev. edition (Harmondsworth: Penguin, 1991 [[1]Doubleday, 1983]).

Findlater, Richard, *Banned! – A Review of Theatrical Censorship in Britain* (London: MacGibbon and Kee, 1967).

Fitzgerald, Percy, *Principles of Comedy and Dramatic Effect* (London: Tinsley, 1870).

Gardner, Erle Stanley, "The Case of the Early Beginning", in: Howard Haycraft (ed.), *The Art of the Mystery Story* (New York: Biblo and Tannen, 1976 [[1]1946]), 203–207.

Gassner, John, *The Theatre in Our Times* (New York: Crown, 1963).

Gillen, Francis, "Anthony Shaffer", in: Stanley Weintraub (ed.), *British Dramatists since World War II* (Detroit: Gale, 1982, Dictionary of Literary Biography 13), vol. 2, 445–450.

Gindin, James, "John Galsworthy", in: Stanley Weintraub (ed.), *Modern British Dramatists, 1900–1945* (Detroit: Gale, 1982, Dictionary of Literary Biography 10), vol. 1, 194–206.

Goodlad, John S., *A Sociology of Popular Drama* (London: Heinemann, 1971).

Gordon, Mel, *The Grand Guignol – Theatre of Fear and Terror* (New York: Amok Press, 1988).

Grella, George, "Murder and Manners: The Formal Detective Novel", in: Robin W. Winks (ed.), *Detective Fiction* (Woodstock: Foul Play Press, 1988), 84–102.

Grella, George, "Murder and the Mean Streets", *Contempora* 1.1 (1970), 6–15.

Groff, Edward, "Point of View in Modern Drama", *Modern Drama* 2:3 (December 1959), 268–282.

Hagen, Ordean A, *Who Done It? A Guide to Detective, Mystery and Suspense Fiction* (New York and London: Bowker, 1969).

Haining, Peter, *Agatha Christie: Murder in Four Acts – A Centenary Celebration of the "Queen of Crime" on Stage, Films, Radio & TV* (London: Virgin, 1990).

Hawkins, Harriet, *Classics and Trash: Traditions and Taboos in High Literature and Popular Modern Genres* (Toronto and Buffalo: U of Toronto P, 1990).

Haycraft, Howard (ed.), *The Art of the Mystery Story* (New York: Biblo and Tannen, 1976).

Heilman, Robert, *Tragedy and Melodrama – Versions of Experience* (Seattle and London: U of Washington P, 1968).

Hernadi, Paul, "Entertaining Commitments: A Reception Theory of Literary Genres", *Poetics* 10 (1981), 195–211.

Hesse, Beatrix, "Dead Bodies on Stage", *Journal of Contemporary Drama in English* 1:1 (2013), 43–53.

Hesse, Beatrix, "From Screen to Stage: The Case of *The 39 Steps*", in: Monika Pietrzak-Fragner and Eckart Voigts-Virchow (eds.), *Adaptations – Performing across Media and Genres* (Trier: WVT, 2009), 143–158.

Hesse, Beatrix, "Representations of the British Legal System in Twentieth-Century Popular Courtroom Drama", in: Monika Fludernik and Greta Olson (eds.), *In the Grip of the Law – Trials, Prisons and the Space Between* (Frankfurt: Peter Lang, 2004), 65–81.

Hesse, Beatrix, "Sister Arts? Alfred Hitchcock und das Londoner Theater der 1920er und 1930er Jahre", in: Henry Keazor (ed.), *Hitchcock und die Künste* (Marburg: Schüren, 2013), 67–82.

Hirsch, E. D., *Validity in Interpretation* (New Haven and London: Yale UP, 1967).

Hodgson, Terry, *The Batsford Dictionary of Drama* (London: Batsford, 1988).

Hubin, Allen J, *Crime Fiction II: A Comprehensive Bibliography, 1749–1990* (New York and London: Garland, 1994).

Hühn, Peter, "The Detective as Reader: Narrativity and Reading Concepts in Detective Fiction", *Modern Fiction Studies* 33:3 (Autumn 1987), 451–66.

Hutchby, Ian, *Conversation and Technology – From the Telephone to the Internet* (Oxford: Blackwell, 2001).

Hutcheon, Linda, *The Poetics of Postmodernism* (New York: Routledge, 1988).

Innes, Christopher, *Modern British Drama 1890–1990* (Cambridge: Cambridge UP, 1992).

Jefferson, Michael, *Criminal Law* (Harlow etc.: Pearson Education, [5]2001 [[1]1992]).

Jesse, F. Tennyson, *Murder and Its Motives* (London: Heinemann, 1924).

Kabatchnik, Amnon, *Blood on the Stage, 1900–1925* (Lanham, MD: Scarecrow, 2008).

Kabatchnik, Amnon, *Blood on the Stage, 1925–1950* (Lanham, MD: Scarecrow, 2009).

Kabatchnik, Amnon, *Blood on the Stage, 1950–1975* (Lanham, MD: Scarecrow, 2011).

Kabatchnik, Amnon, *Blood on the Stage,1975–2000* (Lanham, MD: Scarecrow, 2012).

Kayman, Martin A., "The Short Story from Poe to Chesterton", in: Martin Priestman (ed.), *The Cambridge Companion to Crime Fiction* (Cambridge: Cambridge UP, 2003), 41–58.

King, Daniel P., "A. E. W. Mason", in: John M. Reilly, *Twentieth Century Crime and Mystery Writers*, 2nd ed. (London: St. James, 1985), 608–609.

Klein, Holger, *J. B. Priestley's Plays* (Basingstoke: Macmillan, 1988).

Klein, Kathleen Gregory, *The Woman Detective: Gender and Genre* (Urbana and Chicago: U of Illinois P, 1988).

Knight, Stephen, *Form and Ideology in Crime Fiction* (London: Macmillan, 1980).

Knox, Ronald, "A Detective Story Decalogue", in: Robin W. Winks (ed.), *Detective Fiction* (Woodstock: Foul Play Press, 1988), 200–202.

Kosok, Heinz, *Drama und Theater im England des 20. Jahrhunderts* (Düsseldorf: Bagel; Bern and München: Francke, 1980).

Kosok, Heinz, "Traditionen und Konventionen des 19. Jahrhunderts im englischen Drama und Theater des 20. Jahrhunderts", in: Heinz Kosok (ed.), *Drama und Theater im England des 20. Jahrhunderts* (Düsseldorf: Bagel; Bern and München: Francke, 1980), 9–24.

LaBorde, Charles, *Form and Formula in Detective Drama* (PhD Dissertation: Ohio State University, 1976).

Lacan, Jacques, "Seminar on the Purloined Letter", *Yale French Studies* 48 (1973), 39–72.

Lane, Margaret, *Edgar Wallace: The Biography of a Phenomenon* (London: Hamish Hamilton, 1964 [[1]1938])

Lourdeaux, Stanley, "J. B. Priestley", in: Stanley Weintraub (ed.), *Modern British Dramatists, 1900–1945* (Detroit: Gale, 1982, Dictionary of Literary Biography 10), vol. 2, 110–116.

Maack, Annegret, "Dramatiker der Vorkriegszeit im englischen Theater nach 1945", in: Heinz Kosok (ed.), *Drama und Theater im England des 20. Jahrhunderts* (Düsseldorf: Bagel, Bern and München: Francke, 1980), 105–114.

McLuhan, Marshall, *Understanding Media: The Extension of Man* (London: Routledge and Kegan Paul, 1964).

Marwick, Arthur, *British Society Since 1945* (Harmondsworth: Penguin, 1982).

Merivale, Patricia, "The Flaunting of Artifice in Vladimir Nabokov and Jorge Luis Borges", in: L. S. Dembo (ed.), *Nabokov: The Man and His Work* (Madison and London: U of Wisconsin P, 1967), 209–224.

Most, Glenn W., and William W. Stowe (eds.), *The Poetics of Murder: Detective Fiction and Literary Theory* (New York: Harcourt Brace Jovanovich, 1983).
Muller, John P., and William J. Richardson (eds.), *The Purloined Poe: Lacan, Derrida and Psychoanalytic Reading* (Baltimore and London: Johns Hopkins UP, 1988).
Murfin, Ross, and Supryia M. Ray, *The Bedford Glossary of Critical and Literary Terms*, 2nd ed. (Boston and New York: Bedford/St. Martin's, 2003).
Myszor, Frank, *The Modern Short Story* (Cambridge: Cambridge UP, 2001).
Newton, H. Chance, *Crime and the Drama* (London: Stanley Paul, 1970).
Orwell, George, "Raffles and Miss Blandish", in: *Collected Essays* (London: Secker and Warburg, 1961), 249–263.
Osborne, Charles, *The Life and Crimes of Agatha Christie* (London: Collins, 1982).
Parsons, Hesketh, *Conan Doyle* (London: Unwin, 1987).
Pederson-Krag, Geraldine, "Detective Fiction and the Primal Scene", *Psychoanalytic Quarterly* 18 (1949), 207–214.
Phillips, Gene D., *Alfred Hitchcock* (Boston: Twayne, 1984).
Porter, Dennis, *The Pursuit of Crime: Art and Ideology in Detective Fiction* (New Haven and London: Yale UP, 1981).
Priestley, J. B., *Man and Time* (London: Allen, 1978 [[1]London: Aldus, 1964]).
Priestman, Martin, *Detective Fiction and Literature: The Figure on the Carpet* (Basingstoke, Macmillan, 1990).
Priestman, Martin (ed.), *The Cambridge Companion to Crime Fiction* (Cambridge: Cambridge UP, 2003).
Pyrhönen, Heta, *Murder from an Academic Angle: An Introduction to the Study of the Detective Narrative* (Columbia, SC: Camden House, 1994).
Queen, Ellery, "The Detective Short Story: The First Hundred Years", in: Howard Haycraft (ed.), *The Art of the Mystery Story* (New York: Biblo and Tannen, 1976 [[1]1946]), 476–491.
Rabinowitz, Peter, "Chandler Comes to Harlem: Racial Politics in the Thriller of Chester Himes", in: Barbara A. Rader and Howard G. Zettler (eds.), *The Sleuth and the Scholar: Origins, Evolutions and Current Trends in Detective Fiction* (New York: Greenwood Press, 1988), 19–29.
Rahill, Frank, *The World of Melodrama* (University Park and London: Pennsylvania State UP, 1967).
Ramsey, G. C., *Agatha Christie: Mistress of Mystery* (New York: Dodd, Mead & Company, 1967).
Reddy, Maureen, *Sisters in Crime: Feminism and the Crime Novel* (New York: Ungar, 1988).
Reid, Ian, *The Short Story* (London: Methuen, 1977, The Critical Idiom 37).
Reilly, John M., *Twentieth Century Crime and Mystery Writers* (London: St. James Press, 1985).
Reynolds, William (ed.), *It's a Print! – Detective Fiction from Page to Screen* (Bowling Green: Popular Press, 1994).
Richards, Stanley (ed.), *Best Mystery and Suspense Plays of the Modern Theatre* (New York: Dodd, 1971).
Richards, Stanley (ed.), *10 Classic Mystery and Suspense Plays of the Modern Theatre* (New York: Dodd, 1973).
Rodway, Allan, and Brian Lee, "Coming to Terms", *Essays in Criticism* 14:2 (April 1964), 109–125.
Salter, Denis, "Henry Irving, the 'Dr Freud' of Melodrama", in: James Redmond (ed.), *Melodrama* (Cambridge: Cambridge UP, 1992, Themes in Drama 14), 161–182.
Sayers, Dorothy, "Aristotle on Detective Fiction", in: Robin W. Winks (ed.), *Detective Fiction* (Woodstock: Foul Play Press, 1988), 25–34.

Sayers, Dorothy, "Omnibus of Crime", in: Robin W. Winks (ed.), *Detective Fiction* (Woodstock: Foul Play Press, 1988), 53–83.

Schmidt, Johann N., *Ästhetik des Melodramas* (Heidelberg: Winter, 1986).

Schnierer, Peter Paul, "Crime and Detective Story Structures in Modern British Drama", in: Stephen Kaplan and Will Wright (eds.), *The Image of Crime in Literature, the Media, and Society* (Pueblo, Colorado: Society for the Interdisciplinary Study of Social Imagery, 1991), 53–64.

Schumacher, Claude (ed.), *Artaud on Theatre* (London: Methuen, 1989).

Seeßlen, Georg, *Kino der Angst: Geschichte und Mythologie des Film-Thrillers* (Reinbek: Rowohlt, 1980).

Seeßlen, Georg, and Claudius Weil, *Kino des Phantastischen: Geschichte und Mythologie des Horror-Films* (Reinbek: Rowohlt, 1980).

Sexton, James, "Aldous Huxley's Three Plays, 1931–1948", in: C. C. Barfoot (ed.), *Aldous Huxley between East and West* (Amsterdam: Rodopi, 2001), 59–74.

Shepherd, Simon, and Peter Womack, *English Drama: A Cultural History* (Oxford: Blackwell, 1996).

Sierz, Aleks, *In-Yer-Face Threatre – British Drama Today* (London: Faber and Faber, 2001).

Sinyard, Neil, and Adrian Turner, *Journey Down Sunset Boulevard: The Films of Billy Wilder* (Ryde: BCW Publishing, 1979).

Small, Barbara J., "Ben Travers", in: Stanley Weintraub (ed.), *Modern British Dramatists, 1900–1945* (Detroit: Gale, 1982, Dictionary of Literary Biography 10), vol. 2, 184–188.

Smith, James L., *Melodrama* (London: Methuen, 1973, The Critical Idiom 28).

Stephenson, Anthony, "Frederick Lonsdale", in: Stanley Weintraub (ed.), *Modern British Dramatists, 1900–1945* (Detroit: Gale, 1982, Dictionary of Literary Biography 10), vol. 1, 294–300.

Stephenson, Anthony, "Simon Gray", in: Stanley Weintraub (ed.), *British Dramatists since World War II* (Detroit: Gale, 1982, Dictionary of Literary Biography 13), vol. 1, 199–208.

Stirling, Edward, *Old Drury Lane* (London: Chatto and Windus, 1881).

Stokes, John, "Body Parts: The Success of the Thriller in the Inter-War Years", in: Clive Barker and Maggie B. Gale (eds.), *British Theatre between the Wars, 1918–1939* (Cambridge: Cambridge UP, 2000), 38–62.

Stowe, William W., "From Semiotics to Hermeneutics: Modes of Detection in Doyle and Chandler", in: Glenn W. Most and William W. Stowe (eds.), *The Poetics of Murder: Detective Fiction and Literary Theory* (San Diego, New York and London: Harcourt Brace Jovanovich, 1983), 366–383.

Styan, J. L., *The English Stage – A History of Drama and Performance* (Cambridge: Cambridge UP, 1996).

Suerbaum, Ulrich, *Krimi – Eine Analyse der Gattung* (Stuttgart: Reclam, 1984).

Sweeney, S. E., "Locked Rooms: Detective Fiction, Narrative Theory and Self-Reflexivity", in: Ronald G. Walker and June M. Frazer (eds.), *The Cunning Craft – Original Essays on Detective Fiction and Contemporary Literary Theory* (Macomb, IL: Western Illinois UP, 1990), 1–14.

Symons, Julian, *Bloody Murder – From the Detective Story to the Crime Novel: A History*, rev. edition (Harmondsworth: Penguin, 1974).

Tani, Stefano, *The Doomed Detective: The Contribution of the Detective Novel to Postmodern American and Italian Fiction* (Carbondale and Edwardsville: Southern Illinois UP, 1984).

Taylor, John Russell, *The Rise and Fall of the Well-Made Play* (London: Methuen, 1967).

Thomson, Lynn Marian, "The Crook Play", *Journal of American Drama and Theatre*, 13:1 (Winter 2001), 1–35.

Todorov, Tzvetan, "The Origin of Genres", *New Literary History* 8 (1976–77), 159–170.

Todorov, Tzvetan, "The Typology of Detective Fiction", in: Tzvetan Todorov (ed.), *The Poetics of Prose* (i.e. *La poetique de la prose*, Paris: Seuil, 1971, transl. Richard Howard, Oxford: Blackwell, 1977), 42–52.

Trewin, J. C., "The Great War and the Theater, 1914–1918", in: Stanley Weintraub (ed.), *Modern British Dramatists, 1900–1945* (Detroit: Gale, 1982, Dictionary of Literary Biography 10), vol. 2, 277–284.

Trilling, Lionel, *Sincerity and Authenticity* (Cambridge, MA: Harvard UP, 1972).

Truffaut, François, with the collaboration of Helen Scott, *Hitchcock – Revised Edition* (London: Paladin Grafton Books, 1986).

Trussler, Simon, *The Cambridge Illustrated History of British Theatre* (Cambridge: Cambridge UP, 1994).

Van Dine, S. S., "Twenty Rules for Writing Detective Stories", in: Howard Haycraft (ed.), *The Art of the Mystery Story* (New York: Biblo and Tannen, 1976 [[1]1946]), 189–193.

Vickers, Roy, "Crime on the Stage – The Criminological Illusion", in: Michael Gilbert (ed.), *Crime in Good Company: Essays on Criminals and Crime-writing* (London: Constable, 1959), 179–191.

Voigts-Virchow, Eckart, "'Your Voice Is Around My Neck' – The Media History and Poetics of Theatre Telephony", *Forum Modernes Theater* 19:2 (2004), 131–151.

Walker, Ronald G., and June M. Frazer (eds.), *The Cunning Craft* (Macomb: Western Illinois UP, 1990).

Wearing, J. P., *The London Stage: A Calendar of Plays and Players* (Metuchen, NJ: Scarecrow, 1976–1993).

Weintraub, Stanley (ed.), *British Dramatists Since World War II* (Detroit: Gale, 1982, Dictionary of Literary Biography 13).

Weintraub, Stanley (ed.), *Modern British Dramatists, 1900–1945* (Detroit: Gale, 1982, Dictionary of Literary Biography 10).

Wingate, Nancy, "Getting Away with Murder", *Journal of Popular Culture* 12:4 (Spring 1979), 581–603.

Winks, Robin W. (ed.), *Detective Fiction: A Collection of Critical Essays* (Woodstock: Foul Play Press, 1988).

Winks, Robin W., and Maureen Corrigan (eds.), *Mystery and Suspense: The Literature of Crime, Detection, and Espionage*, 2 vols (New York: Scribner, 1998).

Wright, Willard Huntington, "The Great Detective Stories", in: Howard Haycraft (ed.), *The Art of the Mystery Story* (New York: Biblo and Tannen, 1976 [[1]1946]), 33–70.

Žižek, Slavoj, *Looking Awry: An Introduction to Jacques Lacan through Popular Culture* (Cambridge, MA, and London: The MIT Press, 1991).

Index

www.ingramcontent.com/pod-product-compliance
Lightning Source LLC
Chambersburg PA
CBHW060750150926
51501CB00004B/141
9781137463036